Data Protection

Chaminda Hewage • Lasith Yasakethu
Dushmantha Nalin K. Jayakody
Editors

Data Protection

The Wake of AI and Machine Learning

Springer

Editors
Chaminda Hewage
Cardiff, UK

Lasith Yasakethu
Padukka, Sri Lanka

Dushmantha Nalin K. Jayakody
Lisbon, Portugal

ISBN 978-3-031-76472-1 ISBN 978-3-031-76473-8 (eBook)
https://doi.org/10.1007/978-3-031-76473-8

This Springer imprint is published by the registered company Springer Nature Switzerland AG
The registered company address is: Gewerbestrasse 11, 6330 Cham, Switzerland

Preface

Data protection is a key aspect of information security where personal and business data need to be protected from unauthorized access and modification. The number of emerging digital technologies that collect, transmit, and process privacy-sensitive data is on the rise. These technologies range from blockchain applications that perform private business data transactions on public ledger to Internet of Things (IoT) applications that collect personal healthcare data through wearable devices. The collected personal data without user consent can be used to train and validate Artificial Intelligence (AI) and Machine Learning (ML) models. The protection of personal data received significant attention due to an increased number of violations of subject rights. On the other hand, processing of personal data through emerging AI/ML models and applications can impact individual rights. Therefore, it is important to minimize these risks through reducing the bias of these models, increasing the accuracy, and making the AI/ML models transparent and responsible. This book presents key issues, challenges, opportunities, and solutions associated with data protection in the wake of Artificial Intelligence (AI) and Machine Learning (ML) technologies.

In the wake of the rapid proliferation of Artificial Intelligence (AI) technologies, data protection has emerged as a pressing concern. AI heavily relies on vast amounts of data to train and refine its models, often encompassing personal and sensitive information. Ensuring the privacy of this data has become a complex challenge, necessitating robust frameworks to balance the transformative potential of AI with the fundamental rights of individuals. Concepts such as anonymization, which involves removing personally identifiable information, and differential privacy, which adds noise to data to protect individual identities, are being integrated into AI pipelines to safeguard against data breaches. Additionally, the adoption of federated learning techniques enables model training across decentralized devices, minimizing the need for centralized data repositories and reducing the risk of unauthorized access to sensitive information.

Moreover, the ethical dimensions of data protection in AI extend beyond technical considerations. AI systems are susceptible to inheriting biases present in their training data, which can lead to discriminatory outcomes. Efforts are underway to address algorithmic bias through rigorous auditing and refining of training datasets. Furthermore, concepts such as Responsible AI have received much attention across the globe recently. Stricter regulations, such as the General Data Protection Regulation (GDPR) and AI Act in the European Union, are also influencing the development of AI applications by mandating transparent data usage, informed consent, and the ability for individuals to control how their data is processed. As AI continues to reshape industries, the collaboration between technologists, policy-makers, and ethicists becomes paramount in building a future where data protection is enshrined in the very fabric of AI innovation.

Cardiff, UK
Padukka, Sri Lanka
Lisbon, Portugal
July 2024

Chaminda Hewage
Lasith Yasakethu
Dushmantha Nalin K. Jayakody

About the Book

The purpose of this edited book, *Data Protection: The Wake of AI and Machine Learning*, is to establish the state of the art and chart the course for future research in privacy-preserving AI and ML technologies. The scope of this book includes not only all aspects of data protection challenges with recent advancements in AI and ML technologies but related areas, such as the Metaverse, misinformation and disinformation. The book serves as a central source of reference for data protection research and developments in the wake of AI and ML. The book aims to publish thorough and cohesive overviews on specific topics in data protection in an AI era, as well as works that are larger in scope than survey articles and that will contain more detailed background information. The book also provides a singular coverage of advanced and timely topics and a forum for topics that may not yet have attained the level of maturity to warrant a comprehensive textbook.

Contents

About the Editors

Chaminda Hewage is a Reader (Associate Professor) in Data Security in the Cardiff School of Technologies at Cardiff Metropolitan University, UK, where he is also the Associate Dean Research for the school. He is the Founding Director of the Cybersecurity and Information Networks Centre (CINC) at Cardiff Metropolitan University. He received a B.Sc. Engineering (First Class Honours Degree) in Electrical and Information Engineering from the Faculty of Engineering, University of Ruhuna (Sri Lanka) in 2004 and a Ph.D. in Multimedia Communications from the University of Surrey (UK) in 2009. He was awarded the Gold Medal for best performance in Engineering (2014) by the University of Ruhuna for his achievements in undergraduate studies at the General Convocation held in 2004. After graduation, he joined Sri Lanka Telecom PLC (Sri Lanka) as a Telecommunication Engineer (2004). In September 2005, he was awarded the Overseas Research Scholarships (ORS) scholarship by the Higher Education Funding Council of England (Universities UK) to pursue his Ph.D. at the University of Surrey, UK. Upon completion of his Ph.D., he works as a Researcher at the University of Surrey, UK, and Kingston University, London, UK. In 2015, he joined Cardiff Metropolitan University, UK, as a Senior Lecturer in Networking and Cyber Security. In 2014, he received a Post-Graduate Certification in Higher Education (HE) Teaching and Learning from Kingston University—London. He is a Fellow of the Higher Education Academy (HEA), UK. He is a Senior Member of the IEEE and Member of IET and the Chartered Institute of Information Security (CIISec).

Lasith Yasakethu is a Senior Lecturer in the Department of Applied Electronics, Sri Lanka Technology Campus. He is currently serving as the Dean of Faculty of Technology. Dr. Yasakethu received his B.Sc. Engineering degree (First Class Hons.) in Electrical and Electronic Engineering from the University of Peradeniya, Sri Lanka, in 2007. He was awarded the prize for best performance in Electronic Communication Engineering by the University of Peradeniya for his achievements in undergraduate studies. After graduation, he joined Dialog Telecom (Pvt.) Ltd., Sri Lanka, as a Core-Network Engineer (2007). In October 2007 he was awarded the Overseas Research Scholarships Award by the Higher Education Funding

Council of England to pursue Ph.D. at the University of Surrey, UK. After completing his Ph.D., he worked as a Research Engineer for Technicolor Research and Innovations (formerly known as THOMSON R&D), in Rennes, France (2010–2012). Dr. Yasakethu worked as a Research Fellow and a Senior Research Fellow at the University of Surrey, UK, and Kingston University, London, UK, respectively (2012–2014). Afterwards, he worked as a Senior Lecturer at the Department of Mechatronics, Faculty of Engineering, SAITM, Sri Lanka (2014–2016) and the Department of Electrical and Electronic Engineering, Faculty of Engineering, University of Peradeniya, Sri Lanka (2016–2018). His area of expertise is Video Communication, Perception and Cyber Security. Dr. Yasakethu has worked for several European Commission Framework Program-6 (FP6), Framework Program-7 (FP7) and Horizon 2020 projects. He is a member of IEEE and an author/reviewer of several IEEE/IET journal/conference publications.

Dushmantha Nalin K. Jayakody is a Senior Member of IEEE, Fellow of IET and received the Ph.D. degree in Electronics and Communications Engineering from the University College Dublin, Ireland, in 2014. He received his M.Sc. degree in Electronics and Communications Engineering from Eastern Mediterranean University, Turkey. He was ranked as the top 2% scientist in the world in 2021, 2022 and 2023 by the Stanford Elsevier list. In 2021, he received the Sri Lankan Presidential Award for his outstanding research performance among many other awards and recognitions. Since 2022, he is a Professor at the Lusófona University, Portugal. Since 2023, he is also with the Sri Lanka Institute of Information Technology, Sri Lanka. From 2019 to 2022, he served as the Dean, Faculty of Postgraduate Studies and Research at Sri Lanka Technological Campus, Sri Lanka, from 2021 to 2022 as the Director Postgraduate, Research and Impact at the Research at Sri Lanka Technological Campus, Sri Lanka. He is currently serving as the Vice Chair of IEEE ComSoc (Sri Lanka Chapter). He held visiting positions at the Center for Telecommunications Research, University of Sydney, Australia in 2015, and Texas A&M University in 2018, Qatar Foundation (HBKU). He was a Visiting Professor at the University of Jyvaskyla, Finland, both in 2019 and 2022 within the framework of the Academy of Finland. He also served as a Visiting Professor at the University of Juiz de Fora in Brazil in 2019. From 2019 to 2022, he has been a SPARC Professor at the Department of Electronics and Communication Engineering, National Institute of Tiruchirappalli, India, within the framework of the Ministry of Human Resources in India. From 2014 to 2016, he was a Postdoc Research Fellow at the University of Tartu, Estonia, and the University of Bergen, Norway. From 2016 to 2021, he was a Professor at the School of Computer Science & Robotics, National Research Tomsk Polytechnic University (TPU), Russia. From 2021 to 2022, he was with the Department of Engineering and Computer Science/ Autonoma TechLab, Universidade Autónoma de Lisboa, Portugal. He has supervised over 8 Ph.D. students and many Master and over 50 undergraduate students and 5 Postdoc researchers. In his career, so far, he has attracted nearly 6M $ research funding from many international grant agencies and has published over 230 international peer reviewed journal and conference papers and books. Prof. Jayakody has received the best paper award from the ICCMIT in 2017 and International Conference on Emerging Technologies of Information and Communications (ETIC 2019), March 2019 and IRC 2023. He has received the Sri Lankan President's Award for Research Excellence in 2020. In July 2019, Prof. Jayakody received the Education Leadership Award from the World Academic Congress in 2019. In 2017 and 2018, he received the outstanding faculty award by National Research Tomsk Polytechnic University, Russia. He also received the Distinguished Researcher in Wireless Communications award in Chennai, India 2019. Prof. Jayakody received the Presidential Award for outstanding research performance in the year 2021. He also received "Best Publication Award" at the Sri Lanka Technological Campus, Sri Lanka, in 2019 and 2020.

Chapter 1
Data Protection Challenges and Opportunities Due to Emerging AI and ML Technologies

Chaminda Hewage, Lasith Yasakethu, and Dushantha Nalin K. Jayakody

Abstract Artificial intelligence (AI) and machine learning (ML) research and innovation have created significant advances in how technology is used in diverse sectors. These technologies can create significant impact outside academia, for example, on our economy, society, and industry landscape. However, most of these AI technologies could be biased or inaccurate if they are not developed and validated well. Irresponsible AI applications can pose significant risks to individuals' rights. People have started to resist "black box-" type AI models. They want more transparent, equitable, and responsible AI models. To this end, technological paradigm changes are taking place to make emerging AI models more responsible and useful for individuals and the society. To reinforce the technological efforts, policies and regulations are being enacted across the world to regulate "out-of-control" AI and ML applications. This chapter discusses emerging AI and ML applications, data protection challenges, and privacy-preserving AI and ML algorithms.

1.1 Introduction

In the wake of the rapid proliferation of artificial intelligence (AI) technologies, data protection has emerged as a pressing concern. AI heavily relies on vast amounts of data to train and refine its models, often encompassing personal and sensitive

C. Hewage (✉)
Cybersecurity and Intelligent Networks Centre (CINC), Cardiff School of Technologies, Cardiff Metropolitan University, Cardiff, UK
e-mail: Chewage@cardiffmet.ac.uk

L. Yasakethu
SLTC Research University, Padukka, Sri Lanka
e-mail: lasithy@sltc.ac.lk

D. N. K. Jayakody
Lusofona University, COPELABS, Lisbon, Portugal
e-mail: dushantha.jayakody@ulusofona.pt

C. Hewage et al. (eds.), *Data Protection*,
https://doi.org/10.1007/978-3-031-76473-8_1

information. Ensuring the privacy of this data has become a complex challenge, necessitating robust frameworks to balance the transformative potential of AI with the fundamental rights of individuals. Concepts such as anonymization, which involves removing personally identifiable information, and differential privacy, which adds noise to data to protect individual identities, are being integrated into AI pipelines to safeguard against data breaches. Additionally, the adoption of federated learning techniques enables model training across decentralized devices, minimizing the need for centralized data repositories and reducing the risk of unauthorized access to sensitive information.

Moreover, the ethical dimensions of data protection in AI extend beyond technical considerations. AI systems are susceptible to inheriting biases present in their training data, which can lead to discriminatory outcomes. Efforts are underway to address algorithmic bias through rigorous auditing and refining of training datasets. Furthermore, concepts such as responsible AI have received significant attention across the globe recently. Stricter regulations, such as the General Data Protection Regulation (GDPR) and the AI Act in the European Union (EU), are also influencing the development of AI applications by mandating transparent data usage, informed consent, and the ability of individuals to control how their data is processed. As AI continues to reshape industries, collaboration between technologists, policymakers, and ethicists becomes paramount in building a future where data protection is enshrined in the very fabric of AI innovation.

The main discussions of this chapter are organized as follows. Section 1.2 elaborates on emerging AI and ML algorithms together with the opportunities and challenges. The impact of AI and ML algorithms on data protection is discussed in Sect. 1.3. Section 1.4 analyzes the proposed privacy-preserving solutions for emerging AI and ML technologies.

1.2 Emerging AI and ML Algorithms: Opportunities and Challenges

The sciences of artificial intelligence (AI) and machine learning (ML) are closely related and have transformed many facets of technology and the society. Artificial intelligence (AI) is the creation of intelligent systems that can carry out tasks like learning, thinking, problem-solving, and decision-making, all of which normally need human intelligence. ML, a branch of AI, is the study of building models and algorithms that let computers learn from data, anticipate future events, and gradually get better at what they do without the need for explicit programming.

1.2.1 Introduction

The core concept behind AI is to simulate human-like intelligence in machines, allowing them to analyze complex data, recognize patterns, and make informed decisions. This field has evolved significantly, from early symbolic AI systems that relied on predefined rules to modern AI systems that leverage ML algorithms for data-driven learning.

ML algorithms play a pivotal role in AI development by enabling machines to learn from data. These algorithms are designed to extract meaningful insights and patterns from large datasets, which are then used to make predictions or decisions. There are several types of ML algorithms, each suited for different tasks and data types:

Supervised Learning The system learns from labeled data in supervised learning, where each input has a corresponding output or target. By reducing the discrepancy between expected and actual results, the algorithm gains the ability to map inputs to outputs. Tasks like classification (e.g., spam detection, image recognition) and regression (e.g., sales revenue prediction) frequently use this kind of learning [1, 2].

Unsupervised Learning Unsupervised learning is the process of using unlabeled data to teach an algorithm to find hidden structures or patterns in the data. Unsupervised learning approaches employ clustering algorithms, such as K-means clustering and hierarchical clustering, to group together comparable data points [1].

Semi-supervised Learning With the use of both a greater quantity of unlabeled data and a smaller amount of labeled data, semi-supervised learning integrates aspects of supervised and unsupervised learning. When the process of labeling data is costly or time-consuming, this method works well [1].

Reinforcement Learning Through interactions with an environment, an agent is trained to make successive decisions through reinforcement learning. By making mistakes and learning from them, the agent is either rewarded or penalized for its behavior. Applications like gaming, robots, and self-driving cars frequently use this kind of learning [1].

Numerous industries, including healthcare, finance, retail, and entertainment, have seen radical transformations because of AI and ML algorithms. They help businesses make data-driven decisions, automate repetitive procedures, streamline operations, and customize customer experiences. Personalized therapy recommendations, disease diagnosis, medication discovery, and medical picture analysis are just a few applications of machine learning algorithms in the healthcare industry. However, the quick development of AI and ML has also brought about social and ethical issues, including algorithmic prejudice, data privacy, job displacement, and the influence on social dynamics. A multidisciplinary strategy combining cooperation between engineers, legislators, ethicists, and society at large is needed to address these issues. AI and ML algorithms represent a paradigm shift in how we interact with technology and leverage data to solve complex problems. As these fields continue to evolve, responsible AI development and deployment will be essential to ensure ethical use, transparency, and accountability in AI-powered systems.

1.2.2 Opportunities in AI and ML Algorithms

Algorithms for machine learning (ML) and artificial intelligence (AI) are bringing about revolutionary shifts in a variety of industries, offering a wide range of options from efficiency improvements and automation to personalized experiences and predictive analytics. These technologies have the power to transform workflows, improve judgment, and spur innovation across a range of industries, opening the door to a future that is increasingly intelligent and data-driven.

1.2.2.1 Automation and Efficiency

Across all industries, automation of monotonous jobs and increased productivity are made possible by AI and ML algorithms. AI-enabled robotics in manufacturing automate assembly lines, conduct quality control inspections, and perform predictive maintenance, resulting in higher production rates and lower costs. Similar to this, AI-driven chatbots and virtual assistants in customer service manage standard enquiries, offer tailored advice, and free up human agents for more difficult jobs, all while enhancing response times and service quality [3].

1.2.2.2 Personalization and Customer Experience

Businesses are able to provide clients with individualized experiences owing to AI and ML algorithms. Recommendation engines in e-commerce that are driven by machine learning algorithms examine consumer behavior and preferences to make product recommendations based on unique interests. This promotes client loyalty, raises sales, and improves consumer happiness. Similar to this, in marketing, consumer data is analyzed by AI-driven analytics tools to build tailored campaigns that increase engagement and conversion rates [4].

1.2.2.3 Data Analysis and Insights

Algorithms based on AI and ML are essential for deriving practical insights from massive datasets. While exploratory data analysis (EDA) finds patterns and relationships in data, descriptive analytics techniques summarize data features. Businesses can forecast trends, find growth opportunities, and make data-driven decisions with the use of inferential statistics and predictive modeling. New pathways for analysis and comprehension are made possible by the superior analysis of unstructured data, such as text, audio, and images, by advanced machine learning techniques like deep learning and neural networks [5].

1.2.2.4 Healthcare and Life Sciences

Healthcare and life sciences could undergo revolutionary change because of AI and ML algorithms. Deep learning techniques enable early disease diagnosis and increase diagnostic accuracy in medical picture analysis. Treatment planning, result prediction, and patient risk assessment are all aided by predictive analytics models. AI-driven research tools also expedite drug discovery, pinpoint possible therapeutic targets, and enhance clinical trial efficiency, resulting in advances in medical science and better patient outcomes [6].

1.2.2.5 Finance and Risk Management

Algorithms based on AI and ML are essential for risk management, fraud detection, and investing strategies in the financial industry. Massive volumes of financial data are analyzed by ML models to find trends, abnormalities, and possible dangers. Algorithmic trading systems optimize investment portfolios and maximize profits by executing transactions based on predictive analytics and real-time market data. Furthermore, chatbots and virtual assistants driven by AI improve client interactions, offer financial advice, and expedite banking procedures [7].

1.2.2.6 Supply Chain Optimization

Supply chain management is improved by AI and ML algorithms through demand forecasting, inventory optimization, and predictive analytics. These systems estimate demand, manage production schedules, and optimize inventory levels by analyzing past data, market trends, and outside influences. Predictive maintenance coupled with real-time monitoring increases overall operating efficiency, lowers downtime, and increases equipment uptime [8].

1.2.2.7 Energy and Sustainability

Sustainability initiatives, integration of renewable energy sources, and energy efficiency are all driven by AI and ML algorithms. ML algorithms are used by smart grid technologies to optimize energy distribution, cut waste, and improve grid resilience. Algorithms for predictive maintenance identify equipment problems before they happen, reducing maintenance expenses and downtime. Furthermore, AI-driven models promote sustainable practices in energy management by facilitating resource allocation, assessing environmental effect, and optimizing energy usage [9].

1.2.2.8 Education and Learning

Learning and education are revolutionized by AI and ML algorithms. ML algorithms are used by personalized learning systems to modify learning paths and content according to the preferences and progress of each individual student. Algorithms for natural language processing (NLP) underpin automated grading systems, virtual tutors, and language learning apps. With the help of AI-driven educational analytics, teachers may better understand student performance, engagement levels, and learning outcomes, enabling them to design interventions that will help students succeed [10].

1.2.2.9 Cybersecurity and Threat Detection

Because AI and ML algorithms can identify and mitigate cyber risks instantly, they improve cybersecurity protocols. In order to spot anomalies and possible security breaches, machine learning models examine user behavior, system records, and network traffic patterns. Security tools driven by AI increase defenses against criminal activity, data breaches, and cyberattacks by automating threat detection, response, and remediation [11].

Algorithms using AI and ML offer a wide range of constantly changing options. Numerous advantages are provided by these technologies, such as increased decision-making capacity, automation, efficiency improvements, personalized experiences, and data-driven insights. In the current digital world, businesses and organizations may enhance customer happiness, acquire a competitive edge, and achieve sustainable growth by utilizing the potential of AI and ML.

1.2.3 Challenges in Data Protection

The rapid advancements in artificial intelligence (AI) and machine learning (ML) algorithms have brought about tremendous opportunities, but they also pose significant challenges in terms of data protection and privacy. As these technologies rely heavily on data, ensuring the confidentiality, integrity, and security of data has become paramount. Below are some of the key challenges in data protection associated with AI and ML algorithms.

1.2.3.1 Data Privacy Concerns

Keeping data private is one of the biggest challenges. Large volumes of data must be accessible to AI and ML algorithms in order for them to be trained and learned. Sensitive personal data, including financial transactions, health records, and personal preferences, is frequently included in this data. A key problem is making sure that this data is appropriately kept and utilized in accordance with privacy laws, like the California Consumer Privacy Act (CCPA) or GDPR [12].

1.2.3.2 Data Quality and Bias

Training machine learning algorithms requires high-quality data. Inaccurate forecasts and biased results might result from biased or low-quality data. Data bias can originate from a number of factors, such as unbalanced representations of particular groups, historical biases in training data, or a lack of diversity in the dataset. In order to eliminate biases and guarantee justice in algorithmic decision-making, addressing data quality and bias issues involves meticulous data preprocessing, feature selection, and continuous monitoring [13].

1.2.3.3 Security Risks

Security issues associated with AI and ML algorithms include adversarial attacks, model weaknesses, and data breaches. Adversarial assaults occur when data manipulation is used to trick machine learning algorithms into making inaccurate predictions or judgments. Robust cybersecurity methods, encryption strategies, access limits, and frequent security audits are necessary for safeguarding AI and ML systems in order to identify and address possible risks.

1.2.3.4 Data Governance and Compliance

The procedures and guidelines controlling the gathering, storing, using, and sharing of data inside an organization are referred to as data governance. Making sure that AI and ML adhere to ethical standards and data protection laws is crucial. To prove compliance and responsibility, organizations need to put in place data protection measures, create explicit data governance frameworks, and carry out frequent audits [14].

1.2.3.5 Explainability and Transparency

Because AI and ML algorithms frequently function as "black boxes," it can be difficult to comprehend how they make judgments or forecast outcomes. Algorithmic accountability and bias are raised by this lack of explainability and transparency. In order to eliminate biases, foster human oversight of algorithmic judgments, and increase trust, it is imperative that explainable AI (XAI) and model interpretability be improved in AI and ML models [15].

1.2.3.6 Data Access and Sharing

It might be difficult to strike a balance between data protection and sharing and access. Sharing data raises the possibility of data breaches and unauthorized access, even while it can also foster cooperative research, innovation, and information exchange. Encouraging ethical data sharing habits while reducing risks can be achieved by the implementation of data anonymization techniques, access limits, and secure data sharing protocols.

1.2.3.7 Ethical Considerations

Algorithms using AI and ML bring up moral questions of responsibility, justice, openness, and society influence. Ethical AI concepts and rules are necessary to guarantee that AI systems do not reinforce or magnify preexisting biases, discriminate against specific groups, or violate individuals' rights. The responsible application of AI in delicate fields like healthcare, criminal justice, and surveillance is also subject to ethical problems [16].

1.2.3.8 Regulatory Compliance and Enforcement

Ensuring regulatory compliance and staying up to date with rapidly changing data protection standards are ongoing challenges. Algorithms using AI and ML must abide by industry norms, legal frameworks, and best practices for data protection. Organizations that operate internationally have difficulties since different authorities have different enforcement methods and sanctions for noncompliance.

An all-encompassing strategy for data protection in AI and ML algorithms is needed to resolve these problems. To create strong data protection policies, moral standards, and legal frameworks that encourage innovation while preserving people's rights and privacy, data scientists, cybersecurity specialists, attorneys, legislators, and other stakeholders must all work together. Organizations may fully utilize AI and ML while guaranteeing ethical and responsible data use by taking proactive measures to address these issues.

1.2.4 Summary

In summary, a wide range of opportunities are presented by the development of artificial intelligence (AI) and machine learning (ML) algorithms, spanning from efficiency benefits and automation to personalized experiences and predictive analytics. However, there are a lot of privacy and data protection issues with these options. Key priorities in navigating the complexities of data protection in AI and ML algorithms include addressing privacy concerns, guaranteeing data quality and fairness, mitigating security risks, implementing strong data governance, improving explainability and transparency, addressing ethical considerations, and assuring regulatory compliance. Organizations can ensure that algorithmic systems are trustworthy, can protect individual rights, and can promote fairness while utilizing AI and ML to their full potential by embracing a comprehensive approach that incorporates ethical norms, technical solutions, and legal frameworks. We can address the issues of data protection in the digital age and unlock the transformational power of AI and ML for the good of society through responsible data practices, cooperation, and ongoing innovation.

1.3 Impact of AI and ML Algorithms on Data Protection

1.3.1 Introduction

AI and ML algorithms are creating diverse opportunities and challenges for humanity, which was not anticipated 5–10 years ago. Moreover, these rapid advancements are causing human rights violations, such as privacy and data protection issues, which are widely discussed in the literature [17–24]. As long as emerging AI and ML applications are providing technological advancements and creating wider impact, they will be well received by users. However, if they creep into personal data space, it will impact individual rights and will face natural resistance from policymakers and the public.

1.3.2 Big Data Versus Data Protection

Large amounts of data (big data) are required to train AL and ML algorithms and provide reliable and accurate outcomes. This has led to collecting large amounts of data with or without the consent of the original owners or data owners. These efforts increase the risk of exposing personal data to unauthorized users, violating both confidentiality and integrity [25, 26]. If the collected data contains sensitive information such as biomatrices, any processing of such data could lead to significant consequences to individuals. This was a common practice during COVID-19 due to contact tracing, medical research, and prevention efforts. A number of studies can be found in the literature, which discuss privacy and security challenges during and beyond COVID-19 [22, 27–29]. If consent has been obtained, then there are legal grounds for processing according to current data protection regulations such as GDPR [30, 31]. If not, then it is illegal to process such data. This also highlights the importance of having privacy notices. A privacy notice will dictate the types of data being collected, proposed, and processed. Hence, data collection and processing for training of AI and ML models should be transparent, fair, and lawful [32]. Otherwise, the models' accuracy and lawfulness will be questioned.

Metaverse environments harness large amounts of data through sensors. The collection of sensitive data in this way has come under scrutiny and has raised serious concerns regarding the privacy of the data subjects. The collection of specific types of data on brain wave patterns, facial expressions, eye movements, hand movements, speech and biometric features, and physiological, physical, biometric, and social interactions with the surrounding environment using sensors is going beyond generic data collection practices. The amount and type of data collected by such a vast platform carries substantial privacy and security concerns, with incidental data breaches having a severe impact on the security and privacy of users. Unlike passwords or other types of data, biometric data breaches will have a greater impact on individuals since they cannot be replaced. The content (including users' personal information) stored in a metaverse platform can be leaked and forged [21]. For example, an avatar's information (like audio and video recording) might be hacked while the user is using the platform or an attacker might forge the avatar and misuse it (similar to taking control of a social media account on Facebook). However, it is hard to establish when to correctly check the security actions required to exercise control over personal information since complex services in the metaverse share various types of private information in real time instead of at a particular moment. Detailed privacy and security issues of the metaverse are discussed in Bentotahewa et al. [21] and Wylde et al. [24].

Sensitive Data Processing and Data Protection Another important aspect of AI and ML models or emerging technologies is that they should take necessary steps to protect individual rights. Most of emerging technologies and innovations venture into areas such as:

- Location tracking.
- Behavior tracking.
- Systematic monitoring of a publicly accessible place.
- Processing of sensitive data (racial or ethnic origin, sexual orientation, political opinions, religious or philosophical beliefs, or trade union membership), genetic data, and biometric data.
- Automated decision-making/automated profiling.
- Data processing of minors.

These types of processing of data using emerging AI and ML technologies could cause adverse impact or harm to individuals. A few such example scenarios are listed below.

- A CV-filtering tool based on AI/ML may block a candidate by error or bias who had an opportunity to be shortlisted.
- A GNOME-based sequencing tool, which identifies kids at a young age who have the potential to be high-performing athletes in the future.
- Face recognition software, which fails to identify people of color.
- A credit score check application (based on AI/ML) rejecting a customer with good financial history.
- Rejecting an application for vehicle insurance through an automated tool.
- Unauthorized tracking of individuals or groups by a business or a government.
- Monitoring of behaviors of individuals or groups without consent (e.g., tools to tackle knife crime).
- Social scoring systems by governments.
- Profiling social media accounts/individuals based on interests and behaviors using AI/ML.
- Recommendation systems based on AI/ML (health recommendations).

The above types of processing may have a significant impact on or pose a "high risk" to individuals' privacy and rights, if we do not take measures to reduce the severity of the impact. Data protection regulations such as EU GDPR advise data controllers and processors to carry out data protection impact assessment (DPIA) to assess the impact and come up with mitigation measures to reduce the impact [30]. These regulations also impose fines and penalties for those not carrying out DPIA when launching new applications like AI and ML (including a fine of up to £8.7 million or a 2% global annual turnover if higher—Data Protection Act (DPA), 2018 UK).

1.3.3 Large Language Models (LLMs) Versus Data Protection

Large language model (LLM)-based applications (e.g., ChatGPT) have significantly impacted society and instigated a large debate on ethics and fair usage of such applications. Currently, these models are being trained on publicly available data, and, thus far, personal data has not been used. However, this situation can change rapidly, if these models start to train on personal data, which would start providing recommendations on individuals and groups. Recently, an interview created by Character.ai on behalf of Michael Schumacher has gone viral due to the nature of the data presented [33]. Even though the data looks authentic, it was not an interview that took place. Similar cases can create huge storms in society and pose high risks to related individuals. Therefore, content owners, controllers, and processors should handle data carefully and should not share the same with any other party who could use them for training LLM models on personal data. If such models come into existence, it will violate individual rights and could cause lasting harm to individuals both mentally and physically.

Misinformation and disinformation pose significant risks to individuals and businesses. These campaigns are common during conflicts. For instance, several disinformation campaigns were reported during the ongoing conflict between Ukraine and Russia. These are aimed at taking undue advantage for their missions and creating panic among the public. AI-generated misinformation and disinformation can creep into personal information as well. Furthermore, fake news and fake profiles can be easily created with LLMs like ChatGPT. Fake news can be used in many scenarios, which could have a significant impact on individuals and the society. For example, it can influence election campaigns and election outcomes. However, increasingly, AI and ML techniques are used to detect fake news as well [34]. Fake profiles can be used in bullying, identity theft, and phishing campaigns. Furthermore, LLM-based Chatbots and SMSbots can be used for scamming the public with a high success rate.

Deepfake Generation Versus Data Protection Increasingly, individuals are sharing personal images and features on social media and for other purposes. These widely available images could be used by generative AI algorithms to create face swaps and deepfake videos. Some researchers argue that fake images or videos cannot be considered personal data since it can no longer be attributed to any individual after modification [35]. According to emerging data protection regulations such as GDPR, personal data does not have to be objective. Subjective information such as opinions, judgments, or estimates can be considered as personal data. Therefore, deepfakes could be considered within the broader classification of personal data as defined in GDPR. On a separate issue, data protection regulations such as GDPR only apply to information that relates to an identifiable living individual, and, hence, information relating to a deceased person does not constitute personal data and is therefore not subject to GDPR. This is an issue when deepfakes of deceased individuals, especially popular politicians, celebrities, or spiritual leaders, are created to provide misleading information. To this end, supplementary laws are being enacted by the legislators (e.g., the privacy act of Hungary) to obtain consent from the heirs of such deceased persons [35]. If we can categorize deepfakes as personal data, processing of personal images or videos needs to be subjected to either informed consent or legitimate interest. If the creators (controllers or processors) have not obtained prior consent, they are violating the GDPR regulation. The consent is twofold in this case: it should be obtained from the person in the original video and from the person in the fabricated video. Users need to be vigilant when using apps such as Facebook, TikTok, or any other emerging face apps and read their privacy notices carefully to investigate their data usage for further processing. The acquired face data may not necessarily be used for creating deepfakes, rather as training data for their AI models. Either way, explicit consent should be enforced by these categories of apps to avoid privacy and security infringements.

AI-generated content is creating havoc in the creative arts sector as well [36]. There are many AI tools that can create songs, poems, music, and animations [37].

These technologies may harm creativity in future generations as they will create copyright issues with the original creators [36].

The use of video fingerprinting and marking the content explicitly as fabricated content would be able to solve this problem to a certain extent. Another popular approach is to probe the provenance of the image and video content to find the origin, for example, conduct a reverse image search to find similar content that has appeared before. Recent research has also suggested that distributed technology solutions such as blockchain can provide the much-needed solution to protect from fake news, disinformation, and deepfakes [38].

Deepfake Detection Versus Data Protection AI and ML algorithms such as generative adversarial networks (GANs) are widely deployed to create realistic-looking deepfakes. The collection of sample data through emerging face apps with or without user consent has increased the effectiveness of these algorithms by having a large training dataset to train these algorithms. Most of the proposed detection methods are also based on AI and ML. However, it will take considerable time to train these detection models, and embedding these complex detection algorithms in real-time application will be a challenge. Similar to generation of deepfakes, detection algorithms need to go through a training phase, which will result in collecting more user data, which would, in turn, infringe user privacy.

1.3.4 AI Regulations and Laws

Moses [39] and Schuett [40] argue that current legal definitions are vague and do not cover the current status of technology development. Furthermore, several other studies have highlighted artificial general intelligence (AGI) systems, which exceed human cognitive abilities [41, 42]. To tackle out-of-control AI algorithms, regulators across the world are enacting laws to support transparent and responsible AI technologies [43, 44]. A summary of such laws is presented in Table 1.1.

There are efforts to recognize AI applications and promote safe and transparent AI technologies. These are led by many organizations such as G20 and UNESCO. OECD.AI provides a live repository of more than 1000 AI policy initiatives from 69 countries, territories, and the EU [45]. Within this framework, countries and stakeholder groups join forces to shape trustworthy AI in the world. In 2019, it formulated the OECD (Organization for Economic Cooperation and Development) AI principles, the first intergovernmental standard on AI. The formulated G20 AI principles are based on these OECD AI principles. Furthermore, UNESCO's first-ever global standard on AI ethics—the "Recommendation on the Ethics of Artificial Intelligence"—was adopted by all its 193 member states in November 2021 [46]. These recommendations will help governments and organizations across the world to evaluate, scope, and drive AI technologies toward transparency and safety. The Bletchley Declaration was signed by the countries attending the AI Safety Summit in the United Kingdom on 1–2 November 2023 [47]. In recognition of the transformative positive potential of AI, and as part of ensuring wider

Table 1.1 Selected AI laws across the world [43]

Country	Regulation	Aims/themes	Other relevant laws
China	Algorithmic recommendation management provisions	Collection of individual legislations to tackle different AI trends	Cybersecurity Law
	Interim measures for the Management of generative AI services		Data Security Law
	Deep synthesis management provisions		Personal Information Protection Law
	AI guidelines and summary of regulations		Shenzhen Special Economic Zone Regulations on AI Industry Promotion
	Scientific and technological ethics regulation		
	New generation AI development plan		
EU	The AI Act (adopted by the EU parliament in June 2023)	Creates harmonized rules for placing AI on the EU market	General Data Protection Regulation
		Applies to the EU and any third-country providers and deployers that place AI systems on the EU market	The Digital Services Act
		Centers around a risk-based approach	The Digital Markets Act
		Prohibits use of certain AI systems and provides specific requirements for high-risk systems	The AI Liability Directive
		Creates harmonized transparency rules for certain AI systems	The EU Cyber Resilience Act
			Ethics Guidelines for Trustworthy AI
			The New Product Liability Directive
USA	Executive orders:	Aims to preserve US leadership in AI research and development as well as control government use of AI	The FTC Act, Section 5
	Maintaining American leadership in AI		The Fair Credit Reporting Act
	Promoting the use of trustworthy AI in the federal government		The Equal Credit Opportunity Act
	The safe, secure, and trustworthy development and use of AI		The Title VII of the Civil Rights Act
	Acts and bills:		The Americans with Disabilities Act
	The AI training act		The Age Discrimination in Employment Act
	The national AI initiative act (Division E, Sec. 5001)		The Fair Housing Act
	The AI in government act (Division U, Sec. 101)		The Genetic Information and Nondiscrimination Act

(continued)

Table 1.1 (continued)

Country	Regulation	Aims/themes	Other relevant laws
India	The proposed Digital India Act (replacing the IT Act 2000)	Advocates for a robust, citizen-centric, and inclusive "AI-for-all" environment	The Information Technology Act
		National strategy for AI, India	The Information Technology Rules
		Hopes to become an "AI garage" for emerging and developing economies, where scalable solutions can be easily implemented and designed for global deployment	The Competition Act
			The Motor Vehicles Act
			The Digital Personal Data Protection Act
			The Copyright Act
			The National e-Governance Plan
Australia	Plans to use existing regulations for AI (however, they are in public consultation on safe and responsible AI)	Plans to build AI capability and accelerate the development and adoption of trusted, secure, and responsible AI technologies in Australia	The Patents Act
		An AI ethics framework	The Copyright Act
			The Privacy Act
			The Data Availability and Transparency Act
			The Consumer Data Right
			The Competition and Consumer Act
Brazil	Has a published AI strategy	Prohibits certain "excessive risk" systems	The General Data Protection Act
	Has drafted an AI bill	Has established a regulatory body to enforce the law	The Civil Rights Framework for the Internet
		Creates civil liability for AI providers	The Consumer Protection Code
		Requires reporting obligations for significant security incidents	
		Guarantees various individual rights, such as explanation, nondiscrimination, rectification of identified biases, and due process mechanisms	

(continued)

Table 1.1 (continued)

Country	Regulation	Aims/themes	Other relevant laws
Mauritius	Has published an AI strategy	Sets out a clear vision for development of AI	Financial Services (Robotic and AI Enabled Advisory Services) Rules
	Other initiatives from the Mauritius government include:		The Data Protection Act
	AI society		The National Cyber Security Strategy
	AI for agricultural projects		The Cybersecurity And Cybercrime Act
			The Industrial Property Act
			The Copyright Act
			The Protection Against Unfair Practices (Industrial Property Rights) Act

international cooperation on AI, the Bletchley Declaration resolves to sustain an inclusive global dialogue that engages existing international fora and other relevant initiatives and contributes in an open manner to broader international discussions and to continue research on frontier AI safety to ensure that the benefits of the technology can be harnessed responsibly for good and for all.

1.3.5 Summary

It is evident that emerging AI and ML algorithms are creating data protection issues for individuals. However, increasingly, both technical and policy measures are initiated to tackle such issues. Emerging AI regulations across the world will provide required policy landscapes, hopefully without limiting the innovation in AI and ML. Such privacy-preserving solutions are elaborated in the next section.

1.4 Privacy-Preserving Solutions for Emerging AI and ML Technologies

1.4.1 Introduction

The rapid advancement of artificial intelligence (AI) and machine learning (ML) technologies has revolutionized numerous industries, including healthcare, finance, and autonomous systems [48, 49]. These technologies offer unprecedented capabilities in data analysis, predictive modeling, and automation. However, the widespread adoption of AI and ML also raises significant privacy concerns. The data used to train and deploy these models often contains sensitive personal information, making privacy preservation a critical issue [50]. Ensuring that AI and ML systems can operate effectively while protecting user privacy is paramount for fostering trust and compliance with regulations like the General Data Protection Regulation (GDPR) and the California Consumer Privacy Act (CCPA) [51].

This chapter explores the landscape of privacy-preserving solutions for AI and ML technologies. It reviews the current literature on the subject, evaluates existing methodologies, and provides recommendations for future developments in this field.

1.4.2 Literature Review

This brief review elaborates on diverse privacy-preserving technologies. The approaches covered in this study are depicted in Fig. 1.1.

1.4.2.1 Differential Privacy

Differential privacy has emerged as a robust standard for privacy preservation in data analysis and machine learning. This concept ensures that the removal or addition of a single data point does not significantly affect the output of an algorithm,

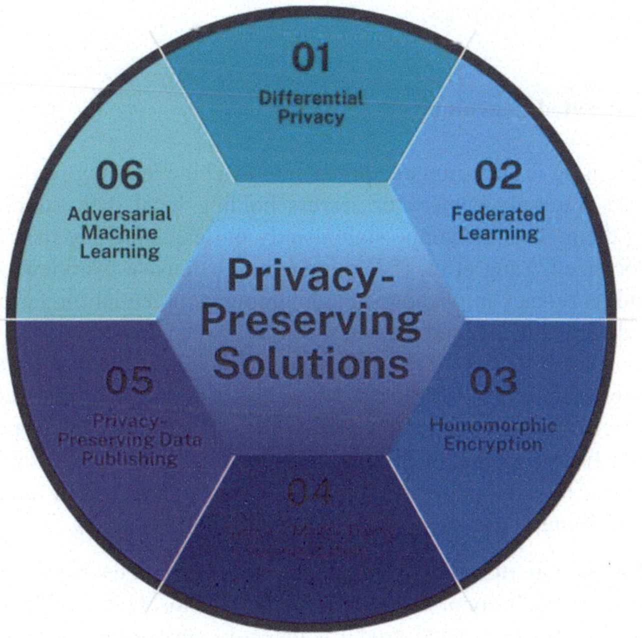

Fig. 1.1 Privacy-preserving solutions

thereby protecting individual privacy. The study by Dwork [52] defined differential privacy and provided a mathematical framework that guarantees a certain level of privacy protection, regardless of the amount of external information an adversary possesses.

Recent studies have focused on improving the practical application of differential privacy in AI and ML. El Ouadrhiri and Abdelhadi [53] discuss a differential privacy framework for federated learning, which introduces noise to the training process, thus preserving user privacy while maintaining model performance. Similarly, Xue et al. [54] developed a method to balance the trade-offs between utility and privacy in differentially private learning algorithms, demonstrating that it is possible to achieve high model accuracy while ensuring privacy.

Applications and Advancements Differential privacy has found applications in various domains, particularly where sensitive data is involved. In healthcare, for example, differential privacy has been employed to enable the sharing of patient data for research without compromising individual privacy. Recent advancements have also focused on reducing the noise added for differential privacy, thereby improving the utility of the data. Phan et al. [55] introduced adaptive noise mechanisms that adjust the amount of noise based on the sensitivity of the query, thus maintaining higher data utility.

Moreover, differentially private synthetic data generation has emerged as a promising area. By generating synthetic datasets that preserve the statistical properties of the original data, organizations can share data more freely while ensuring privacy. Torfi et al. [56] developed a method for generating differentially private synthetic data using GANs, which has shown to be effective in maintaining data utility while protecting privacy.

1.4.2.2 Federated Learning

Federated learning is a distributed approach to machine learning where models are trained across multiple devices or servers holding local data samples, without exchanging them. This technique significantly mitigates privacy risks by keeping raw data localized. Yang et al. [57] provided an extensive overview of federated learning and its privacy implications, highlighting its potential for various applications, including healthcare and finance.

Advancements in federated learning have focused on enhancing privacy guarantees and improving efficiency. For instance, Truex et al. [58] proposed a privacy-preserving federated learning framework that incorporates differential privacy techniques to further secure client data. Additionally, Kang et al. [59] examined the use of federated learning in mobile networks, demonstrating how it can protect user data while enabling the development of robust predictive models.

Security Enhancements and Scalability Recent research has aimed at improving the security and scalability of federated learning systems. One significant challenge is ensuring that the aggregated model updates do not leak sensitive information.

Bonawitz et al. [60] introduced a secure aggregation protocol that prevents the server from learning individual updates, thus enhancing privacy.

Scalability is another critical area. As federated learning systems scale to millions of devices, efficient communication becomes paramount. Advances in communication-efficient federated learning algorithms, such as those proposed by Konečný et al. [61], have significantly reduced the communication overhead, making federated learning feasible for large-scale applications.

1.4.2.3 Homomorphic Encryption

Homomorphic encryption allows computations to be performed on encrypted data without decrypting it, thus preserving privacy throughout the processing pipeline. This technique is crucial for scenarios where sensitive data must be processed without exposing it to potential breaches. Recent advancements have made homomorphic encryption more practical for real-world applications.

In the context of AI and ML, homomorphic encryption enables privacy-preserving training and inference. Mohassel and Zhang [62] demonstrated how linear regression models could be trained on encrypted data, ensuring that sensitive information remains protected. Similarly, Vizitiu et al. [63] applied homomorphic encryption to deep learning, enabling encrypted neural network training and prediction.

Efficiency Improvements One of the major challenges of homomorphic encryption is its computational overhead. Recent studies have focused on optimizing these computations to make homomorphic encryption more practical. Lee et al. [64] proposed a new scheme that significantly reduces the computational complexity of homomorphic operations, making homomorphic encryption feasible for use in deep learning models.

Another promising approach is the use of hybrid schemes that combine homomorphic encryption with other cryptographic techniques to improve efficiency. Das [65] introduced a hybrid approach that combines homomorphic encryption with secure multiparty computation (SMPC), providing enhanced privacy while reducing the computational burden.

1.4.2.4 Secure Multiparty Computation

Secure multiparty computation (SMPC) allows multiple parties to jointly compute a function over their inputs while keeping those inputs private. This method has found applications in privacy-preserving machine learning, where collaborative training is required without sharing raw data. Bonawitz et al. [60] developed an SMPC protocol for federated learning, ensuring that model updates remain confidential. This approach enables collaborative training while preserving data privacy. Moreover, Tran et al. [66] proposed a framework for training machine learning models using SMPC techniques, demonstrating SMPC's feasibility for practical applications in scenarios where data privacy is paramount.

Practical Implementations Practical implementations of SMPC in machine learning have focused on reducing communication overhead and improving computation efficiency. Mohassel and Rindal [67] developed ABY3, a high-performance framework for SMPC that enables efficient training of machine learning models. This framework uses a combination of arithmetic, Boolean, and Yao sharing to optimize performance.

Another practical implementation is the use of SMPC in federated learning systems. Kanagavelu et al. [68] proposed an SMPC-based federated learning framework that ensures privacy-preserving model training across multiple parties, demonstrating its application in financial data analysis.

1.4.2.5 Privacy-Preserving Data Publishing

Privacy-preserving data publishing focuses on releasing data in a way that balances utility and privacy. Techniques such as k-anonymity, l-diversity, and t-closeness have been developed to anonymize datasets, reducing the risk of reidentification. Recent studies have explored new methods to enhance privacy while maintaining data utility.

In AI and ML, these techniques are crucial for creating public datasets that can be used for training without compromising individual privacy. Majeed and Lee [69] provided a comprehensive survey of anonymization methods and their application to data publishing, highlighting the trade-offs and challenges involved. Additionally, Gao [70] proposed a new anonymization technique that improves the balance between data utility and privacy.

1.4.2.6 Adversarial Machine Learning

Adversarial machine learning explores the vulnerabilities of AI and ML models to adversarial attacks and aims to develop robust defenses. Liu et al. [71] reviewed the field, discussing various attack vectors and defense mechanisms. Privacy-preserving adversarial training involves creating models that are resilient to attacks while maintaining privacy guarantees. Bu et al. [72] proposed a technique that combines differential privacy with adversarial training to protect models from both privacy breaches and adversarial examples. This approach enhances the robustness and security of AI systems, making them more reliable in practical applications.

1.4.3 Recommendations and Predictions

1.4.3.1 Enhancing Privacy Techniques

While significant progress has been made in developing privacy-preserving techniques for AI and ML, further enhancements are necessary. Future research should focus on improving the efficiency and scalability of these methods. For instance, optimizing homomorphic encryption schemes to reduce computational overhead will make them more practical for large-scale applications.

Integrating multiple privacy-preserving techniques could offer stronger guarantees. For example, combining differential privacy with federated learning and homomorphic encryption can provide a multilayered approach to privacy protection. These hybrid approaches can address the limitations of individual methods and offer comprehensive privacy solutions.

Research methods to provide explanations for AI models' decisions without compromising data privacy should be implemented. This can involve privacy-preserving techniques for feature importance analysis, interpretable model training, and secure visualizations.

1.4.3.2 Regulatory Compliance and Standardization

As privacy concerns grow, regulatory frameworks like GDPR and CCPA will continue to evolve. AI and ML systems must comply with these regulations, necessitating the development of standardized privacy-preserving techniques. Collaboration between researchers, industry stakeholders, and policymakers will be crucial for establishing these standards.

Standardization will also facilitate interoperability between different systems and technologies, promoting widespread adoption of privacy-preserving methods. Developing benchmarks and evaluation metrics for privacy-preserving AI and ML will enable consistent assessment and comparison of different approaches.

1.4.3.3 Privacy by Design

Incorporating privacy by design principles into AI and ML development processes will ensure that privacy is considered from the outset. This approach involves embedding privacy-preserving techniques into the system architecture, rather than retrofitting them as an afterthought.

Designing AI and ML models with privacy in mind will require interdisciplinary collaboration, bringing together experts in machine learning, cryptography, and data privacy. By fostering a culture of privacy awareness and integrating best practices into the development lifecycle, organizations can build more secure and trustworthy systems.

1.4.3.4 User Empowerment and Transparency

Empowering users with greater control over their data is essential for building trust in AI and ML technologies. Developing user-friendly interfaces that allow individuals to manage their privacy settings and understand how their data is used will enhance transparency.

Transparent AI systems should provide explanations of their decision-making processes, enabling users to understand how their data influences outcomes. This transparency can be achieved through techniques like explainable AI (XAI), which aims to make complex models more interpretable.

1.4.3.5 Quantum-Resistant Privacy Techniques

It is important to explore cryptographic methods that are resistant to quantum attacks to ensure the long-term security of privacy-preserving techniques. This involves researching new algorithms and protocols that can withstand the computational power of quantum computers.

Moreover, investigating the potential of quantum homomorphic encryption for performing computations on encrypted data is essential to provide enhanced privacy for future quantum computing environments.

Future research should explore novel privacy-preserving techniques and their applications to emerging AI and ML technologies. For instance, quantum computing offers potential for developing new cryptographic methods that could enhance privacy protection.

1.4.3.6 Privacy-Preserving Data Sharing and Collaboration

Development of frameworks for secure data marketplaces where data owners can share data with buyers under strict privacy guarantees is an interesting avenue. This includes mechanisms for differential privacy, secure multiparty computation, and blockchain for transparent and secure transactions.

1.4.3.7 Privacy-Preserving AI for Internet of Things (IoT) and Edge Computing

Here, the focus is on privacy-preserving techniques for AI models deployed on edge devices, where data is processed locally. This includes lightweight encryption methods, differential privacy for edge data, and federated learning for edge AI.

Moreover, addressing the privacy challenges posed by new AI paradigms, such as reinforcement learning and generative adversarial networks (GANs), will be important. Investigating how privacy-preserving techniques can be adapted to these contexts will ensure that emerging technologies remain secure and trustworthy.

1.4.4 Summary

Privacy-preserving solutions are essential for the continued growth and acceptance of AI and ML technologies. Techniques such as differential privacy, federated learning, homomorphic encryption, and secure multiparty computation offer robust frameworks for protecting sensitive data while enabling advanced analytics.

Despite significant progress, challenges remain in optimizing these methods for practical use and ensuring compliance with evolving regulatory standards. By adopting privacy by design principles, enhancing transparency, and empowering users, the AI and ML community can build systems that respect privacy and foster trust.

1.5 Conclusions

Emerging AI and ML algorithms offer ways to improve privacy compliance, strengthen encryption, and enable strong threat detection—all of which can improve data protection. These technologies can assist organizations in better protecting confidential data and upholding privacy laws. On the other hand, issues include protecting federated learning systems, guaranteeing differential privacy, and resolving biases in AI models. To effectively use AI and ML in data protection, it is still necessary to strike a balance between innovation and strict data privacy regulations.

Everyone is welcoming the innovation in AI and ML and anticipates their wider impact on the society and economy. However, they do not anticipate that these algorithms would sneak into personal data space and violate individual rights and pose significant risks. Privacy-preserving solutions are crucial for the growth of AI and ML, employing techniques such as differential privacy and federated learning to protect sensitive data. Despite advancements, optimizing these methods for practical use and regulatory compliance remains challenging. Adaptation of privacy by design, enhancing transparency, and empowering users can help build trustworthy AI and ML systems.

The chapters in this book address the challenges and opportunities discussed in this chapter in more detail. Furthermore, they also discuss specific data protection use cases such as smart cities, marketing, and social media content publishing. The authors believe that the remaining chapters of this book will provide a detailed literature review on data protection challenges in the emerging AI era, thus making a valuable contribution to both our knowledge and understanding.

References

1. Sarker, I. H. (2021). Machine learning: Algorithms, real-world applications and research directions. *SN Computer Science, 2*(3), 160.
2. Song, C., Ristenpart, T., & Shmatikov, V. (2017, October). Machine learning models that remember too much. In *Proceedings of the 2017 ACM SIGSAC conference on computer and communications security* (pp. 587–601).
3. Yaseen, A. (2022). Accelerating the SOC: Achieve greater efficiency with AI-driven automation. *International Journal of Responsible Artificial Intelligence, 12*(1), 1–19.
4. Ifekanandu, C. C., Anene, J. N., Iloka, C. B., & Ewuzie, C. O. (2023). Influence of artificial intelligence (AI) on customer experience and loyalty: Mediating role of personalization. *Journal of Data Acquisition and Processing, 38*(3), 1936.
5. Qin, S. J., & Chiang, L. H. (2019). Advances and opportunities in machine learning for process data analytics. *Computers & Chemical Engineering, 126*, 465–473.
6. Leite, M. L., de Loiola Costa, L. S., Cunha, V. A., Kreniski, V., de Oliveira Braga Filho, M., da Cunha, N. B., & Costa, F. F. (2021). Artificial intelligence and the future of life sciences. *Drug Discovery Today, 26*(11), 2515–2526.
7. Fritz-Morgenthal, S., Hein, B., & Papenbrock, J. (2022). Financial risk management and explainable, trustworthy, responsible AI. *Frontiers in Artificial Intelligence, 5*, 779799.
8. Abaku, E. A., Edunjobi, T. E., & Odimarha, A. C. (2024). Theoretical approaches to AI in supply chain optimization: Pathways to efficiency and resilience. *International Journal of Science and Technology Research Archive, 6*(1), 092–107.
9. Ahmad, T., Zhang, D., Huang, C., Zhang, H., Dai, N., Song, Y., & Chen, H. (2021). Artificial intelligence in sustainable energy industry: Status Quo, challenges and opportunities. *Journal of Cleaner Production, 289*, 125834.
10. Pedro, F., Subosa, M., Rivas, A., & Valverde, P. (2019). *Artificial intelligence in education: Challenges and opportunities for sustainable development.*
11. Shah, V. (2021). Machine learning algorithms for cybersecurity: Detecting and preventing threats. *Revista Espanola de Documentacion Cientifica, 15*(4), 42–66.
12. Murdoch, B. (2021). Privacy and artificial intelligence: Challenges for protecting health information in a new era. *BMC Medical Ethics, 22*, 1–5.
13. Ntoutsi, E., Fafalios, P., Gadiraju, U., Iosifidis, V., Nejdl, W., Vidal, M. E., & Staab, S. (2020). Bias in data-driven artificial intelligence systems – An introductory survey. *Wiley Interdisciplinary Reviews: Data Mining and Knowledge Discovery, 10*(3), e1356.
14. Fischer, B., & Piskorz-Ryń, A. (2021). Artificial intelligence in the context of data governance. *International Review of Law, Computers & Technology, 35*(3), 419–428.
15. Waltl, B. (2021). Explainability and transparency of machine learning in ADM systems. In *Research handbook on big data law* (pp. 341–356). Edward Elgar Publishing.
16. Safdar, N. M., Banja, J. D., & Meltzer, C. C. (2020). Ethical considerations in artificial intelligence. *European Journal of Radiology, 122*, 108768.
17. Meurisch, C., & Mühlhäuser, M. (2021). Data protection in AI services: A survey. *ACM Computing Surveys (CSUR), 54*(2), 1–38.
18. Wachter, S., & Mittelstadt, B. (2019). A right to reasonable inferences: Re-thinking data protection law in the age of big data and AI. *Columbia Business Law Review*, 494.
19. Hacker, P. (2023). Manipulation by algorithms. Exploring the triangle of unfair commercial practice, data protection, and privacy law. *European Law Journal, 29*(1–2), 142–175.
20. Ukwandu, E., Hewage, C., & Hindy, H. (2024). Editorial: Cyber security in the wake of fourth industrial revolution: Opportunities and challenges. *Frontiers in Big Data, 7.*
21. Bentotahewa, V., Khattak, S., Hewage, C., Sengar, S. S., & Jenkins, P. (2024). Privacy and security landscape of Metaverse. In N. Naik, P. Jenkins, P. Grace, L. Yang, & S. Prajapat (Eds.), *Advances in computational intelligence systems. UKCI 2023. Advances in intelligent systems and computing* (Vol. 1453). Springer. https://doi.org/10.1007/978-3-031-47508-5_32

22. Hewage, C., Rahulamathavan, Y., & Ratnayake, D. (Eds.). (2023). *Data protection in a post-pandemic society: Laws, regulations, best practices and recent solutions*. Springer.
23. Hewage, C. T., Khattak, S. K., Ahmad, A., Mallikarachchi, T., Ukwandu, E., & Bentotahewa, V. (2022). *Multimedia privacy and security landscape in the wake of AI/ML* (pp. 203–228). Social Media Analytics, Strategies and Governance.
24. Wylde, V., Prakash, E., Hewage, C., & Platts, J. (2023). Post-Covid-19 Metaverse cybersecurity and data privacy: Present and future challenges. In C. Hewage, Y. Rahulamathavan, & D. Ratnayake (Eds.), *Data protection in a post-pandemic society*. Springer. https://doi.org/10.1007/978-3-031-34006-2_1
25. Wylde, V., Prakash, E., Hewage, C., & Platts, J. (2023). Ethical challenges in the use of digital technologies: AI and big data. In R. Montasari, V. Carpenter, & A. J. Masys (Eds.), *Digital transformation in policing: The promise, perils and solutions. Advanced sciences and technologies for security applications*. Springer. https://doi.org/10.1007/978-3-031-09691-4_3
26. Wylde, V., Prakash, E., Hewage, C., & Platts, J. (2022). The use of AI in managing big data analysis demands: Status and future directions. In R. Montasari (Ed.), *Artificial intelligence and National Security*. Springer. https://doi.org/10.1007/978-3-031-06709-9_3
27. Bentotahewa, V., Yousif, M., Hewage, C., Nawaf, L., & Williams, J. (2022). Privacy and security challenges and opportunities for IoT technologies during and beyond COVID-19. In R. Montasari, F. Carroll, I. Mitchell, S. Hara, & R. Bolton-King (Eds.), *Privacy, security and forensics in the internet of things (IoT)*. Springer. https://doi.org/10.1007/978-3-030-91218-5_3
28. Bentotahewa, V., Hewage, C., & Williams, J. (2021). Solutions to big data privacy and security challenges associated with COVID-19 surveillance systems. *Frontiers in Big Data, 4*.
29. Bentotahewa, V., Hewage, C., & Williams, J. (2021). Security and privacy issues associated with coronavirus diagnosis and prognosis. In S. Paiva, S. I. Lopes, R. Zitouni, N. Gupta, S. F. Lopes, & T. Yonezawa (Eds.), *Science and technologies for smart cities. Smart City 360° 2020* (Lecture Notes of the Institute for Computer Sciences, Social Informatics and Telecommunications Engineering) (Vol. 372). Springer. https://doi.org/10.1007/978-3-030-76063-2_8
30. GDPR. (2018). *General data protection regulation (GDPR). [online]*. General Data Protection Regulation (GDPR). Available at: https://gdpr-info.eu/
31. Bentotahewa, V., Hewage, C., & Williams, J. (2022). The normative power of the GDPR: A case study of data protection laws of South Asian countries. *SN Computer Science, 3*, 183. https://doi.org/10.1007/s42979-022-01079-z
32. Wylde, V., Rawindaran, N., Lawrence, J., et al. (2022). Cybersecurity, data privacy and Blockchain: A review. *SN Computer Science, 3*, 127. https://doi.org/10.1007/s42979-022-01020-4
33. Magazine editor sacked over Schumacher AI 'interview'. (n.d.). *BBC Sport* [online]. Available at: https://www.bbc.co.uk/sport/formula1/65361193
34. Zhou, X., & Zafarani, R. (2020). A survey of fake news: Fundamental theories, detection methods, and opportunities. *ACM Computing Surveys (CSUR), 53*(5), 1–40.
35. Hewage, C., Infosecurity Magazine. (n.d.). *Data protection in the wake of deepfakes* [online] Available at: https://www.infosecurity-magazine.com/next-gen-infosec/data-protection-wake-deepfakes/
36. Abbott, R., & Rothman, E. (2023). Disrupting creativity: Copyright law in the age of generative artificial intelligence. *Florida Law Review, 75*, 1141.
37. Wang, L., Zhao, Z., Liu, H., Pang, J., Qin, Y., & Wu, Q. (2024). A review of intelligent music generation systems. *Neural Computing and Applications*, 1–21.
38. Heidari, A., Navimipour, N. J., Dag, H., Talebi, S., & Unal, M. (2024). A novel blockchain-based deepfake detection method using federated and deep learning models. *Cognitive Computation*, 1–19.
39. Moses, L. B. (2007). Recurring dilemmas: The law's race to keep up with technological change. *University of Illinois Journal of Law, Technology, & Policy*, 239.
40. Schuett, J. (2023). Defining the scope of AI regulations. *Law, Innovation and Technology, 15*(1), 60–82.

41. Cichocki, A., & Kuleshov, A. P. (2021). Future trends for human-AI collaboration: A comprehensive taxonomy of AI/AGI using multiple intelligences and learning styles. *Computational Intelligence and Neuroscience, 2021*(1), 8893795.
42. Korteling, J. H., van de Boer-Visschedijk, G. C., Blankendaal, R. A., Boonekamp, R. C., & Eikelboom, A. R. (2021). Human-versus artificial intelligence. *Frontiers in Artificial Intelligence, 4*, 622364.
43. Anon. (n.d.). *Global AI Legislation Tracker* [online]. Available at: https://iapp.org/resources/article/global-ai-legislation-tracker/
44. Sheehan, M. (2023). China's AI regulations and how they get made. *Horizons: Journal of International Relations and Sustainable Development, 24*, 108–125.
45. oecd.ai. (n.d.). *The OECD Artificial Intelligence Policy Observatory* [online]. Available at: https://oecd.ai/
46. www.unesco.org. (n.d.). Recommendation on the Ethics of Artificial Intelligence | UNESCO [online]. Available at: https://www.unesco.org/en/articles/recommendation-ethics-artificial-intelligence
47. GOV.UK. (2023). The Bletchley Declaration by Countries Attending the AI Safety Summit, 1–2 November 2023 [online]. GOV.UK. Available at: https://www.gov.uk/government/publications/ai-safety-summit-2023-the-bletchley-declaration/the-bletchley-declaration-by-countries-attending-the-ai-safety-summit-1-2-november-2023
48. Cioffi, R., Travaglioni, M., Piscitelli, G., Petrillo, A., & De Felice, F. (2020). Artificial intelligence and machine learning applications in smart production: Progress, trends, and directions. *Sustainability, 12*(2), 492.
49. Sahu, M., Gupta, R., Ambasta, R. K., & Kumar, P. (2022). Artificial intelligence and machine learning in precision medicine: A paradigm shift in big data analysis. *Progress in Molecular Biology and Translational Science, 190*(1), 57–100.
50. Sun, Y., Liu, J., Wang, J., Cao, Y., & Kato, N. (2020). When machine learning meets privacy in 6g: A survey. *IEEE Communications Surveys & Tutorials, 22*(4), 2694–2724.
51. Voigt, P., & Von dem Bussche, A. (2017). *The EU general data protection regulation (GDPR): A practical guide* (Vol. 10(3152676), 1st ed., pp. 10–5555). Springer.
52. Dwork, C. (2006). Differential privacy. In *International colloquium on automata, languages, and programming* (pp. 1–12). Springer.
53. El Ouadrhiri, A., & Abdelhadi, A. (2022). Differential privacy for deep and federated learning: A survey. *IEEE Access, 10*, 22359–22380.
54. Xue, R., Xue, K., Zhu, B., Luo, X., Zhang, T., Sun, Q., & Lu, J. (2023). Differentially private federated learning with an adaptive noise mechanism. *IEEE Transactions on Information Forensics and Security, 19*, 74.
55. Phan, N., Wu, X., Hu, H., & Dou, D. (2017). Adaptive Laplace mechanism: Differential privacy preservation in deep learning. In *In 2017 IEEE international conference on data mining (ICDM)* (pp. 385–394). IEEE.
56. Torfi, A., Fox, E. A., & Reddy, C. K. (2022). Differentially private synthetic medical data generation using convolutional GANS. *Information Sciences, 586*, 485–500.
57. Yang, Q., Liu, Y., Chen, T., & Tong, Y. (2019). Federated machine learning: Concept and applications. *ACM Transactions on Intelligent Systems and Technology (TIST), 10*(2), 1–19.
58. Truex, S., Baracaldo, N., Anwar, A., Steinke, T., Ludwig, H., Zhang, R., & Zhou, Y. (2019). A hybrid approach to privacy-preserving federated learning. In *Proceedings of the 12th ACM workshop on artificial intelligence and security* (pp. 1–11).
59. Kang, J., Xiong, Z., Niyato, D., Zou, Y., Zhang, Y., & Guizani, M. (2020). Reliable federated learning for mobile networks. *IEEE Wireless Communications, 27*(2), 72–80.
60. Bonawitz, K., Ivanov, V., Kreuter, B., Marcedone, A., McMahan, H. B., Patel, S., Ramage, D., Segal, A., & Seth, K. (2017). Practical secure aggregation for privacy-preserving machine learning. In *Proceedings of the 2017 ACM SIGSAC conference on computer and communications security* (pp. 1175–1191).

61. Konečný, J., McMahan, H. B., Ramage, D., & Richtárik, P. (2016). Federated optimization: Distributed machine learning for on-device intelligence. *arXiv preprint arXiv*, 1610.02527.
62. Mohassel, P., & Zhang, Y. (2017). SecureML: A system for scalable privacy-preserving machine learning. In *2017 IEEE symposium on security and privacy (SP)* (pp. 19–38). IEEE.
63. Vizitiu, A., Niţă, C. I., Puiu, A., Suciu, C., & Itu, L. M. (2020). Applying deep neural networks over homomorphic encrypted medical data. *Computational and Mathematical Methods in Medicine, 2020*(1), 3910250.
64. Lee, S., Lee, G., Kim, J. W., Shin, J., & Lee, M. K. (2023). Hetal: Efficient privacy-preserving transfer learning with homomorphic encryption. In *International conference on machine learning* (pp. 19010–19035). PMLR.
65. Das, D. (2018). Secure cloud computing algorithm using homomorphic encryption and multi-party computation. In *2018 international conference on information networking (ICOIN)* (pp. 391–396). IEEE.
66. Tran, A. T., Luong, T. D., Karnjana, J., & Huynh, V.-N. (2021). An efficient approach for privacy preserving decentralized deep learning models based on secure multi-party computation. *Neurocomputing, 422*, 245–262.
67. Mohassel, P., & Rindal, P. (2018). ABY3: A mixed protocol framework for machine learning. In *Proceedings of the 2018 ACM SIGSAC conference on computer and communications security* (pp. 35–52). ACM.
68. Kanagavelu, Z., Li, Z., Samsudin, J., Yang, Y., Yang, F., Goh, R. S. M., & Wang, S. (2020). Two-phase multi-party computation enabled privacy-preserving federated learning. In *2020 20th IEEE/ACM international symposium on cluster, cloud and internet computing (CCGRID)* (pp. 410–419). IEEE.
69. Majeed, A., & Lee, S. (2020). Anonymization techniques for privacy preserving data publishing: A comprehensive survey. *IEEE Access, 9*, 8512–8545.
70. Gao, S., Ma, J., Sun, C., & Li, X. (2014). Balancing trajectory privacy and data utility using a personalized anonymization model. *Journal of Network and Computer Applications, 38*, 125–134.
71. Liu, J., Nogueira, M., Fernandes, J., & Kantarci, B. (2021). Adversarial machine learning: A multilayer review of the state-of-the-art and challenges for wireless and mobile systems. *IEEE Communications Surveys & Tutorials, 24*(1), 123–159.
72. Bu, Z., Li, P., & Zhao, W. (2021). Practical adversarial training with differential privacy for deep learning. *arXiv preprint arXiv*, 2103.09298.

Chapter 2
Death Becomes Data

Nisha Rawindaran and Vibhushinie Bentotahewa

Abstract Traditionally, data protection safeguards have always revolved around the living. However, as our digital footprints extend beyond mortality, a new frontier has emerged to protect the personal information of deceased individuals. This chapter investigates this critical, multifaceted need, moving beyond legal mandates to explore the ethical, social, and technological implications. Protecting the deceased's data is not just about compliance, it is the ability to highlight respect above all. The departed are dutifully owed their right to privacy, preventing their digital selves from being exploited or manipulated. Their online presence, often imbued with memories and personal details, is an extension of their identity, deserving protection even after their passing. Safeguarding this data also preserves legacies, ensuring that the deceased's values and contributions are faithfully represented. For grieving families, the unauthorised use of a loved one's data can be deeply hurtful. Intrusions can exacerbate their pain of loss, hindering their journey of grief. Conversely, responsible data management can offer solace, allowing access to cherished memories and digital fragments of their loved one's life. However, safeguarding posthumous data presents complex challenges. The rise of artificial intelligence (AI), machine learning (ML), and the metaverse (MV) raises questions about ownership, consent, and control over digital representations. Existing legal frameworks often struggle to keep pace, demanding revisions to ensure robust data protection for the deceased. Navigating this digital afterlife requires a comprehensive approach. Clear legal frameworks with enforcement mechanisms in place are vital. However, legal solutions alone are insufficient. Ethical awareness of and respect for the deceased's digital legacy are equally crucial. Technological developers must prioritise adopting privacy-centric design principles and transparency, providing responsible control

N. Rawindaran (✉)
University of South Wales, Newport, UK
e-mail: nisha.rawindaran@southwales.ac.uk

V. Bentotahewa
Cybersecurity and Information Networks Centre (CINC), Cardiff School of Technologies, Cardiff Metropolitan University, Cardiff, UK
e-mail: vibentotahewa@cardiffmet.ac.uk

C. Hewage et al. (eds.), *Data Protection*,
https://doi.org/10.1007/978-3-031-76473-8_2

over data even after life. Protecting the deceased's data in the digital age is not simply a technical or legal issue. It is about upholding human dignity, respecting legacies, and acknowledging the delicate balance between privacy and the evolving online world. As this chapter delves deeper into the digital afterlife, ensuring responsible data management of the deceased's data, this remains an essential step towards a more respectful and ethically grounded online reality, for both the living and the departed.

2.1 Introduction

In the digital age, where personal information often permeates the online environment, data protection has become a fundamental human right. Yet, traditional safeguards primarily focus on the living, leaving the personal data of deceased individuals in a precarious and often unregulated space. However, it is essential to examine the importance of managing data of deceased individuals in order to uphold their dignity. Additionally, measures aimed at safeguarding the data of deceased individuals are crucial for preventing identity theft and fraudulent activities. Overall, governmental efforts to protect the digital legacy of deceased individuals will demonstrate the state's responsibility to protect and ensure ethical treatment of the data of its citizens both during their lifetime and after their passing. The importance of safeguarding the privacy of deceased individuals has been highlighted in an article written by Buitelaar, J.C. titled *Post-mortem privacy and informational self-determination*. This article focuses on the privacy of Internet users and discusses the significance of post-mortem privacy protection [1].

This article argues that data protection for the deceased is not simply a legal imperative but a multifaceted necessity informed by ethical considerations, the delicate balance between honouring legacies and respecting private lives, the emotional well-being of grieving families, and the complexity of evolving technology and privacy rights. While traditional data protection mechanisms revolve around living individuals, data protection of deceased people has begun to gain prominence, as societies have increased their digitalisation and personal information has become more accessible online, and goes beyond more than just a legal obligation. It encompasses ethical considerations, preservation of legacy, the emotional impact on grieving families, technological challenges, and the need for accountability. Striking a balance between respecting the privacy of the deceased and addressing the evolving nature of digital technologies is crucial for navigating this complex digital landscape. As society grapples with privacy challenges, a thoughtful and comprehensive approach is essential for upholding the principles of dignity, respect, and empathy for the deceased and their loved ones. In this chapter, the researchers will delve into the significance of safeguarding the personal data of deceased people, of existing legal frameworks, of exploring the social, legal, ethical, political, and technical implications, and of the evolving role of the data protection mechanisms in managing the privacy of deceased people.

2.2 Literature Review

As our lives increasingly intertwine with the digital world, the question of what happens to our online presence after death becomes ever more pressing. The realm of 'post-mortem privacy', coupled with complex ethical and legal dilemmas, challenges us to reimagine the boundaries of privacy and control beyond the final curtain. At the forefront of this digital frontier lies the burgeoning field of legal scholarship. In their seminal work, Schafer et al. [2] argue that protecting the deceased's privacy is not solely a matter of individual dignity but also crucial for fostering intergenerational trust. Unchecked digital footprints, they warn, can become harmful 'data breadcrumbs' that expose past generations to unintended scrutiny and judgement. This echoes concerns raised by Harbinja et al. [3] who highlight the unsettling rise of autonomous artificial intelligence (AI) agents programmed to mimic the deceased on social media. These 'ghostbots' blur the lines between remembrance and appropriation, raising questions about consent, agency, and the ethical implications of simulating a digital doppelganger.

Kohl [4] expands this perspective, suggesting that grappling with the complexities of digital afterlife illuminates the fragility of privacy rights even for the living. The lessons learned, he argues, can inform broader discourses on data ownership, control, and the need for robust legal frameworks to safeguard our digital selves. Beyond purely legal concerns, philosophers also weigh in on the ethical complexities. Fabre [5] explores the concept of 'posthumous rights', challenging the traditional focus on individual rights tied to physical embodiment. He argues for a nuanced understanding of rights that extends beyond the mortal coil, potentially granting deceased individuals some degree of control over their digital legacy. For popular media, Burkeman [6] offers a practical guide to empower individuals to take proactive steps towards managing their online afterlife, outlining strategies for data deletion, account closure, and designation of digital executors.

However, navigating this uncharted territory is not without its challenges. Articles like *What Happens to Your Digital Footprint When You Die?* [7], published by online resource platforms such as IFLScience, provide accessible overviews but often lack the depth needed for informed decision-making. Similarly, sensational headlines like O′Grady's 'Metaverse: Zuckerberg says technology could bring back the dead' [8] risk blurring the lines between speculative fiction and technological feasibility, further complicating public understanding. While the legal landscape remains in flux, earlier works of Danziger et al. [9] provide valuable groundwork. This study identifies the ethical and legal vacuum surrounding post-mortem privacy, urging policymakers to actively engage in crafting legislation that protects the deceased without infringing on the rights of the living. Livingstone [10] echoes this sentiment for people going to court over dead family members' Facebook pages to apply for post-mortem privacy, which, in turn, highlights real-world legal battles emerging from digital afterlife disputes. She emphasises the need for clear regulatory frameworks that empower families to manage the deceased's online presence while respecting individual privacy rights. Looking ahead, Choi [11] explores options using blockchain to suggest the technology's ability to simplify the complex process of navigating the digital afterlife for families by offering a glimpse into potential technological solutions and to manage digital inheritances and enforce post-mortem data deletion protocols.

In conclusion, the question of post-mortem privacy stands at the crossroads of technology, ethics, and law. With research moving forward, a multifaceted approach is crucial. Robust legal frameworks, informed public discourse, and responsible technological development are all essential for protecting the digital legacies of the deceased while safeguarding the rights of the living. Only then can navigation of the intricate landscapes of the digital afterlife be completed with respect, clarity, and justice.

2.3 Legal Context

The recognition and protection of the privacy rights of deceased individuals plays a crucial role in maintaining social and emotional well-being. Acknowledging and respecting the same is essential for cultural and historical preservation. Rituals and traditions are highly valued in many societies, and upholding these customs contributes to the preservation of a community's cultural identity and legacy. By protecting the right to privacy of the deceased, communities make sure that their cultural heritage lives on, preserving traditions and ideals for future generations. Furthermore, it is imperative to guarantee that individuals receive dignified and equitable treatment even after their demise. Legal frameworks should be established to prevent any mistreatment or abuse, acknowledge and safeguard privacy rights, and ensure the dignified handling of the data of deceased individuals.

The rights of deceased individuals sometimes intersect with broader societal issues. In the wake of a loved one's passing, families often grapple with grief while navigating the complexities of handling the deceased person's digital assets. Accessing and managing these assets, from sentimental emails to financial records, can be emotionally challenging. Therefore, careful consideration and legal clarity are required to strike a balance between granting bereaved families the access they need and safeguarding the deceased's privacy. Beyond the cold grip of legal frameworks lies family, grief, and technology. This depicts the impact of data protection on the deceased, not merely as a matter of rules but as a vital element for safeguarding cherished memories, navigating the complexities of digital evolution, and alleviating the anguish of grieving families. For families struggling with the raw ache of loss, unauthorised access or misuse of a deceased loved one's data can be akin to tearing at the very fabric of their grief. Intimate details, precious memories, and even financial or medical information, when exposed or manipulated, can exacerbate their already profound pain of loss.

Imagine, for a moment, the chilling prospect of a deceased parent's social media account being impersonated or their online shopping habits used to target grieving family members with intrusive advertising. Such invasions not only violate the deceased's privacy but also inflict fresh wounds on those already struggling to heal. Responsible data management can offer solace and hope, even amidst the darkness of grief. Access to a loved one's digital memories such as shared photos, heartfelt messages, or even home videos can provide a vital link to the past, fostering connection and remembrance, as clearly outlined by Keyes et al. [12]. Family members may also be frustrated by the exposure of the deceased person's privacy and their inability to stop information from being shared. This may also damage people's faith in organisations and systems designed to protect data. This damage to trust might go beyond the data breach/leak and increase worries about the security and privacy of living individuals. Family members may start worrying about who has access to data and how they might use it. This may include concerns such as potential identity theft or using personal information for malicious purposes. Concerns over susceptibility and privacy may become more acute when financial or medical records are among the information that has been compromised.

Meanwhile, Taylor [13] noted that having the data rights of the deceased can sometimes 'facilitate the grieving process' of the surviving family as they can access cherished memories and maintain a connection with the deceased if they were given prior consent as per the deceased's last wishes. A carefully curated digital memorial can sometimes provide comfort and serve as a reminder of the departed's life and legacy. As the digital landscape spins into ever-unimaginable forms with artificial intelligence (AI) weaving personalised avatars, machine learning (ML) shaping digital legacies, and the metaverse (MV) promising immersive, virtual encounters, data ownership and control in the context of the deceased is becoming increasingly complex.

Consequently, this leads to questions such as Who owns the digital representations of the departed? Under what conditions and with whom can the data of deceased individuals be shared? Who holds the right to consent to their usage? How is the data of deceased individuals stored, processed, and secured to prevent unauthorised access or leaks? What are the possible long-term repercussions of improper management of the data of deceased people? How can we ensure respect and authenticity in a realm where boundaries between reality and simulation blur? Here, existing legal frameworks often struggle to keep pace with the breakneck speed of technological innovation and lead to inadequate answers.

Floridi [14], in dealing with the ethical and legal issues raised by the emerging possibilities of digital afterlives, much explored the legal system. Gaps and ambiguities leave families and loved ones vulnerable to the potential misuse of the deceased's digital identities. Navigating this multifaceted challenge requires a comprehensive approach, weaving together legal, ethical, technological, and social considerations. Robust legal frameworks with clear guidelines, consent mechanisms, and strong enforcement are crucial. Yet, legal solutions alone cannot heal the delicate threads of grief or guarantee responsible data management, as expressed by Floridi in their research.

Ethical awareness of and a culture of respect for the deceased's digital legacy are equally vital. Developers must prioritise privacy-centric design principles and transparency, ensuring clear ownership rights and control over data even beyond life.

Moreover, public education and social discourse are essential for fostering a collective understanding of the importance of protecting the data of the deceased and the potential repercussions of misuse. This is not simply a matter of technical challenge or legal compliance; it is about bringing together respect, empathy, and technological responsibility. It is also about honouring the memories of the departed, safeguarding their legacies, and offering solace to grieving families as they navigate the complex terrain of the digital afterlife. Taking this approach ensures that the stories, identities, and legacies of the deceased remain protected and valued, even after the final thread of life has been spun.

The legal landscape regarding data protection for deceased individuals varies globally. Some authorities treat the deceased person's privacy as an extension of the individual's living rights, while others consider it a matter of inheritance law [15]. Therefore, establishing clear legal frameworks is important to provide guidance on handling personal data after death and prevent unauthorised access and misuse of the data. Mishandling or unauthorised use of a deceased person's data can exacerbate the emotional burden on the grieving family. Therefore, respecting data protection in this context is a legal, ethical, and compassionate approach towards supporting individuals coping with loss.

In line with the current data protection mechanisms, establishing clear lines of accountability and transparency to ensure that personal data is handled according to the deceased person's wishes and legal frameworks is important. Integrating transparent practices and secure storage to prevent unauthorised access, obtaining explicit consent for addressing disputes, and maintaining data integrity would help stop people from stealing deceased people's information to create fake profiles and IDs.

It is imperative that the guidelines are clear and offer transparency to be able to identify the gaps in the legal framework for accessing the data of the deceased. As it currently stands, the General Data Protection Regulation (GDPR) and a variety of national laws governing post-mortem data are covered and discussed in this chapter. While the GDPR does not explicitly regulate post-mortem data, Recital 27 [16] states that it 'does not apply to the personal data of deceased persons'. However, member states can introduce national laws governing the processing of deceased persons' data under Recital 27 of the GDPR,[1] Article 9.2(j). Cultures have different norms, beliefs, and traditions about the afterlife, privacy, and family rights. Therefore, encouraging member states to enact national laws to govern the processing and storing of deceased people's data would allow them to create customised laws in line with their national legislations. This also ensures uniformity, transparency, and clarity across all jurisdictions, guaranteeing that the data of deceased people is managed and protected according to established policies and procedures to reduce potential uncertainty and conflicts. In addition, this also helps individual nations create and update their national laws, their legal framework, and their policies to accommodate evolving technological changes and address any privacy

[1] https://gdpr-info.eu/art-27-gdpr/

concerns effectively over time. In order to allow for the processing for archiving purposes in the public interest, scientific or historical research purposes or statistical purposes), Guidelines 06/2020 on Article 6(4) of the GDPR[2] specifically mentions deceased persons in relation to the 'balancing of interests' provision).

Some European Union (EU) member states have taken steps to extend the definitions of personal data and ensure the privacy rights of deceased individuals. The Swedish Data Protection Act defines personal data as all kinds of information that, directly or indirectly, relate to a natural person, specifically mentioning the personal data of living people (DLA Piper global law firm. ND. Definition, Definition in Sweden). Likewise, data that relates to a living individual is defined as personal data under the United Kingdom (UK) Data Protection Act (DLA Piper global law firm. ND. Definition, Definition in UK). Likewise, the Estonian Data Protection Act[3] (Article 44, Section 2) grants heirs the right to access, rectify, and restrict the processing of the deceased's personal data. Unless the data subject specifies otherwise, the consent is legitimate for the duration of the data subject's life and for 10 years following the data subject's passing. The consent of the data subject is valid for 20 years following his or her death if the subject passed away at an early age. With the approval of one of the data subject's heirs, processing the data subject's personal information after their death is permitted (DLA Piper global law firm. ND. Collection & Processing, Collection & Processing in Estonia).

Reflecting on France's Digital Republic Act, this allows individuals to designate a digital executor to manage their online accounts and data after death[3] (Article 37). Equally, Catalonia's Data Protection Act recognises the right to digital wills, enabling individuals to specify how their data should be handled posthumously.[4] Italy's Data Protection Code formally specifies that personal data of deceased persons can only be processed for specific purposes, such as historical, statistical, or scientific research.[5]

The courts in Germany, for instance, have occasionally upheld the deceased's non-commercial and commercial interests within the EU [17]. However, the Court of Cassation in France decided that 'the right to act in respect of privacy disappears when the person in question, the sole holder of that right, dies' [17]. In addition, there is a lack of evidence to support post-mortem privacy protection in Article 8 of the European Convention on Human Rights; instead, it grants protection only to living individuals [18]. The Court has often declined to acknowledge this right for the deceased unless there is a connection between their privacy and that of the living [18].

State and federal laws in the United States restrict access to patients' medical records, whether they are alive or deceased [19]. According to the Health Insurance Portability and Accountability Act of 1996 (HIPAA) [20], a living individual's

---

[2] https://eur-lex.europa.eu/eli/reg/2016/679/oj

[3] France's Digital Republic Act: https://www.legifrance.gouv.fr/jorf/id/JORFTEXT000046113741

[4] Catalonia's Data Protection Act: https://www.boe.es/buscar/pdf/2006/BOE-A-2006-1769-consolidado.pdf (Article 30)

[5] Italy's Data Protection Code: https://www.garanteprivacy.it/web/garante-privacy-en (Article 84).

medical records can be accessed by authorised people only if there is a legitimate reason to do so (US Department and Health Services. ND. Summary of the HIPAA Privacy Rule). There is very little guidance about post-mortem confidentiality in HIPPA. However, the act protects the personally identifiable information for 50 years following the date of the individual's death (US Department and Health Services. ND. Health Information of Deceased Individuals). During that period, the medical records of the deceased can usually be obtained by the personal representative of the decedent (US Department and Health Services. ND. Health Information of Deceased Individuals).

Contractual frameworks can also differ in this context. Terms of service (ToS) and privacy policies of online platforms often include provisions for addressing data handling after death, such as account memorialisation or deletion options. They can offer analysis of specific examples from major platforms like Facebook, Google, and Twitter; however, these can vary significantly from one social media platform to the next. Digital inheritance platforms could also offer tools to individuals to create digital wills or designate data custodians to manage their online assets after death. Everplans, Legacy Locker, and Entrust are just some of the major commercial digital inheritance platforms that could offer their insights into what tools are currently offered or can be offered.

Whilst contracts are bounded by law, ethical frameworks are driven by social boundaries. Guidelines from data protection authorities (DPAs) offer ethical considerations for handling deceased persons' data, such as balancing privacy rights with the interests of family members and researchers. Specific DPA guidelines review documents from the Article 29 Working Party or individual DPAs like the UK Information Commissioner's Office (ICO) and the Spanish Agency for Data Protection (AEPD) to paint a picture that offers better clarity and explanation where necessary Additionally, professional codes of conduct for data processors may also include provisions for respecting the privacy of deceased persons. These professional codes of conduct will examine relevant codes from external organisations like the International Association of Privacy Professionals (IAPP) or national data protection associations for formulating conclusions on the matter.

Additional considerations for specific individuals and groups such as children and other vulnerable groups pose unique challenges, such as balancing parental rights with the child's right to privacy and providing special protection to vulnerable groups (e.g. minorities, victims of crime) to prevent their data from being misused. Children and Deceased Data Protection act refer to resources from United Nations International Children's Emergency Fund (UNICEF), the United Nations (UN) Committee on the Rights of the Child, or child-focused data protection advocacy groups. Likewise, vulnerable groups will need consultation from human rights organisations, digital divide studies, and research specific to their data concerns.

2.4 Ethical Context

In an increasingly digital age, personal data may include basic information (name, email addresses, IP addresses, etc.), memories, photographs, and communications. Therefore, people should be treated with respect even after they pass away, and their private information should be handled carefully. This helps maintain the deceased person's legacy and memory and the faith they had in institutions during their lifetime. Beyond what is required by law, ethical considerations highlight the importance of treating the privacy of deceased people with the same respect as the living.

The ethical considerations surrounding the privacy rights of the deceased people are further heightened by the rise of technologies like deepfakes. The potential use of a deceased person's images, biometrics, and voice raises significant concerns. For example, the director of the film *Roadrunner* used artificial intelligence to effectively bring late celebrity chef Anthony Bourdain's voice back to life, and there have been some critics over the unannounced use of a deepfake voice to say sentences that Bourdain never spoke [21]. To mitigate these risks, ethical guidelines must be established, emphasising the responsible use of technologies involving deceased people. Striking a balance between innovation and ethical considerations is imperative to prevent the exploitation of the deceased for unauthorised purposes.

On the other hand, Meta chief Mark Zuckerberg has said that the company's metaverse platform may one day help people interact with the virtual avatars of their deceased loved ones [22]. The metaverse is a digital universe where users can interact with computer-generated environments and other users in real time [23]. This concept involves using dozens of datasets from people's lives. The re-creation of deceased people in virtual environments raises profound ethical and moral questions about privacy, consent, and the boundaries of technology.

A major concern is what would happen if the deceased individuals had not provided explicit consent for the use of their digital existence after their death. In such a context, the use of the deceased individual's personal information, without proper safeguards, could violate their privacy even after death. This might not directly impact the deceased person; however, there could be detrimental impacts on the surviving family members and relatives. Moreover, while some people may find comfort in interacting with a virtual image of a loved one who has passed away, for others, it could hinder the natural progression of their grief. Additionally, interacting with these digital avatars may have long-term emotional and psychological repercussions. Therefore, conducting a thorough analysis to identify the pros and cons and having proper mechanisms in place to strike a balance between preserving memories and respecting the dignity of the departed are crucial.

The digital realm stretches like a vast canvas, capturing the intricate threads of our lives in pixels and algorithms. However, its reach extends beyond the living, weaving its embrace around the memories and identities of those who have passed. With this digital immortality comes a pressing question:

> What ethical duties do we owe the deceased when it comes to protecting their data, their online selves, their digital legacies?

This is not merely a matter of legal compliance, a checkbox ticked on a bureaucratic form. It is a conversation stemming from diverse ethical strands, each demanding our attention and weaving a compelling case for the responsible, respectful protection of the departed's digital footprints. At the heart of this lies the golden thread of respect. We owe the deceased the same right to privacy and autonomy they enjoyed in life. Their personal information, often imbued with intimate details and revealing experiences, is an extension of their very being, a digital tapestry deserving protection even after the thread of life has been snipped. Imagine the chilling prospect of a deceased individual's data—their social media posts, medical records, or even financial information—being exploited for financial gain, political manipulation, or even malicious impersonation. This violates the very essence of autonomy, robbing them of the right to control their own narrative and potentially tarnishing their online legacy. The digital realm serves as a crucial repository of memories, reflections, and contributions. It becomes a canvas upon which the deceased paint their stories, share their experiences, and leave their mark on the world. Safeguarding the data of the deceased is not just about respecting their privacy; it is about preserving their posthumous reputation and legacy. Uncontrolled access or manipulation of their online persona can distort their values, misrepresent their beliefs, and erase their authentic selves from the digital landscape.

Taylor [13] observed the rights of the deceased to have the ability to tell their story accurately and respectfully, both now and in the future. However, this ethical obligation extends beyond the deceased, coupled with compassion and empathy towards the living, particularly grieving families. Unauthorised access or misuse of intimate information can exacerbate the pain of loss, reopening wounds and hindering the delicate process of grieving. Conversely, responsible data management can offer solace and comfort, allowing families to access cherished memories and preserve precious fragments of their loved one's life. Mayer-Schönberger [24], in his research, has highlighted the importance of protecting the data of the deceased, leading to the empowerment of the living to navigate remembering and letting go.

Beyond the cold dictates of legal frameworks, protecting the data of the deceased is a profound ethical and humanistic obligation. It is about honouring the privacy of the departed, preventing the exploitation of their digital selves, and safeguarding their legacies for future generations. It is about acknowledging the deep emotional impact on grieving families and fostering a culture of respect and empathy as we navigate the complex terrain of the digital afterlife. This is not a matter of ticking boxes but about ensuring that the memories, contributions, and identities of the departed remain protected and valued even in the ever-evolving digital landscape.

2.5 Technical Context (AI, ML, Metaverse, Digital Footprint)

The importance of safeguarding the rights of deceased individuals in the face of evolving technologies cannot be overstated. The rapidly advancing technological world has made deceased people's rights more important than ever before. One of the fundamental reasons for prioritising the rights of deceased individuals lies in the preservation of their dignity and respect. Our lives are increasingly integrated into Internet platforms in the digital age, leaving a significant digital trail from email accounts to social media profiles. Digital footprints, social media profiles, and online accounts persist long after an individual's demise. Therefore, ensuring that their digital presence is in line with their desires and values should be a part of honouring their autonomy in life and in death. However, due to the lack of existing regulations, balancing the need to close these accounts to preserve the privacy of the deceased person needs innovative solutions.

Evolving technologies introduce new concerns about data ownership and control after someone's death. If not adequately protected, the enormous amount of personal data kept online can be altered, mishandled, or exploited. Thus, protecting the rights of the departed necessitates creating precise guidelines for data handling and making sure that people continue to have control over what happens to their personal data after they pass away. This is especially important as more machine learning and artificial intelligence systems are using personal data for various purposes. In order to achieve this, some platforms offer features for managing accounts after death, allowing users to specify their preferences regarding data retention, memorialisation, or deletion.

We can designate people as 'legacy contacts' on social media sites like Facebook (Facebook, ND), and they will be able to manage the main profile in the event that it is memorialised after someone's death. Once the primary profile has been memorialised, the person added as a legacy contact will have the authority to make changes to it. Otherwise, the deceased person's profile will turn up as people one may know or prompt to recall happy memories that happened a year/s ago. Outlook's parent company, Microsoft (Microsoft, ND), has also clarified what happens to users' accounts in the event of their death. A family member has the authority to terminate an account on their own if they are aware of the login details. However, in the worst-case scenario, if they do not have the login credentials, according to Microsoft, Outlook.com and OneDrive accounts will be frozen after 1 year and any email messages and files stored on OneDrive will be deleted shortly after. Microsoft accounts will also expire after 2 years of inactivity (Microsoft, ND). On the other hand, sites like Instagram usually reserve the right to use our photos to advertise the platform even if they do not have ownership of our content (Instagram, ND). This could technically continue after someone dies. That is why, after the death of one's loved ones, it is important to delete the content on their account or the account as a whole.

2.6 Discussion

The digital age has extended the scope of personal data beyond the living, leaving the protection of the deceased's online information in uncharted territory. This report explores the ethical, legal, and technological challenges surrounding post-mortem data protection, drawing insights from various studies and articles. Whilst the emphasis has been on the significant emotional consequences of post-mortem data leaks, potentially hindering the grieving process and causing further distress, Taylor [13] highlights data rights after death as a means to facilitate the grieving process by allowing families access cherished memories and maintain a connection with the deceased. Floridi [14] underscores the inadequacy of existing legal systems to handle the ethical and legal issues raised by emerging possibilities of digital afterlives. Mayer-Schönberger [24] reminds us that controlled forgetting can be a virtue, allowing us to move on from past experiences and make room for new ones.

2.6.1 Deceased People's Data and Its Impact on Different Fields

2.6.1.1 Ethical Context

With regard to the information of deceased individuals, there are numerous ethical considerations that must be taken into account. One of the primary concerns is the issue of consent, as deceased individuals are not able to provide consent to organisations to use their data. It is important to establish clear guidelines for handling this sensitive information of the deceased. Implementing regulations that specify how the government or organisations should handle the data of deceased individuals may be necessary. This could involve allowing a relative to provide consent on behalf of the deceased or requiring individuals to indicate their preferences regarding the use of their data after their passing.

Misusing or mishandling the data of deceased individuals can have negative consequences for both the deceased and their family members. Therefore, it is crucial to have safeguards in place to protect the privacy of deceased individuals as a way of honouring their dignity and memory. Failing to handle their data in accordance with their wishes could be seen as disrespectful. Additionally, using the data of deceased individuals without explicit consent from the deceased or their loved ones can heighten grief and distress for their families.

Furthermore, in order to honour the memory and legacy of the deceased, it is vital to establish protocols for the appropriate use of their data. It is important to note that the legal frameworks governing the use of deceased individuals' data may vary by jurisdiction, but it is essential to comply with these laws to ensure ethical behaviour. Furthermore, when handling the data of deceased individuals, it is important to be culturally sensitive and aware, as different cultures have unique practices for commemorating and managing the information of their deceased loved ones.

2.6.1.2 Legal Context

The data of deceased individuals can have a significant impact on legal matters as it is still subject to privacy and security risks, even after the individual has passed away. While some countries have implemented regulations to protect the data of deceased individuals, others are still in the process of establishing their stance on this issue.

Deceased individuals often leave behind a digital footprint, including social media profiles, email accounts, and data shared with marketing companies. Without proper mechanisms in place to protect this digital legacy, there is a risk of identity theft and fraud if the data falls into the wrong hands. Managing this digital legacy requires considerations such as privacy protection, access rights, data retention periods, and preserving personal memories.

Although technology companies have begun implementing measures to protect online data, there are challenges to managing the digital data of deceased individuals. It is crucial to develop and implement clear policies and procedures at national, regional, and global levels to govern the handling of deceased users' data, with companies needing to adjust their practices accordingly. Countries should also emphasise the consequences of failing to comply with legal requirements to encourage entities collecting and processing the data of deceased individuals to adhere to the regulations, as non-compliance could damage their reputation.

2.6.1.3 Technological Context

Information regarding deceased individuals can be a valuable asset for research and innovation in diverse fields, as long as it is used in accordance with ethical standards. Before utilising data related to deceased individuals, it is essential to ensure that proper consent is obtained and that the information is both accurate and current. Using outdated or inaccurate data of deceased individuals can introduce bias into algorithms, which may jeopardise the fairness and transparency of data results. Therefore, it is imperative to thoroughly assess the accuracy and timeliness of data when choosing information to train algorithms and to take proactive steps to mitigate and reduce algorithmic bias.

Considering all the above factors, a legal framework must be developed to address the security and privacy issues pertaining to the use of deceased people's data. A multifaceted framework is proposed, emphasising the need for robust legal frameworks, ethical awareness, social discourse, and technological advancements, to ensure respectful and responsible data management for the deceased, as discussed in this chapter. It is also imperative for legal frameworks to keep up with

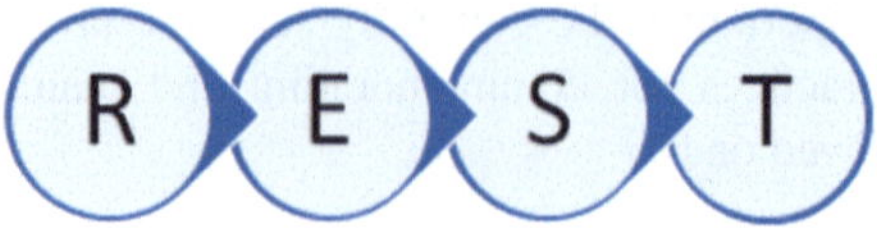

Fig. 2.1 The REST framework [25]

technological advancements to address potential issues adequately and in a timely manner (Fig. 2.1).

A comprehensive and robust legal framework with ethical, technological, and social aspects is crucial for effective post-mortem data protection. Key elements from the research's academic review have acknowledged gaps in the system to include:

Robust Legal Frameworks A clear and comprehensive legal framework should be able to implement well-defined rights, responsibilities, and enforcement mechanisms for post-mortem data management by addressing the limitations identified by Floridi [14]. Issues of data protection, ownership, consent, and privacy must be addressed for developing a strong legal framework for post-mortem data. Guidelines for the collection, processing, storage, access, and the use of such information must be clearly set in the framework, considering the rights of the deceased and the interests of the surviving family members. Accountability, transparency, and potential liability must also be considered when data is used for legal or research purposes.

Ethical Awareness A culture of respect for the deceased's digital legacy can be fostered through public education, setting out ethical guidelines for developers and technology companies, and implementing responsible data practices by aligning with the recommendations of Taylor [13]. Raising awareness about respecting a deceased person's digital legacy is paramount. This includes protecting their privacy, obtaining permission from their loved ones before accessing, using, and sharing their information, and treating the deceased with dignity and respect.

Social Discourse Encouraging public dialogue and debate about the ethical dilemmas related to death is essential. The social dilemma of sharing the personal information of a deceased person centres on the balance between the potential benefits of sharing such information and the right to privacy of the deceased person. On the other hand, there is a need for accountability and transparency if the person who passed away was a public figure. However, there are concerns about the potential impact of sharing the personal information of the deceased on their surviving loved ones. Ultimately, there is no robust solution or choice about sharing the deceased's personal information; therefore, this requires serious ethical consideration and consideration of the interests of all parties.

Technological Advancements With the advancements of technology, new methods and tools are developed to protect the rights and privacy of living and deceased individuals. Some social media platforms offer their users the option to designate a legacy contact to manage their account after they pass away [26]. This allows a designated person to respond on behalf of the deceased. In addition, technology companies are adopting processors to delete user data once the individuals pass away [27]. This makes it possible to prevent sensitive data from being kept indefinitely/longer than necessary following the death of the data subject. Developing privacy-enhancing technologies with strong data security measures, clear ownership models for digital representations, and control mechanisms for authorised access and usage is important.

The digital legacy of the deceased deserves respect and protection. Ethical considerations underpinning this stance is included in the design of the REST framework. Personal data, often intertwined with intimate details and experiences, extends the deceased's being into the digital realm. Protecting this information safeguards their right to control their narrative and prevents potential exploitation or manipulation. In preserving legacies, the digital landscape acts as a repository of memories, stories, and contributions. Safeguarding the deceased's data honours their posthumous reputation and ensures authentic representation, preventing distortion or erasure. This extends compassion to grieving families as uncontrolled access or misuse of post-mortem data can exacerbate their pain of loss and hinder their grieving process. Responsible data management, conversely, empowers families to access cherished memories and navigate remembrance with respect, aligning with Mayer-Schönberger's [24] notion of controlled forgetting as a tool for coping and moving forward.

From a legal and regulatory standpoint, the question of what happens to the deceased's data and individual privacy after death is far from resolved. Currently, most data protection laws do not accept this concept of post-mortem privacy on a legal basis and there is a lack of measures to safeguard the personal information of deceased individuals. Thus, it is important to note that securing personal information during life and after death is essential. Existing legal frameworks struggle to keep pace with the evolving digital landscape, leaving gaps and ambiguities in post-mortem data protection. Key shortcomings include the limited scope within the legal frameworks that often prioritise the living, leaving the data of the deceased in a grey area or subject to fragmented regulations. Consent mechanisms lead to challenges in obtaining consent for post-mortem data usage, as the deceased can no longer provide explicit authorisation. This then leads to implementing and enforcing data protection laws through robust mechanisms and dedicated resources.

The rapid evolution of technology has further complicated data protection in the digital afterlife, leading to emerging technologies such as AI, ML, and MV and raising new questions about ownership, consent, and control of digital representations of the deceased, as evidenced by Zuckerberg's (2022) vision for the MV. Data access and control mechanisms are needed to ensure authorised access to and control over the deceased's data by designated individuals or entities. Developers of digital platforms and services must prioritise privacy in their centric design principles and transparency in their data management practices.

2.7 Conclusions

The increasing amount of personal information, memories, and interactions that people keep online has led to a growing concern about their privacy and security after their death. It is apparent that the majority of the current data protection legislations have not been able to guarantee the security and privacy of deceased individuals. Thus, it is time to unite nations to establish robust data protection laws that safeguard the privacy of living and deceased people alike. When drafting these mechanisms, giving people more power over their data and privacy choices would enable them to safeguard personal information in the afterlife. This could entail allowing people to indicate how they would like their data handled after their passing. Furthermore, by putting together a thorough strategy for handling and getting rid of their digital accounts and assets after death, individuals can make proactive decisions to safeguard their legacy after death. On the other hand, protection of personal data in the digital afterlife is not simply about technical challenges or legal obligations. It is about upholding human dignity, respecting legacies, and acknowledging the delicate balance between privacy and the evolving nature of our online existence. As we delve deeper into the digital realm, ensuring responsible data management for the deceased remains an essential step towards fostering a more respectful and ethically grounded online world, both for the living and the departed.

Acknowledgement The authors extend their appreciation to the editors and reviewers for their valuable feedback and insightful comments, which contributed to enhancing the quality of this chapter.

References

1. Buitelaar, J. C. (2017). Post-mortem privacy and informational self-determination. *Ethics and Information Technology, 19*(2), 129–142. Available at:. https://doi.org/10.1007/s10676-017-9421-9
2. Schafer, B., Briggs, J., Moncur, W., Nicol, E., & Azzopardi, L. (2023). What the Dickens: Post-mortem privacy and intergenerational trust. *Computer Law & Security Review, 49*, 105800.
3. Harbinja, E., Edwards, L., & McVey, M. (2023). Governing ghostbots. *Computer Law & Security Review, 48*, 105791.
4. Kohl, U. (2022). What post-mortem privacy may teach us about privacy. *Computer Law & Security Review, 47*, 105737.
5. Fabre, C. (2009). Posthumous rights. In M. H. Kramer et al. (Eds.), *The legacy of H.L.A. hart: Legal, political, and moral philosophy*. (Oxford, 2008; online edn, Oxford Academic,1 Jan. 2009), https://doi.org/10.1093/acprof:oso/9780199542895.003.0012. Accessed 4 January 2024.
6. Burkeman, O. (2022, April 4). Upon my death, delete how to plan your digital legacy. *The Guardian*. Retrieved from https://www.theguardian.com/society/2022/apr/04/upon-my-death-delete-how-to-plan-your-digital-legacy
7. IFLScience. (2022). *What Happens to Your Digital Footprint When You Die?* [online] Available at: https://www.iflscience.com/what-happens-to-your-digital-footprint-when-you-die-66262. Accessed 4 January 2024.

8. O'Grady, S. (2022). Metaverse: Zuckerberg says technology could bring back the dead. *The Independent*. Retrieved from https://www.independent.co.uk/tech/metaverse-bringing-back-the-dead-zuckerberg-b2422876.html:
9. Danziger, J., Cummings, M. E., & Taylor, L. A. M. (2017). Postmortem privacy: An ethical and legal lacuna. *International Review of Law, Computers & Technology, 31*(4), 387–406. https://doi.org/10.1080/13600869.2017.1275116
10. Livingstone, S. (2018). People are going to court over dead family members' Facebook pages: it is time for post-mortem privacy. *The Conversation*. Retrieved from https://theconversation.com/people-are-going-to-court-over-dead-family-members-facebook-pages-its-time-for-post-mortem-privacy-78375
11. Choi, J. M. (2016). How blockchain could make life & death easier for families. *Hacker Noon*. Retrieved from https://medium.com/hackernoon/how-blockchain-could-make-life-death-easier-for-families-9c14ca4e3415.
12. Keyes, K. M., Pratt, C., Galea, S., McLaughlin, K. A., Koenen, K. C., & Shear, M. K. (2014). The burden of loss: Unexpected death of a loved one and psychiatric disorders across the life course in a national study. *American Journal of Psychiatry, 171*(8), 864–871.
13. Taylor, S. W. (2019). Data rights after death: An ethical and legal analysis. *Washington Journal of Law, Technology & Intellectual Property, 28*(1), 1.
14. Floridi, L. (2018). The philosophy of digital afterlives. *International Journal of Philosophical Studies, 26*(8), 868–884.
15. Malgieri, G. (2019). R.I.P.: Rest in privacy or rest in (Quasi-) Property?, *Data Protection and Privacy* [Preprint]. https://doi.org/10.5040/9781509926237.ch-006.
16. General Data Protection Regulation (GDPR). (n.d.). *Recital 27 - not applicable to data of deceased persons*. [online] Available at: https://gdpr-info.eu/recitals/no-27/
17. Harbinja, E. (2001). Does the EU data protection regime protect Post-Mortem privacy and what could be the potential alternatives?. *A Journal of Law, Technology and Society*. Available at: https://doi.org/10.2966/scrip.00
18. European Court of Human Rights. (2022). *Guide on Article 8 of the European Convention on Human Rights*. Available at: https://www.echr.coe.int/documents/d/echr/guide_art_8_eng. Accessed 18 February 2024.
19. Smolensky, K. (2009). *Rights of the dead*. Arizona Legal Studies Discussion Paper No. 06-27, Available at SSRN: https://ssrn.com/abstract=924499 or https://doi.org/10.2139/ssrn.924499.
20. www.garanteprivacy.it. (n.d.). Home - Garante privacy en. [online] Available at: https://www.garanteprivacy.it/web/garante-privacy-en
21. Yang, M. (2021). Anthony Bourdain documentary sparks backlash for using AI to fake voice. *The Guardian*, Available at: https://www.theguardian.com/food/2021/jul/16/anthony-bourdain-documentary-ai-voiceover-roadrunner. Accessed 20 December 2023.
22. Sankaran, V. (2023). Zuckerberg says Metaverse can bring back the dead – virtually. *The Independent*, Available at https://www.independent.co.uk/tech/metaverse-bringing-back-the-dead-zuckerberg-b2422876.html. Accessed 20 December 2023.
23. Ofcom. (2023). *Future technology and media literacy*. Available at: https://www.ofcom.org.uk/__data/assets/pdf_file/0024/263427/Future-technology-and-media-literacy.pdf. Accessed: 18 December 2023.
24. Mayer-Schönberger, V. (2010). *Delete: The virtue of forgetting in the digital age*. Princeton University Press.
25. Rawindaran, N. (2023). Impact of Cyber Security Awareness in Small, Medium Enterprises (SMEs) in Wales, Ph. D. Thesis. Cardiff Metropolitan University. Available at: https://figshare.cardiffmet.ac.uk/articles/thesis/Impact_of_cyber_security_awareness_in_small_medium_enterprises_SMEs_in_Wales/23599497/1?file=41412072 (Accessed: 20 November 2024).
26. Legal and General. (2024). What happens to someone's social media when they die? Available at https://www.legalandgeneral.com/insurance/life-insurance/lifestyle/digital-legacy/. Accessed 10 February 2024.
27. Ryan-Mosley, T. (2023). Your digital life isn't as permanent as you think it is. *MIT Technology Review*, 19 May. https://www.technologyreview.com/2023/05/19/1073367/digital-life-isnt-permanent-google-twitter-inactive-accounts/. Accessed 10 February 2024.

Chapter 3
Redefining Reality in Political Propaganda: Exploring the Impact of Superimposed Deepfakes in Misinformation Campaigns

Harpith Pandian and Nisha Rawindaran

Abstract In an era characterized by the rampant dissemination of misinformation, the advent of superimposed deepfakes marks a troubling development in media manipulation. The realms of social media, news outlets, and political arenas have transformed into battlegrounds where the lines between fact and fiction are increasingly blurred. This research endeavors to delve into the far-reaching social, technological, and political implications of deepfakes within misinformation campaigns, elucidating the potential adverse aspects of this emerging technology. Employing a comprehensive mixed-methods methodological approach, this study conducts a primary survey targeting individuals susceptible to disinformation. This chapter sheds light on the challenges associated with recognizing superimposed deepfakes and explores opportunities to implement regulatory measures that can mitigate the detrimental repercussions of such media. In addition to the survey, a systematic literature review is meticulously conducted to gain crucial insights into the contextual impact of superimposed deepfakes. This review also delves into the individualistic motives behind the stark reality of these deceptive media forms. Furthermore, this research includes an elaborate discussion of the criteria guiding the systematic literature review methodology. By amalgamating the diverse perspectives garnered from the survey with findings from the literature review, the analysis identifies trends and patterns, providing nuanced suggestions. Consequently, this chapter aims to not only analyze and evaluate the current discourse on digital literacy, media ethics, and technology but also to construct a robust policy framework. This framework seeks to address and mitigate the risks associated with the weaponization of deepfake technology in both political and social contexts.

H. Pandian
Honors College – Rutgers University New Brunswick, New Jersey, USA
e-mail: harpith.p@rutgers.edu

N. Rawindaran (✉)
Cybersecurity and Information Networks Centre (CINC), Cardiff School of Technologies, Cardiff Metropolitan University, Cardiff, UK
e-mail: nrawindaran@cardiffmet.ac.uk

C. Hewage et al. (eds.), *Data Protection*,
https://doi.org/10.1007/978-3-031-76473-8_3

3.1 Introduction

Superimposed deepfakes, readily created with the capabilities of generative artificial intelligence (AI) to manipulate videos, audio recordings, and speech, have the potential to influence public perception. Political campaigns worldwide rely on digital platforms and the ubiquitous media to disseminate information, and, as such, deepfakes can alter public opinion, fuel offensive or rebellious actions against the public, and greatly damage the reputation of world leaders. Therefore, exploring this rapidly evolving topic will emphasize the importance of constructing ethical boundaries to govern uneventful incidents in political discourse. The specific targeting of influential politicians has now become more prominent in this digital age. In 2020, a deceptive video, utilizing synthetic image altering, of Joe Biden garnered more than one million views on Twitter. This fake video portrayed Joe Biden in a way where the US president forgot which state he was in, which indeed is a cruel act of disseminating misinformation as a means of dethroning powerful political personalities. For misinformation-prone individuals, this type of deception could redefine reality and lower their trust in news media across the Internet. Likewise, in March of 2022, a deepfake video extensively surfaced across the Internet supposedly depicting Ukrainian president Volodymyr Zelenskyy ordering Ukrainian soldiers to surrender to Russia amidst the raging weeks-old war. In its entirety, the morphed video seemed eerily real with the president positioned in front of a presidential backdrop with a podium and traditional Ukrainian symbols to further prove the deceptive message. This video was subsequently proven to be deceptive. Such adverse targeting of high-profile figures can undermine public trust and alter voting perceptions. Several other malicious occurrences have been reported in the last few years, accompanied by the rise of superimposed deepfake technology.

By comprehending superimposed deepfakes and their potential to redefine reality, policymakers can formulate informed strategies to mitigate the deleterious effects on public trust. Preemptively, media literacy programs can be tailored to address the specific challenges posed by manipulated content, empowering citizens to discern fact from fiction in an era where synthetic media is synonymous with misinformation. Additionally, an exploration of evolving deepfake technology can set the foundational framework for ethical guidelines and legal responsibilities, urging responsible creation and technological governance.

3.2 Literature Review

3.2.1 Social and Political

Vaccari and Chadwick carried out a comprehensive study to investigate the implications of deepfakes in the realms of deception, uncertainty, and trust within the news media. It encompassed social perception and the impact of political misinformation.

Through a combination of content analyses and surveys, their research analyzed the extent to which deepfake videos infiltrated news reporting and their impact on the credibility of news sources, public trust, and the viewers' ability to discern truth from deception. The results of the study revealed a concerning trend: the presence of deepfake videos in the news ecosystem has led to heightened levels of uncertainty among consumers, decreased trust in media outlets, and an increased susceptibility to deceptive content. The study also highlighted the capacity of deepfakes to generate uncertainty and erode trust in news sources and underscored the importance of raising media literacy levels and fact-checking to mitigate the potentially harmful impact of synthetic political videos on public perception. The results called for a multidisciplinary approach, integrating technology, media literacy, and policy interventions, to mitigate the negative effects of deepfakes on political information dissemination [1].

In his research study, Ahmed explored the factors influencing the inadvertent sharing of deepfake content, its profound impact on the society, and the role of social network size. Employing a mixed-methods approach, the study surveyed a diverse sample of participants to gather quantitative data on their political interest, cognitive abilities, and social network size. Qualitative insights were obtained through interviews and focus groups. This approach comprehensively examined multiple variables, providing a nuanced understanding of the complexities surrounding deepfake sharing. However, there was indeed the challenge of accurately assessing cognitive abilities and the inherent subjectivity of self-reported political interest. The results reveal correlations between high political interest and increased deepfake sharing, suggesting a potential avenue to target a specific sector. Cognitive ability and social network size also played significant roles, with participants possessing larger networks and varied cognitive abilities demonstrating heightened vulnerability to inadvertent sharing. Furthermore, the study contributed valuable insights for developing targeted awareness campaigns and technological solutions to mitigate the inadvertent dissemination of deepfake content [2].

van der Sloot and Wagensveld discussed the legal and social implications of the creation and dissemination of deepfake content. They explored existing legal frameworks, the limitations of the current legislation, and the challenges of attributing responsibility to deepfake content. Additionally, their study investigated the ethical considerations surrounding the use of deepfake technology, including issues related to consent, privacy, and the potential for harm. Their study also delved into broader societal implications, such as the impact of deepfakes on democratic processes and public discourse [3].

Pawelec delved into the powerful intersection of evolving deepfake technology and its impact on democratic procedures. The research specifically investigated the methods used to create and spread synthetic audiovisual content. Through a holistic approach that combined both qualitative and quantitative methods, the study analyzed the scope of deepfake production and its profound influence on crucial democratic functions. The research incorporated case studies, data extraction methods, and content analysis to examine instances where deepfakes had been employed to spread misinformation. The study also included interviews of experts from the field of technology in order to gain insights into the shifting nature of fabricated media. The research brought attention to the adverse disruption of trust in political organizations, the manipulation of public opinion, and the potential of deepfakes to impact the election process.

Furthermore, it emphasized the widespread consequences of synthetic media, particularly in amplifying disinformation and hate speech, posing a direct threat to the fundamental principles of democratic governance. By focusing more on the methods used to take advantage of deepfake technology, the study contributed to a nuanced understanding of the threats posed to democratic processes such as elections. However, the rapidly evolving nature of deepfake technology made it challenging to keep up with the pace of recent developments, making risk management difficult. To enhance the study's practical implications through a holistic overview, a more in-depth exploration of potential countermeasures and policy interventions is required [4].

Mustak et al., in their research study, analyzed scientific papers related to deepfakes, focusing on their implications for firms and consumers across various disciplines such as business, communications, computer science, information science, journalism, and social sciences. Their study highlighted the potential risks of deepfake-induced marketplace deceptions, the strategies and mechanisms for protection against harmful effects, and the opportunities presented by deepfake technology. It also emphasized the challenges posed by the rapid advancement of deepfake technology. The study identified the fact that erosion of trust is a significant concern, addressing the impact of consumer faith in the marketplace. The authors argued that addressing deepfake-induced marketplace deception requires a multifaceted approach, combining legal, technical, market, and educational responses, thus suggesting that the rise of deepfake-induced deception may lead to further disruption of consumer trust in marketing [5].

Nanjundaswamy and Krishnan conducted an integrative literature review on deepfakes, addressing the topic across various disciplines, and proposed a holistic understanding of the social impacts and implications of this harmful technology. The authors analyzed empirical and theoretical research on deepfakes, identifying annual trends, geographic coverage, and frameworks. Their study aimed to answer two research questions, and did do effectively with an integrative approach, going beyond previous reviews to offer a more comprehensive overview. The impact of deepfakes, driven by advancements in AI, was highlighted, with a focus on the increasing concern in corporate and government sectors. The authors emphasized the scarcity of empirical research and the lack of a strong theoretical framework, indicating a need for future studies to guide progress in developing regulatory measures. Their study concluded with recommendations for scholars and practitioners, providing a state-of-the-art framework for deepfake research. Although the integrative approach can enrich our understanding of the topic, the study acknowledged limitations and suggested avenues for improvement, contributing to the evolving discourse on deepfakes [6].

3.2.2 Technological

Nguyen et al. illustrated the rapid advancement of deepfake technology and its implications for various applications, including entertainment, political manipulation, and cybersecurity threats. An extensive review of the existing literature and research developments in deepfake creation and detection was conducted, categorizing them into different approaches and methodologies. A diverse range of generative models were used for deepfake creation, such as variational autoencoders (VAEs) and generative adversarial networks (GANs), to understand how these models have evolved over time. The research also explored the various strategies and techniques employed in deepfake detection, including the use of convolutional neural networks (CNNs) and recurrent neural networks (RNNs). This survey critically analyzed the strengths and limitations of these methods, providing insights into the ongoing challenges in the arms race between deepfake creators and detectors [7].

Agarwal et al. delved into a meticulous examination of the technology used to detect deepfake videos, focusing on the fusion of appearance and behavioral cues as the key indicators. By employing advanced computer vision techniques, this research scrutinized the anomalies present in deepfake videos that may have been overlooked. The technology incorporates not only facial features but also considers the nuances of movement and behavior, acknowledging that convincing deepfakes must extend beyond static visual elements. The methodology involved a comprehensive dataset of authentic and deepfake videos, allowing for a thorough analysis of distinguishing factors. By combining appearance and behavior analysis, the research contributed to the development of more robust and sophisticated detection mechanisms to counter the growing threat of deceptive and malicious deepfake content [8]. Additionally, research conducted by Rawindaran et al. (2022) showed that the growing sophistication of deepfakes creates a unique cybersecurity issue. Unlike larger corporations with robust security infrastructures, smaller businesses often have only limited resources to combat these threats. Deepfake technology allows attackers to create disturbingly realistic audios and videos to seemingly impersonate a business owner receiving a seemingly genuine video call from their customers, complete with voice and mannerisms, requesting an urgent money transfer or a payment of an invoice through a supply chain mechanism. The ease with which deepfakes can bypass traditional security measures makes them particularly dangerous for various small businesses and individuals. To mitigate such risks, there is a requirement to leverage technological advancements themselves. Employee training programs can help staff identify suspicious communications, while incorporating multifactor authentication (MFA) protocols adds an extra layer of security beyond passwords. By understanding the technological impact of deepfakes and implementing appropriate solutions, businesses and individuals can strengthen their defenses and protect themselves from financial losses and operational disruptions [9].

3.2.3 Legal and Ethical Frameworks

Whittaker et al. presented a deepfake framework, which primarily focused on exploring the implications of deepfakes, with an emphasis on their potential benefits and risks for both businesses and customers. While this framework was robust in dissecting the applications of deepfakes in advertising, there were some notable gaps. One major limitation was the relatively vague attention given to the social, technological, and legal dimensions surrounding deepfakes. This framework largely neglected a thorough examination of the societal impact, ethical considerations, and potential legal attributes associated with the proliferation of deepfakes.

To bridge this gap, a more comprehensive framework should be developed, incorporating a multifaceted analysis of social acceptance, the technological evolution of deepfakes, and the legal and ethical frameworks governing their use. Additionally, the proposed framework could delve deeper into societal perceptions, considering how different demographics may respond to deepfake-enhanced videos. This study proposed a framework aiming to bridge these identified gaps by adopting a more comprehensive and integrated approach and ensuring a thorough exploration of the multifaceted factors surrounding the adoption of deepfakes, particularly in the political landscape [10].

3.3 Insights

3.3.1 Comparative Insights

While Vaccari and Chadwick [1] and Ahmed [2] primarily concentrated on the impact of deepfakes on public perception and sharing behavior, van der Sloot and Wagensveld [3] take a broader approach by delving into legal and ethical dimensions. These studies have collectively highlighted the multifaceted nature of the deepfake phenomenon, indicating that a comprehensive understanding must consider not only the societal impact but also the legal and ethical frameworks.

Pawelec [4], Mustak et al. [5], and Nanjundaswamy and Krishnan [6] collectively shed light on the extensive consequences of deepfake technology. While Pawelec [4] focuses on its impact on democratic processes, Mustak et al. [5] and Nanjundaswamy and Krishnan [6] provide insights into marketplace deceptions and the need for a holistic understanding across various disciplines. The integration of these perspectives underscores the far-reaching implications of deepfake technology in both political and commercial domains.

Both Nguyen et al. [7] and Agarwal et al. [8], while approaching the deepfake issue from different angles, contribute to the ongoing arms race between creators and detectors. Nguyen et al. [7] provide a broad overview of deepfake creation methods, while Agarwal et al. [8] emphasize the importance of considering behavioral cues for effective detection. The integration of these insights can inform the development of more robust detection mechanisms.

3.3.2 Cross-Study Insights

The studies collectively highlight that the impact of deepfakes extends beyond technological aspects, encompassing legal, ethical, societal, and democratic dimensions. A multidisciplinary approach resonates across the studies. Whether addressing public trust, political manipulation, or marketplace deceptions, the findings consistently advocate for holistic strategies that combine legal, technological, and educational interventions. The rapidly evolving nature of deepfake technology is a common challenge acknowledged in several studies, and an urgent need for ongoing research, empirical evidence, and regulatory measures is emphasized, pointing toward future directions for scholars and practitioners.

Moreover, the literature review offers profound implications for understanding and mitigating the multifaceted challenges posed by this evolving technology. First, the unanimous recognition of the multifaceted nature of deepfake impact underscores the necessity for a comprehensive, interdisciplinary approach to address the issue.

From legal frameworks and societal perceptions to the erosion of public trust and democratic processes, the convergence of findings emphasizes that any effective strategy must consider the intricate interplay of technological, ethical, legal, and societal dimensions. A consistent call for holistic approaches across the studies reinforces the idea that singular interventions are insufficient. Instead, a multifaceted strategy that amalgamates legal interventions, technological advancements, media literacy efforts, and policy considerations is imperative to confront the diverse challenges posed by deepfakes. Additionally, the shared acknowledgment of the challenges posed by the rapid evolution of deepfake technology underscores the need for ongoing research and adaptive strategies. These insights emphasize the urgency of developing robust countermeasures that not only address the current landscape but also remain agile in the face of the dynamic and unpredictable nature of deepfake advancements.

The research landscape on deepfakes consistently demonstrates the pervasive influence of synthetic media on public and political opinions. Vaccari and Chadwick [1], Pawelec [4], and Mustak et al. [5] converge on a disconcerting trend that the presence of deepfake videos in news media and political discourse has led to heightened uncertainty among consumers, decreased trust in media outlets, and increased susceptibility to deceptive content. In accordance, the study by Pawelec [4] is centered on the adverse disruption of trust in political organizations and the manipulation of public opinion through the use of synthetic audiovisual content, especially those posing a risk to democratic processes. Similarly, Mustak et al. [5] emphasize the potential risks of deepfake-induced marketplace deceptions, specifically highlighting the erosion of consumer trust.

The convergence of these insights leads to an understanding of the profound impact of deepfakes on public and political arenas, in an era where the lines between truth and deception are starkly blurred. The implications are far-reaching, necessitating urgent attention to the development of comprehensive countermeasures. The multidisciplinary approach advocated in Vaccari and Chadwick [1], Pawelec [4], and Mustak et al. [5] becomes imperative, emphasizing not only technological solutions but also the importance of legal, educational, and policy interventions. However, as highlighted by Nanjundaswamy and Krishnan [6], the scarcity of empirical research and the lack of a strong theoretical framework prompts the need for more nuanced investigations into the intricate ways in which deepfakes shape public and political opinions.

A comprehensive plan for further research should focus on understanding the cognitive, emotional, and societal factors that mediate the impact of deepfakes on individuals and communities. This includes exploring the effectiveness of media literacy programs, evaluating the legal frameworks governing deepfake creation and dissemination, and developing adaptive technological solutions. Such a holistic research plan will be instrumental in informing evidence-based policies and interventions that safeguard the integrity of public and political discourse in the face of evolving synthetic media threats.

3.4 Methodology

This research study uses a systematic literature review, which is a rigorous method for screening existing research. This methodology follows a predefined set of steps in order to minimize bias and ensure transparency in the methods used to gain results. Each step focuses on the research question in synthesizing its findings and is then documented and tested for its reproducibility, allowing for a stronger foundation in this research's analysis. This methodology allows for mitigating the risk of selecting convenient studies and ensures a comprehensive evaluation of the current knowledge base. To complement this process, this research also incorporates primary research, which uses a mixed-methods approach, utilizing a survey questionnaire to gather primary data from a target sampled population. This combination of secondary and primary research, with its qualitative and quantitative approaches, offers a richer and more nuanced understanding of the research questions.

3.5 Primary Survey

In developing these research questions, the main goal of the research was to thoroughly explore the connections between deepfakes, political misinformation, and media literacy. Concerned about the rising use of deepfakes in politics, the research intended to capture the diversified sentiments of individuals in society by uncovering their personal insights. These questions act as tools to understand not only what people think but also how they feel about the influence of deepfakes. This approach brings a more relatable and meaningful dimension to the academic investigation, enhancing its significance within scholarly discussions. The survey questions played a pivotal role in shaping the depth of this research study, significantly impacting both the findings and personal perspective on the subject matter. Moreover, the methodology not only amplified the data but also enriched the overall analysis by providing a qualitative depth to the quantitative findings. The responses to questions assessing confidence in discerning deepfakes, perceptions of social media platforms, and proposed solutions offered an expansive view of individual perspectives.

3.5.1 Qualitative Survey Questions

The understanding of deepfakes' societal impact is crucial to this study, and, to achieve this, the survey was designed based on four key questions.

1. "What are the negative effects of deepfakes in your view?"
2. "How often do you use social media platforms as a source of political news?"
3. "Do you think social media platforms are doing enough to combat the spread of deepfake-based misinformation?"
4. "How would you tackle the issue of political deepfakes?"

The first question, "What are the negative effects of deepfakes in your view?" assesses the participants' understanding of the adverse effects of superimposed deepfakes. The second question, "How often do you use social media platforms as a source of political news?" analyzes the trend between politics and its spread through social media platforms. The third question, "Do you think social media platforms are doing enough to combat the spread of deepfake-based misinformation?" examines potential societal perception toward the adverse effects that deepfakes exert in today's fast-paced world. Finally, the question "How would you tackle the issue of political deepfakes?" explores the various solutions proposed by each individual to mitigate the spread of deepfakes and the need for unity. This section of the survey aims to capture the public's perception on deepfakes and their impact on the political landscape.

The next section of the survey was to understand the news and media literacy levels of the participants. The below questions were administered to them. These questions evaluated the extent to which respondents were actively engaged in daily occurrences. The questions also showed how individuals would respond to misinformation online and analyzed the spread of such misinformation. It also showed the participants' response to different news and media sources, including their sentiments about them.

3.5.1.1 News and Media Literacy (Questions)

1. I follow news from different media sources (scale 1–5): 1 = disagree, 5 = strongly agree
2. I would warn other individuals on misinformation (scale 1–5): 1 = disagree, 5 = strongly agree
3. I would exchange news information with my family and friends (scale 1–5): 1 = disagree, 5 = strongly agree
4. I comment and share my feedback on various types of news and articles (scale 1–5): 1 = disagree, 5 = strongly agree

This survey employed non-probability sampling, a method in which participants are chosen based on specific criteria rather than random selection [11]. Unlike probability sampling, which aims for a statistically representative sample of the entire population, non-probability sampling focuses on gathering data from a relevant subgroup. In this case, this research leveraged professional platforms such as LinkedIn and industry contacts to target participants. While this approach may not guarantee that every member of the target population has an equal chance to participate, it allows the research to gather insights from a group with specific industry knowledge or professional experience relevant to the research questions. This sample size represents a percentage of the total target population, and generalizing the results may depend on several factors.

3.6 Results and Analysis

3.6.1 Specific Survey Insights

Most respondents had heard of the term "deepfake" before the survey. Figure 3.1 shows that a significant number of them had encountered political deepfake videos, but some had not.

With regard to perception, the majority of respondents expressed low confidence in discerning whether a video is real or a deepfake. As shown in Fig. 3.2, many individuals acknowledged that deepfakes can be used to spread misinformation and propaganda and manipulate political opinions.

Respondents generally follow news from various sources, but there is a mix in their inclination to share news and information with others. Figure 3.3 shows that their willingness to warn others about misinformation varies.

We identified numerous negative effects, including swaying opinions, spreading false information, privacy invasion, political manipulation, trust issues, and cybersecurity risks. Respondents have varied levels of reliance on social media platforms a source of political news. Some express skepticism about the reliability of social media for political information. A common sentiment is that social media platforms are not doing enough to combat the spread of deepfake-based misinformation.

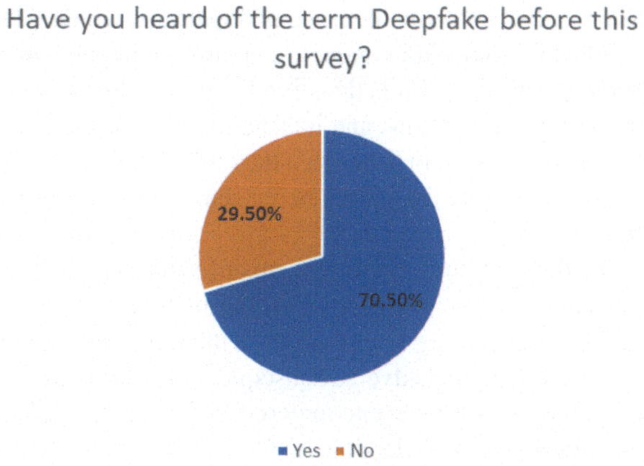

Fig. 3.1 The term "deepfake"

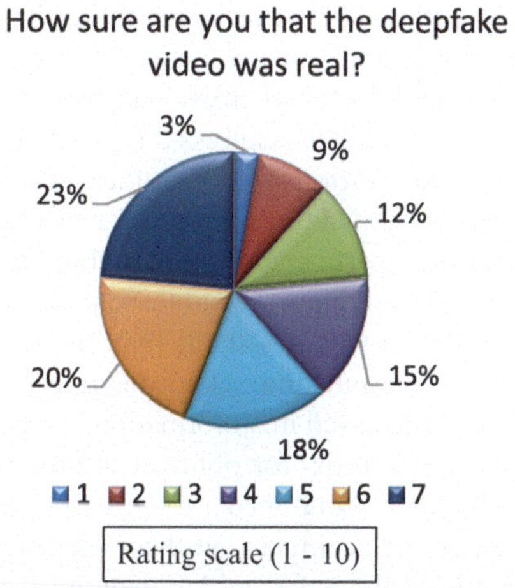

Fig. 3.2 Are deepfakes real?

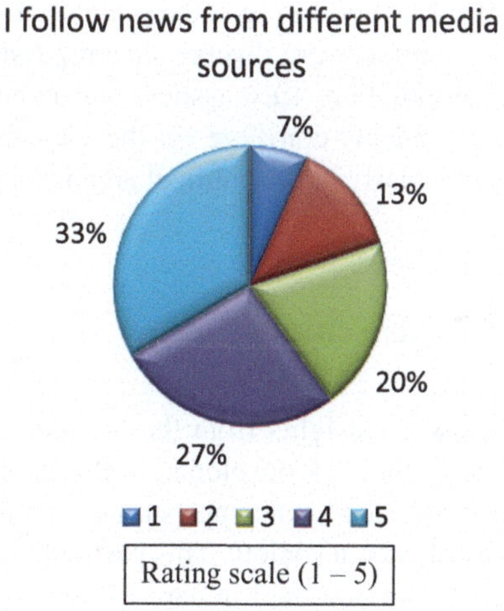

Fig. 3.3 Follow news

Suggestions to tackle the issue of superimposed deepfakes include promoting independent fact-checking agencies, creating independent bodies, developing advanced detection systems, and educating the public. Some respondents propose implementing stricter regulations and international bodies to address deepfake use in politics. Respondents vary in how often they use social media platforms as a source of political news, despite the survey being sent out to a majority of individuals in the education sector.

In the domain of political misinformation, the results of the deepfake survey present a diverse perspective on the impact of deepfakes. The awareness of deepfakes varies among respondents, with some individuals acknowledging their existence and potential adverse consequences, while others are less informed. A majority of respondents have encountered political deepfake videos, regardless of their awareness level of deepfakes. Conversely, there is a significant difference in the perception of truth of these videos, as some respondents express uncertainty about the authenticity of the content they have encountered.

Intriguingly, the respondents' level of media literacy and critical thinking skills is a major factor when evaluating the social and political impact of deepfakes. While some individuals express a high level of skepticism and actively seek out information from various sources, others seem more susceptible to the persuasive nature of deepfakes. One respondent, for example, noted that deepfakes can sway opinions toward a particular agenda, highlighting the potential danger of these manipulated videos in influencing public sentiment.

The survey also sheds light on the role of social media as a source of political news. Despite the acknowledgment of social media's limitations in combating deepfake-based misinformation, a significant portion of survey respondents rely on these platforms for political information. This reliance, accompanied by the frequency of social media use, establishes the need for effective measures to curb the spread of deepfakes on these platforms. Individuals propose various solutions to mitigate the unfavorable impact of deepfakes, with suggestions ranging from promoting independent fact-checking agencies to creating governmental bodies dedicated to monitoring and regulating deepfake content. The negative effects attributed to deepfakes are diverse, encompassing misinformation, privacy invasion, political manipulation, trust issues, and cybersecurity risks. It is significant to note that respondents converge on the idea that deepfakes can have severe consequences, particularly in the political arena, despite their varied backgrounds and experiences.

3.7 Discussion

Drawing insights from the literature review and the primary survey on deepfakes, the framework developed in this research intends to encompass the social, technological, and legal aspects related to mitigating the adverse effects of superimposed deepfakes in society. This particular framework additionally aims to provide a general overview with regard to how high-authority organizations, governments, and the public can take measures to comprehend the devastating effects of deepfakes in terms of breaching data privacy. Subsequently, necessary security changes can be put into effect to limit the dissemination of misinformation, which stems from superimposed deepfakes.

D - Digital Literacy and Education
Enhancing digital literacy and empowering individuals with the knowledge and skills needed to discern authentic from manipulated content is crucial. This involves integrating media literacy programs into educational curricula, conducting awareness campaigns, and promoting critical thinking skills. By fostering a digitally literate society, individuals can become more resilient to the influence of deepfakes.

E - Ethical Deepfake Policies
Establishing ethical deepfake policies is essential for the political and social sectors. Governments and social media platforms should collaborate to create and enforce regulations that govern the creation, distribution, and detection of deepfakes. These policies should prioritize the protection of individuals' privacy, prevent malicious use in political campaigns, and impose consequences for those who violate these standards. Ethical guidelines can act as a deterrent and provide a legal basis for prosecuting those who misuse deepfake technology.

F - Fact-Checking and Verification Mechanisms
The third pillar focuses on encouraging the use of reliable fact-checking agencies and promoting the integration of advanced technologies, such as blockchain or cryptographic verification, which can enhance the authenticity of digital content. Social media platforms should invest in robust algorithms and collaborations with independent fact-checkers to swiftly identify and label deepfake content, providing users with accurate information and building trust in online sources.

E - Engagement of Technology
Leveraging technology involves the development and deployment of advanced detection algorithms, artificial intelligence (AI), and machine learning (ML) models, specifically designed to identify deepfake content. Collaboration between technology companies, research institutions, and government bodies is essential to stay ahead of evolving deepfake techniques. Additionally, the integration of watermarking or digital signatures in media content can assist in verifying its authenticity.

N - Network of Collaboration
The fifth pillar emphasizes the importance of a collaborative network involving governments, technology companies, academia, and civil society. Building a united front against deepfake threats requires joint efforts in research, information sharing, and policy development. Governments can establish international collaborations to address cross-border challenges, while technology companies can pool resources to develop and implement standardized deepfake detection tools. Civil society and academia can contribute by raising awareness and conducting research on emerging threats and countermeasures.

D - Data Privacy and Protection
As deepfakes often involve the unauthorized use of individuals' personal information, stringent data protection measures are the need of the hour. Governments should enact and enforce comprehensive data privacy laws and companies must prioritize user privacy in the development of deepfake detection tools. This ensures that individuals are safeguarded against potential abuses of their personal data in the creation and dissemination of deepfakes.

3.8 Conclusions

In conclusion, this research study has illuminated the multifaceted challenges posed by superimposed deepfakes in an era marked by the viral spread of misinformation. Major findings emphasize the profound impact of deepfakes on public perception, political manipulation, and democratic processes, with evidence from deceptive videos targeting high-profile political figures. The primary survey revealed a diverse range of perspectives among respondents, highlighting varying levels of awareness, confidence in discerning deepfakes, and reliance on social media platforms for political news. The proposed D.E.F.E.N.D. framework provides a comprehensive and integrated strategy, encompassing digital literacy, ethical policies, fact-checking mechanisms, technological engagement, collaborative networks, and data privacy protection. However, it is worth noting that the research has limitations, including the dynamic nature of deepfake technology and the need for ongoing research to keep pace with advancements. The scarcity of empirical research, as identified in some of the literature, prompts the need for more investigations into how deepfakes shape public and political opinions. The potential for further research lies in understanding the cognitive, emotional, and societal factors mediating the impact of deepfakes as well as in evaluating the effectiveness of media literacy programs and the development of adaptive technological solutions. In navigating the evolving landscape of synthetic media, this research provides a foundation for researchers, policymakers, and practitioners to collaboratively develop evidence-based strategies that will safeguard the integrity of information dissemination and uphold democratic values.

References

1. Vaccari, C., & Chadwick, A. (2020). Deepfakes and disinformation: Exploring the impact of synthetic political video on deception, uncertainty, and trust in news. *Social Media + Society, 6*, 205630512090340.
2. Ahmed, S. (2021). Who inadvertently shares deepfakes? Analyzing the role of political interest, cognitive ability, and social network size. *Telematics and Informatics, 57*, 101508.
3. van der Sloot, B., & Wagensveld, Y. (2022). Deepfakes: Regulatory challenges for the synthetic society. *Computer Law and Security Review, 46*, 105716.

4. Pawelec, M. (2022). Deepfakes and democracy (theory): How synthetic audio-visual media for disinformation and hate speech threaten core democratic functions. *DISO, 1*, 19.
5. Mustak, M., et al. (2023). Deepfakes: Deceptions, mitigations, and opportunities. *Journal of Business Research, 154*, 113368.
6. Vasist, P. N., & Krishnan, S. (2022). Deepfakes: An integrative review of the literature and an agenda for future research. In *Communications of the Association for information systems*.
7. Nguyen, T. T., et al. (2022). Deep learning for Deepfakes creation and detection: A survey. *Computer Vision and Image Understanding, 223*, 103525.
8. Agarwal, S., et al. (2020). *Detecting deep-fake videos from appearance and behavior*. IEEE.
9. Rawindaran, N., Jayal, A., & Prakash, E. (2022). Exploration of the impact of cybersecurity awareness on small and medium enterprises (SMEs) in Wales using intelligent software to combat cybercrime. *Computers, 11*(12), 174.
10. Whittaker, L., et al. (2021). The rise of Deepfakes: A conceptual framework and research agenda for marketing. *Australasian Marketing Journal, 29*, 183933492199947.
11. Qualtrics. (2022). *What is Non-probability sampling? definition & examples*. [online] Qualtrics. Available at: https://www.qualtrics.com/uk/experience-management/research/non-probability-sampling/.

Chapter 4
Profiling and Privacy: The Role of Data Privacy in Emerging Technologies

Vibhushinie Bentotahewa, Joel Pinney, and Matthew Tomlinson

Abstract The privacy landscape in relation to digital marketing is dynamic and evolving with the advancements of technology. Digital marketing relies heavily on data, enabling businesses to target people and deliver customised experiences. It also allows organisations to communicate with potential customers through digital means (i.e. social media). The collecting, processing, and storing of personal data has become crucial to marketing strategies. To grow their digital marketing operations effectively, marketers want to comprehend customer behaviour, including their browsing habits and interests. The question is where to draw the line when it comes to excessive monitoring, tracking, and customised marketing. Digital marketers have to obtain users' explicit consent, maintain transparency, and take precautions to safeguard users' privacy. This highlights the necessity of having regulations in place to protect customers against misuse of and unauthorised access to their data. Whilst advertisers need to ensure that their marketing content aligns with customer expectations, protection of customer data privacy also cannot be ignored. The advancements of technology have also sparked the need to develop and implement data privacy regulations to protect customer data. Some governments already have such mechanisms in place, whilst some are in the process of drafting stringent data protection legislation similar to the General Data Protection Regulation (GDPR) in the European Union (EU)/European Economic Area (EEA) region. This has enabled people to have more control over their personal information. This study will explore the current landscape of privacy in digital marketing and touch upon technological advancements in artificial intelligence (AI) and machine learning (ML) that are pioneering these changes. Moreover, this study

V. Bentotahewa · M. Tomlinson (✉)
Cybersecurity and Information Networks Centre (CINC), Cardiff School of Technologies, Cardiff Metropolitan University, Cardiff, UK
e-mail: vibentotahewa@cardiffmet.ac.uk; mdtomlinson@cardiffmet.ac.uk

J. Pinney
Creative Computing Research Centre (CCRC), Cardiff School of Technologies, Cardiff Metropolitan University, Cardiff, UK
e-mail: jpinney2@cardiffmet.ac.uk

C. Hewage et al. (eds.), *Data Protection*,
https://doi.org/10.1007/978-3-031-76473-8_4

provides an overview of the current and future challenges and opportunities to help advertisers apply emerging technologies in an ethically correct manner.

4.1 Introduction

The digital marketing landscape has changed significantly with the enhancement of technology. These changes have allowed marketers to widen the scope of their campaigns, allocate resources more efficiently, and make accurate and swift decisions. The latest advancements in data collection, automation of processes, and campaign optimisation have been made possible by artificial intelligence (AI) and machine learning (ML) technologies. With enormous datasets, ML algorithms enable marketers to find patterns, anticipate customer behaviour, and gain useful insights. Furthermore, creativity has gained prominence as AI and ML algorithms are reshaping digital marketing strategies. Reports suggest that many marketing companies are currently using AI and ML or are planning to do so in the future [30]. Among the well-known brands, Coca-Cola, a renowned worldwide beverage company, is also using AI for advertising purposes [30].

Despite all their advantages, the extensive application of AI and ML in digital marketing has raised significant ethical and privacy concerns. Obtaining consent, algorithmic biases, and data privacy are the major concerns. Online tracking and profiling have also prompted issues related to data misuse, manipulation, and violation of individual privacy rights. This has also blurred the boundary between personalisation and surveillance. However, it is important to note that technological advancements and digital marketing are like two sides of the same coin. Therefore, the best way forward is to use AI and ML to create meaningful customer interactions while adhering to fairness, transparency, integrity, and privacy standards. In this chapter, the researchers investigate the moral problems that marketers face and look into the existing laws, policies, and frameworks to reduce potential risks and promote the responsible use of AI and ML.

4.2 Current Data Privacy Regulations

The European Union (EU) implemented the General Data Protection Regulation (GDPR) as a comprehensive data protection regulation to protect its citizens' personal information and privacy. Its extraterritorial scope means that organisations outside the EU also need to comply if they process the personal data of EU residents, even though it primarily applies to businesses operating within the EU. Globally, the GDPR has a significant impact on digital marketing strategies.

According to the GDPR, ‘personal data’ refers to any information that can be used to identify or contact a natural person (Official Legal Text. [33] Art. 4 GDPR—Definitions). This covers both online and basic identifiers such as names, email addresses, IP addresses, and cookies in the context of digital marketing. Marketers need to be cautious when gathering and using these kinds of data. They should collect only the necessary information, need a legitimate reason to process data, and should not retain collected data for longer than necessary (Official Legal Text. [34] Art. 6 GDPR—Lawfulness of processing). In addition, it is crucial that obtaining explicit consent before collecting, processing, and using an individual’s data for marketing purposes is in line with the GDPR principles (Official Legal Text. [35] Art. 7 GDPR—Conditions for consent). It must also be considered that the data subject has the right to withdraw their agreement from both their data being collected and their previously collected data being stored. It is for these reasons that digital marketers must ensure that the collected data is maintained in a format that allows an individual’s data to be transferred (i.e. to other departments) or deleted upon their request [54]. After gathering data, digital marketers must ensure that proper organisational and technical safeguards are put in place to guard against unauthorised access, disclosure, alteration, and destruction. Additional safeguarding measures such as annual data protection impact assessments (DPIAs) must also be conducted to minimise any risks and remain compliant with the GDPR [41]. To safeguard personal data and foster client trust, organisations must adhere to the provisions of the GDPR. Failure to do so may result in breach of data protection regulations and result in an investigation by the relevant data protection authority. It can also result in fines of up to \$20 million (£17.5 million) or 4% of the annual revenue for severe violations, whichever is greater [41]. For minor/less severe infringements, Article 83(4) of the General Data Protection Regulation (GDPR) stipulates potential fines of up to €10 million or up to 2% of the annual worldwide turnover of the previous fiscal year for businesses, whichever amount is greater (GDPR. EU. [36] Art. 83 GDPR—General conditions for imposing administrative fines).

The GDPR states that the data subject shall have the right not to be subjected to a decision based solely on automated processing, including profiling (Official Legal Text. [32] Art. 22 GDPR—Automated individual decision-making, including profiling). In the context of digital marketing, this relates to targeted advertising and personalised content delivery using the information that organisations collect by deploying AI and ML algorithms. From the marketers’ point of view, automated processing is vital for them to understand their consumer base and purchasing habits, personalise customer interactions, allocate resources more effectively, spot market gaps and opportunities, and detect fraud. As a result, if marketers use automated processing to achieve the aforementioned benefits, they have to give people the option to opt out and be open about the processes in place. This helps organisations build trust.

4.3 Challenges Facing Digital Marketers in Light of Privacy Regulations

Digital marketers face challenges due to evolving data protection and privacy regulations at national and regional levels. In line with the GDPR (GDPR. EU [18] What are the GDPR consent requirements?), marketers are required to obtain affirmative consent from digital platform users before collecting, processing, and using their data. The literature highlights that consent is considered valid if several conditions are met. In particular, the data subject must be competent enough to provide consent and do so freely and must have sufficient knowledge of what they are agreeing to [10]. The question is whether the online consent meets the conditions mentioned above. On the other hand, sometimes, the data subject does not read privacy rules; or even if they do, they do not comprehend them/they lack the background knowledge to make an educated choice [10]. This can make the explicit consent requirement confusing to the users.

In terms of digital marketing, the data collection processes are reliant on the data subject acknowledging and accepting the terms and conditions in order to remain complaint. However, as stated, data subjects often overlook these important documents in order to proceed straight to the website/platform they intend to use. Research by Remsen [45] has identified a number of reasons that a data subject may disregard reading these documents, including not wanting to read a long and confusing legal document, not knowing any bad experiences from their friends and relatives, or the fact that millions of others are actively using the same site. Because of these reasons, digital marketers face dilemmas and need to overcome this challenge by having a standardised system across the different platforms. This standardised system may include simplifying such terms of service (as done by Google in 2012) so that they are understandable to the reader whilst retaining the legal security surrounding the documents [50]. However, a later study conducted by Robinson and Zhu [48] found that efforts to simplify legal documents, such as their terms of service by Google, did not make the legal terminology more salient for users to understand. The closing remarks of Robinson and Zhu do provide an interesting insight into the potential of making privacy regulations and legal information more accessible to end users by stating that policymakers must consider those terminologies and legal issues that end users find most pressing to make content salient to them. In this instance, long legal documents are accompanied by easily digestible snippets of information that can resonate with the end users. In the context of digital marketing, essential information relating to a user's personal information such as how it will be used and how it will influence marketing campaigns can be expressed clearly, as opposed to being lost in long legal documents that contain streams of information and complex legal terminology. If done correctly, digital marketers can be confident that the end user has consented to the processing of their data in accordance with the GDPR.

Another critical issue is the limitations on data usage. The GDPR states that the data collected for specified, explicit, and legitimate purposes should not be further processed in an incompatible manner (GDPR. EU. [17] Art. 5 GDPR—Principles relating to processing of personal data). It is clear that privacy regulations restrict the ways in which marketers can use consumer data. This impacts personalised marketing strategies, as marketers must carefully navigate the boundaries of data usage outlined in these regulations. A study by Han et al. underscores the complexity of ensuring compliance while delivering personalised and targeted content to users and suggests some technological solutions to overcome this. The authors claim that by utilising deep learning techniques for privacy-preserving user modelling in digital marketing platforms, users would be able to retain control over their personal data while marketers would be able to discover appropriate target audiences for their campaigns [21]. However, it must be considered that the limitations imposed by the GDPR, such as data minimisation and purpose limitation, limit the ability to utilise certain business approaches. Moreover, the GDPR purposefully generates certain barriers that limit businesses to utilise data through certain big data and AI approaches [41]. The limitations imposed by the GDPR have sparked many conversations surrounding the future development of technology and approaches across many disciplines. Moreover, the GDPR will continue to play a vital role in impacting the future of technology development, and those who do not comply with the GDPR requirements will eventually fail [60]. Concerns surrounding the implications of the GDPR will additionally play a significant role in the future development of digital marketing. The GDPR will continue to substantially influence nearly all areas of digital marketing from data collection methods, targeting, and segmentation strategies to the final campaign execution and analysis of its impact [55].

Cross-border compliance is another significant hurdle. Digital marketing often involves global campaigns that target audiences across various jurisdictions. Navigating the diverse landscape of privacy regulations worldwide requires marketers to stay informed about the legal requirements in each region and develop strategies to ensure compliance on a global scale. For example, in order to mirror some of the provisions of the EU's GDPR, Australia [6] and New Zealand [12] have revised their privacy laws for their developing digital economies. The Chinese government, on the other hand, uses a restrictive regulatory approach. According to the Chinese government, cross-border data transfers might seriously jeopardise its national interests [5]. To create a centrally regulated data governance system, the Chinese government has implemented many new laws over the last 5 years, including the Cybersecurity Law (CSL), the Personal Information Protection Law (PIPL), and the Data Security Law (DSL). In addition, the most recent measures for cross-border data transfers have established a centrally controlled data governance framework based on national security and general welfare concerns [5].

In order for digital marketers to collect data on their audiences, many companies rely on third-party cookies. It is estimated that around 83% of marketers rely on the data collected by these third-party cookies in order to drive their marketing campaigns [61]. A third-party cookie works by tracking a user and their behaviours when interacting with website content. Unlike first-party cookies that are used to enhance user experience, a third-party cookie is primarily used for marketing purposes, and they outnumber first-party cookies by a factor of two [9]. In a study conducted by Altaweel et al. [3], when tested, the 100 most popular websites were found to collect more than 6000 cookies of which 83% were third-party. Therefore, because of the increased use of third-party cookies, it comes as no surprise that for many, privacy concerns have come into question. The question surrounding the collection of data for online marketing has sparked many users to begin considering how they share information when online [1].

The rising connection between social media and digital marketing is a topic explored by many, most notably the influence that social media platforms have on consumer purchasing habits as well as how the social impact of engaging with social media content may go on to influence consumer choices. It is for these reasons that social media data has become an essential component in building reliable customer profiles. A key case study in this field is the data collected and processed by the popular social networking site Facebook. In this instance, users provide personal information ranging from identifiable content (i.e. name, address) to preferences (i.e. favourite sports team, music), which provides substantial value to marketeers and researchers58. Once processed, this information provides marketers with a better understanding of the background characteristics of the potential customer base, which can be utilised to better improve the targeting of their marketing efforts [26]. However, over the years, Facebook has gained a large volume of bad publicity surrounding privacy concerns about the vast amounts of data the company collects from its platform [15]. Through applying the gathered data on each user, Facebook is able to determine what content is both popular and relevant to each use via their algorithm. Facebook applies their algorithm known as EdgeRank to determine popular content by assessing the connection between the user and content creator, the users' interactions (i.e. commenting, liking posts), and the times in which they interacted [14]. Whilst in most instances, Facebook provides its users with the ability to take control of their data and manage its use, there are examples where Facebook's vast data storages have been compromised. A key example of where Facebook's privacy concerns came into play is the case of Cambridge Analytica. In 2014, Cambridge Analytica began offering audience change behaviour services through inappropriately harvesting the personal information and opinions of Facebook users. This, in turn, prompted an increase in concerns surrounding the privacy of users' data [56]. When exploring the case of Cambridge Analytica, it is important to acknowledge Facebook's claims of not being directly involved in the scandal. However, considerations surrounding the persuasive power of the information held by Facebook comes into question.

4.4 Impact of AI and ML on Digital Marketing

Businesses are adopting AI technologies backed by data analytics now more than ever, and AI and ML algorithms enable marketers to analyse vast amounts of data to understand consumer behaviour and preferences. By leveraging this information, digital marketers can create highly personalised user experiences. Many firms have already personalised their websites, emails, social media posts, videos, and other materials to better respond to customer demands. This ranges from tailored content recommendations to individualised product suggestions, ultimately enhancing customer engagement and satisfaction. In addition, this strategy helps improve the relevance of advertisements and the overall effectiveness of advertising campaigns. However, on the contrary, the literature suggests that there are ambiguities around the use of AL and ML, including security, reliability, and ethics [4]. According to one of the studies, out of the significant challenges users of AI-enabled services face, the loss of privacy and control is alarming [4].

The integration of AI-driven chatbots has revolutionised customer service in the digital marketing realm. Chatbots use natural language processing and machine learning to effectively understand and respond to customer queries [29]. This provides immediate assistance and contributes to building positive customer experiences. From the organisation's perspective, they do not have to invest money to train humans to deal with customers and the staff can be assigned tasks that cannot be automated. However, chatbots pose serious security challenges that need urgent solutions. Modern chatbots employ modern natural language and machine learning techniques [22]. This means that chatbots learn from previous conversations, which can contain personal information, and reuse words/phrases when communicating with others. So, if a user shares any personal information with the chatbot, there is a possibility of data breach from the third parties involved. However, in order to overcome this, data can be encrypted, but ML algorithms cannot compute directly on encrypted data. In addition, some applications may not use encryption by default and may allow unauthorised third parties access to messages exchanged between an individual and the chatbot. Furthermore, transported data can be tampered with by perpetrators if organisations fail to use end-to-end encryption.

There could also be some instances where it may be difficult for users to exercise their rights to erase, restrict, or object to the processing of their data when interacting with a chatbot [22], which contradicts the data subjects' rights highlighted in the GDPR. On the other hand, spam chatbots have been created to spam chatbot users [39]. This might potentially result in the exploitation of trust, infection of computers with malware, and theft of personal data. Similarly, malicious chatbots on cross-platform instant messaging platforms can disguise themselves, contact a user, pretend to be a legitimate chatbot, and steal sensitive information.

AI and ML play a crucial role in marketing automation, the foundation of which is the automatic personalisation/customisation of marketing initiatives [23]. Automated marketing heavily relies on extensive data collected from various places, such as social media accounts, menus, online reviews, and websites [20] to create targeted and personalised campaigns. In addition, automated systems powered by these technologies can handle routine tasks using the collected data. It saves time and allows marketers to focus on the more strategic aspects of their campaigns. However, on the other hand, while the impact of AI and ML on digital marketing is largely positive, it is important to acknowledge challenges and ethical considerations. Issues related to data privacy, algorithmic biases [51], and the responsible use of AI technologies need careful attention to ensure that the benefits are maximised without compromising user trust.

Whilst artificial intelligence has been revolutionary in the field of digital marketing, enabling marketers to collect and analyse vast amounts of information, the benefits have also been passed on to customers. Through analysing the vast amounts of information collected, AI can provide customers with real-time personalised and predictive recommendations based on the customer's previous purchases or viewing habits [42], in addition to offering them targeted advertisements and suggestions based on items that may be of interest to them. However, whilst some see this as an enhancement of their experience, others have investigated how the personalisation of advertisements through analysing data is a breach of their personal information [8]. Despite this, the utilisation of this information is not only beneficial to the customer's personalised experience but also to general business operations. Through leveraging the same, data businesses are able to not only develop targeted marketing campaigns but also assist in forecasting demand and identifying new market opportunities [52].

The application of AI and ML techniques in digital marketing has a twofold benefit, offering both front- and back-end opportunities. Businesses may choose to apply ML and AI techniques for back-end tasks such as forecasting and profiling or front-facing tasks to enhance the users' experience and boost brand recognition [42]. Most commonly, AI is applied in digital marketing owing to its ability to analyse vast amounts of data that needs processing. Because of its ability to rapidly recognise trends and patterns in data, AI enables marketers to save time on resources and rely more heavily on ML and AI techniques [40].

Applying AI techniques to digital marketing enables marketers to begin exploring the hidden insights into consumer-generated content. As reported, a popular outdoor clothing brand utilises AI in order to attempt to predict which jackets a consumer may be interested in. The system begins by exploring where, when, why, and what activities the consumer will be conducting before narrowing down the relevant options/matches for the customer [46]. However, it is vital to consider the origin of the data that enables businesses to apply such techniques. Factors that influence the decisions to purchase products may eventually come down to varying psychographics and personality traits [44]. Therefore, the AI resorts to analysing comments posted on people's social media posts in order to reveal personality traits and values [46]. This data will then eventually contribute towards developing customer profiles that have been enhanced with unique insights from the psychographic data.

Artificial intelligence can also be applied in email marketing development and analysis, through examining which types of emails perform best in terms of generating value for a business [2]. Through developing an enhanced understanding of customer preferences, marketers are empowered to develop email marketing campaigns that not only distribute relevant content but also does so at the optimal time. Sterne [52] discusses how the capabilities of AI will enable marketers to determine the days on which emails are mostly likely to be read and to design/layout preferences and any other relevant materials that have been discovered through their profiling and analysis.

Privacy regulations, such as the General Data Protection Regulation (GDPR) (Official Legal Text. [38] General Data Protection Regulation (GDPR), impose legal obligations on businesses regarding user data protection. Non-compliance can result in severe penalties. As a result, automated marketing systems have to comply with existing data protection rules to keep up with the changing threat landscape. However, utilising AI in business has now become a necessity as opposed to an option. Moreover, businesses utilising AI are able to generate smarter and sharper insights as opposed to those not using AI, resulting in them falling behind [46]. To maintain a balance between use of AI and right to privacy, users should be fully informed about the data collection procedures to solve the privacy concerns related to using AI and ML algorithms. Transparent communication and obtaining informed consent are also essential elements in addressing privacy concerns and moving forward. In addition, bias and discrimination may unintentionally be sustained by automated marketing systems. Automated methods have the potential to reproduce and intensify biases present in the previous data that was utilised to train algorithms. This highlights the significance of routine audits and openness in algorithmic decision-making processes.

4.5 Future Privacy Concerns in Light of AI and ML in Digital Marketing

Personal information has become a valuable commodity in the current digital age. Businesses, governments, and organisations, including digital marketers, have been able to collect data by using AI technologies to make predictions and better decisions. However, the data collected using AI and ML might contain sensitive information that people may not want organisations to use without their explicit consent. This is where policymakers will have to ensure a proper balance between deploying new technologies and personal data privacy. Privacy is the right to keep personal information safe from unauthorised access, and, today, it is more important than ever.

Even if AI and ML present enormous potential for advancing and expanding the digital marketing sector, they also put people's right to privacy in danger. Data privacy is one of the main issues associated with the use of AI and ML by digital marketers. The researchers note that since AI depends heavily on user data, there are ethical issues that businesses must address [27]. Mishandling of personal data, illegal access, and data breaches are serious issues that may erode customer faith in digital marketing strategies. In addition, Elon Musk has also warned the world that AI may be even more hazardous than nuclear weapons [28]. In light of that, whether digital marketing companies are prepared enough to address their customer concerns about privacy and security is questionable. However, achieving a balance between commercial interests and data privacy, security, and ethics is crucial for digital marketing companies and digital regulators to overcome future challenges.

Some countries have implemented robust data protection regulations, such as the General Data Protection Regulation (GDPR). Nevertheless, the absence of consistent legislation leaves a big vacuum for businesses looking to expand and deploy AI [53]. To fill the gap and guarantee the ethical and legal processing of data, businesses using AI and ML must support the development of robust mechanisms to protect their customers' data. In addition, defining categories of data as those that digital marketing companies are either permitted or prohibited to collect will also help them operate consistently worldwide. If there are any ambiguities, customers might not use AI-based apps. Therefore, being transparent about the processors would enable digital marketing organisations to gain customer trust and effectively use AI technologies.

AI and ML algorithms use collected data to build comprehensive user profiles and forecast customer behaviour. Even though this enables organisations to create personalised advertisements and improve their customer experiences, there are privacy and security concerns about profiling and targeted advertisement campaigns. The literature highlights that only 25% of global companies understand the true value of AI, and it is predicted to grow as marketers understand the benefits of using AI and ML technologies [53]. This means that the use of AI would increase over time, and not having robust mechanisms in place would put the privacy and security of the customer in danger.

In 2017, a survey was conducted among more than 5000 people from different nations, and the results showed that 63% of the respondents valued privacy more than a positive customer experience [24]. Additionally, the survey found that respondents wanted businesses to refrain from using AI if it could invade an individual's privacy [24]. In addition, the participants raised concerns about how AI is utilised to obtain personal information, what kinds of information it can access, and how serious a privacy violation could be [24]. This clearly suggests that people are conscious of their privacy. Therefore, policymakers and digital marketers must work together to address future privacy and security concerns.

To address the potential challenges, digital marketing organisations should be transparent about their processes and the use of AI and ML technologies. Companies should provide clear insights into the data they will use for profiling and creating targeted marketing advertisements to promote trust and transparency. This entails revealing the categories of information gathered, the goals of data processing, and the standards applied across the organisation. By doing this, companies help customers make informed decisions regarding their data, and this also helps users have great control and responsibility over their data. In addition, it is clear that with the ability to analyse large amounts of data, AI can be deployed to monitor the behaviour of individuals in ways that were impossible before. This means that with the complex algorithms used in AI, it has the potential to make robust decisions by analysing the existing patterns in data [47]. This also means that individuals may not even know how much of their personal data is being used by AI technologies [47], and this may have the potential to pose new challenges to their privacy.

Therefore, AI algorithms should be designed to minimise the collection and processing of personal data and ensure that it is kept secure and in a confidential manner [16].

The use of AI and ML in digital marketing can also increase the possibilities of cybersecurity risks and data breaches. The amount of data collected and stored by digital marketing companies attracts perpetrators who would get unauthorised access to personal information if organisations do not have robust security mechanisms in place. Considering the amount of data being collected and processed, if the data falls into the wrong hands through hacking or other security breaches [16], there may be severe consequences, including data breaches, financial losses, and loss of trust. Furthermore, the data can be utilised to initiate phishing attempts and disseminate misleading information [47]. These may also have detrimental effects on people's reputations, their privacy, and even their mental health.

To mitigate these potential risks, businesses should have certain measures, such as encryption, access controls, regular security audits, and incident management plans, in place. In addition, raising awareness and providing training on how to handle data, the importance of data, and the consequences of data breaches are important. Furthermore, adhering to GDPR principles such as data minimisation, purpose limitation, storage limitations, and accountability (Official Legal Text. [37] Chapter 1—General provisions) would also reduce the potential impact of data breaches. Another approach would be to utilise blockchain technology to enhance transparency [19]. The decentralised architecture of blockchain ensures that data remains secure.

4.6 Ethical Implications of Profiling for Privacy

Using advanced emerging technologies for profiling poses significant threats to privacy and raises ethical concerns. While profiling offers a range of benefits, such as personalised recommendations and targeted advertising, its ethical impact must not be overlooked. One of the ethical concerns is the violation of user privacy. Recently, there has been a case where 'Target', a retail corporation, used their algorithms to analyse purchasing trends to predict a girl's pregnancy and provide them with targeted promotional offers [7]. While algorithms are commonly utilised to improve customer experience, they overstep their boundaries when they invade their customers' privacy without obtaining explicit consent. This incident has sparked concerns regarding consumer privacy and the appropriate limits for companies conducting personal data analysis as part of their marketing strategies. In order to prevent future occurrences, it is important to emphasise the ethical obligation of companies to safeguard the privacy of their customers.

Organisations use AI and ML technologies to gather data and create detailed profiles of individuals, often without their explicit consent. This unlawful collection, processing, and use of personal data goes against established data protection laws like the GDPR, raising concerns about individuals' privacy rights. In addition, collection of vast amounts of data also increases security risks, attracting hackers and malicious actors. This raises concerns about unauthorised access to or misuse of personal information, which could lead to identity theft, financial fraud, and other harmful consequences. Therefore, digital marketing organisations that rely on advanced technologies like AI and ML have an ethical responsibility to communicate with users about how they collect, process, handle, and share data.

Addressing the legal and ethical implications of profiling for privacy requires a comprehensive and strategic approach. It is crucial to inform individuals of how their data is collected, processed, and utilised and educate them on their rights to help them challenge any improper handling of their data. Implementing regulatory frameworks such as data protection mechanisms can ensure that data is ethically processed for marketing purposes. Furthermore, when designing and implementing algorithms, creators of new technologies need to prioritise fairness, transparency, and respect for individuals' privacy rights. This approach will help maintain legal and ethical standards and protect privacy in the digital age.

4.7 Future Application of Data in Digital Marketing

The application and utilisation of digital platforms such as social media show no signs of slowing in the near future. In 2019, around 3.51 billion people were reported to be actively engaged in using social media platforms worldwide, and this number grew to 4.89 billion in 2023 [13]. Dixon [13] states that if the same trend continues, then, by 2027, nearly six billion people will be active users of social media platforms. As the number of social media users continues to increase, the amount of data generated by each user will continue to climb accordingly. Therefore, it is seems logical to ask what the future holds for digital marketers.

Whilst the future will continue to see the increasing use of AI to help enhance the optimisation of customer communication, Niininen [31] discusses how the use of AI will become more commonplace within organisations looking to automate and personalise communication with their customers. However, AI is built on big data, and the more data an organisation can gather on an individual, the better it performs and the more consumers it attracts. This raises concerns then that a small handful of organisations will hold an ever-increasing amount of data on society as a whole and continue feeding this into large-scale AI models and hold the monopoly over user data.

The growth of Internet of Things (IoT) brings about both advantages and privacy issues. There is a chance that information will be misused, even though businesses can use gathered data to provide individualised services. IoT companies use a variety of sources, such as web browsing, purchase history, and even smart appliances, to collect comprehensive customer data [59]. Targeted advertising is made possible by this data in conjunction with information from data brokers, but questions concerning customer awareness and consent are raised. The extent of data collection, which may include private information like age, health, and spending patterns, may not be disclosed to customers. Because of this, there is an imbalance of power where businesses know a great deal more about their clients than they would feel comfortable sharing. Additionally, the quest for a 'perfect' IoT system may cause businesses to gather far more data than is necessary, posing moral questions regarding user consent and transparency. Furthermore, there is a chance that this could be abused to the point where it crosses the line into personalised services and irritates users with invasive advertising tactics [11].

Other fields such as augmented reality (AR) and virtual reality (VR) will also begin to become tools for digital marketing [49]. AR relies on users to always share their location with the application, often forcing this as a condition of use. King et al. [25] explain how geotracking has the potential to identify where the user works and lives, their potential income, their daily routines, and where they holiday. Questions have been raised about the identity of an individual being exposed from location information alone, a situation created when *The New York Times* reported that they had successfully identified an individual based on anonymous location information sold to them by location marketing companies [57], thus raising concerns that anonymity cannot be guaranteed regardless of user preference.

However, with the development of these new channels of communicating and collecting data from customers, new challenges have begun to arise. New considerations surrounding data protection and privacy need to be taken into account. The advancements of these technologies have the power to completely change the way companies are interacting and communicating with their customers [43], but it seems vital to ask whether the current privacy laws are sufficient to deal with new technologies.

4.8 Conclusions and Recommendations

Current data privacy laws, especially the GDPR in the European Union, have had a significant impact on digital marketing strategies. The laws put in place require digital marketers to obtain express consent, limit the use of data, and ensure cross-border compliance. Obtaining valid consent and managing data usage restrictions pose problems for digital marketers, particularly when it comes to using third-party cookies and social media data. Furthermore, the impact of AI and ML has revolutionised digital marketing by creating new avenues for personalisation, chatbot integration, and marketing automation. However, ethical concerns, privacy concerns, and potential effects of AI and ML algorithms on data privacy are what will likely cause future privacy issues in digital marketing. Future data applications in digital marketing point to the continued rise in influence of social media platforms and the increased use of AI in customer communication. As the Internet of Things (IoT) devices and augmented reality become more embedded in our daily lives, questions still need to be answered as to how the data gathered impacts an individual's privacy. The future trends mentioned raise questions about user consent, transparency, and the applicability of current privacy laws given the challenges posed by developing technologies. In conclusion, despite the unprecedented opportunities that artificial intelligence and machine learning (AI and ML, respectively) offer digital marketers, it is imperative to address growing privacy concerns through transparency, responsible data use, and robust security measures to maintain user confidence, enhance trust, and ensure compliance with data protection laws. Striking a balance between user privacy and commercial interests will be necessary for the sustainable growth of the digital marketing industry.

The authors provide the following recommendations to address the legal and ethical considerations surrounding the utilisation of advanced technologies in digital marketing and profiling.

- Given the amount of data that marketing companies collect, they must proactively manage and mitigate privacy risks. It is essential that organisations consistently inform users about the data collection processes, storage methods, and data sharing practices. Additionally, they have the onus of collecting and processing data in a responsible manner in alignment with existing data protection mechanisms.
- Digital content has gone beyond borders, so countries' ability to control their data can be beneficial. To this end, implementing data localisation policies would be a great way to ensure limited sharing of collected data.
- Countries should also review their privacy laws and consider data localisation as part of their strategies to protect the privacy of their citizens.
- In addition, the increased number of data breaches and misuse emphasises the need for solid data protection measures. Companies must adhere to stringent security protocols and comply with evolving regulatory frameworks like the GDPR. Therefore, having the necessary resources to establish a privacy program is essential, and countries should find ways to support these organisations. This will enable organisations to train staff and implement measures to safeguard user privacy.
- While the GDPR is a solid data protection mechanism, there is a need to update existing mechanisms to address the changing privacy landscape. Organisations should consider adopting privacy-enhancing techniques, prioritising transparency, and building trust to tackle market profiling and data privacy challenges effectively.
- Data protection laws must continually evolve to keep pace with the rapid advancements in ML and AI technologies. As these technologies become increasingly sophisticated, the potential for misuse and infringement upon individuals' privacy rights grows exponentially. Without up-to-date regulations, there is a heightened risk of data breaches.

Acknowledgement The authors thank the editors and reviewers for their feedback and comments, which have helped improve the quality of this chapter.

References

1. Acquisti, A., John, L. K., & Loewenstein, G. (2012). The impact of relative standards on the propensity to disclose. *Journal of Marketing Research, 49*(2), 160–174.
2. Alkhayyat, A., & Ahmed, A. (2022). *The impact of artificial intelligence in digital marketing*. Maladalens University.
3. Altaweel, I., Good, N., & Hoofnagle, C. (2015). Web privacy census. *Technology Science, 1*(1), 1–28.
4. Ameen, N., et al. (2021). Customer experiences in the age of artificial intelligence. *Computers in Human Behavior, 114*. https://doi.org/10.1016/j.chb.2020.106548
5. Arroyo, V., et al. (2023). What specific measures could the US, the EU and China take in order to foster and facilitate cross-border data flows? *Comparative Approach to Big Tech Regulation*. Available at: https://www.sciencespo.fr/public/chaire-numerique/wp-content/uploads/2023/09/policy-brief-data-flows.pdf. Accessed 14 Dec 2023
6. Australian Government. (2019). *Chapter 8: app 8 cross-border disclosure of personal information*. Office of the Australian Information Commissioner. Available at: https://www.oaic.gov.au/privacy/australian-privacy-principles/australian-privacy-principles-guidelines/chapter-8-app-8-cross-border-disclosure-of-personal-information. Accessed 14 Dec 2023
7. Bayes Consulting. (2023). *Target's AI-powered insight: How target figured out a teenage girl was pregnant before her father did!* LinkedIn. Available at https://www.linkedin.com/pulse/targets-ai-powered-insight-how-target-figured-out-teenage
8. Boerman, S. C., Kruikemeier, S., & Bol, N. (2021). When is personalized advertising crossing personal boundaries? How type of information, data sharing, and personalized pricing influence consumer perceptions of personalized advertising. *Computers in Human Behavior Reports, 4*(1), 100–144. https://doi.org/10.1016/j.chbr.2021.100144
9. Cahn, A., et al. (2016). An empirical study of web cookies. In *WWW '16: Proceedings of the 25th international conference on World Wide Web* (pp. 891–901).
10. Chomanski, B., & Lauwaert, L. (2023). *Online consent: How much do we need to know?* Springer. https://doi.org/10.1007/s00146-023-01790-2
11. De Cremer, D., Nguyen, B., & Simkin, L. (2017). The integrity challenge of the Internet-of-Things (IoT): On understanding its dark side. *Journal of Marketing Management, 33*(1–2), 145–158. https://doi.org/10.1080/0267257X.2016.1247517
12. De Silva, S. (2023). New Zealand – Data protection overview. *Data guidance*. Available at: https://www.dataguidance.com/notes/new-zealand-data-protection-overview. Accessed 14 Dec 2023.
13. Dixon, S. (2023). Number of social media users worldwide from 2017 to 2027. *Statista*. Available at: https://www.statista.com/statistics/278414/number-of-worldwide-social-network-users/. Accessed 11 Feb 2024.
14. Djurica, M., et al. (2013). Facebook marketing. In *32nd international conference on organizational science development* (pp. 143–151).
15. Drummond, M. (2024). Should you delete your Facebook? Here's what a privacy expert thinks after recording all Zuckerberg's remarks. *SkyNews*. Available at: https://news.sky.com/story/should-you-delete-your-facebook-account-heres-what-the-man-who-records-all-zuckerbergs-remarks-thinks-13042612. Accessed 11 Feb 2024.
16. Economic Times. (2023). AI and privacy: The privacy concerns surrounding AI, its potential impact on personal data. *The Economic Times*. Available at: https://economictimes.indiatimes.com/news/how-to/ai-and-privacy-the-privacy-concerns-surrounding-ai-its-potential-impact-on-personal-data/articleshow/99738234.cms?from=mdr. Accessed 15 Feb 2024.
17. GDPR. EU. (n.d.-a). Art. 5 GDPR – Principles relating to processing of personal data. *GDPR. EU*. Available at: https://gdpr.eu/article-5-how-to-process-personal-data/. Accessed 14 Dec 2023.
18. GDPR. EU. (n.d.-b). What are the GDPR consent requirements? *GDPR. EU*. Available at: https://gdpr.eu/gdpr-consent-requirements/. Accessed 14 Dec 2023.

19. Ginzburg, J. (2023). Data privacy, AI, and more: What's next for digital marketers? *LinkedIn*. Available at: https://www.linkedin.com/pulse/data-privacy-ai-more-whats-next-digital-marketersjean-ginzburg. Accessed 16 Feb 2024.
20. Haleem, A., et al. (2022). Artificial Intelligence (AI) applications for marketing: A literature-based study. *International Journal of Intelligent Networks, 3*, 119–132. https://doi.org/10.1016/j.ijin.2022.08.005
21. Han, Q., et al. (2023). Towards privacy-preserving digital marketing: An integrated framework for user modeling using deep learning on a data monetisation platform. *Electronic Commerce Research, 23*, 1701–1730. https://doi.org/10.1007/s10660-023-09713-5
22. Hasal, M., et al. (2021). Chatbots: Security, privacy, data protection, and social aspects. *Concurrency and Computation: Practice and Experience, 33*(19). https://doi.org/10.1002/cpe.6426
23. Heimbach, I., Kostyra, D. S., & Hinz, O. (2015). Marketing automation. *Business & Information Systems Engineering, 57*(2), 129–133. https://doi.org/10.1007/s12599-015-0370-8
24. Idziniak, S. (2023). AI and data privacy: protecting information in a new era. *Technology Magazine*. Available at: https://technologymagazine.com/articles/ai-and-data-privacy-protecting-information-in-a-new-era. Accessed 16 Feb 2024.
25. King, A., Kaleem, F., & Rabieh, K. (2020). A survey on privacy issues of augmented reality applications. In *2020 IEEE conference on Application, Information and Network Security (AINS). 2020 IEEE Conference on Application, Information and Network Security (AINS)* (pp. 32–40). IEEE. https://doi.org/10.1109/AINS50155.2020.9315127
26. Li, Y., & Shiu, Y. (2012). A diffusion mechanism for social advertising over microblogs. *Decision Support Systems, 54*(1), 9–22.
27. Masłowska, N. (2023). Navigating the future: The Impact of AI in Marketing Strategy. *LinkedIn*. Available at: https://www.linkedin.com/pulse/navigating-future-impact-ai-marketing-strategy-nicol-mas%C5%82owska-h12jf#:~:text=Data%20Privacy%20Concerns%3A%20There%20is,that%20companies%20need%20to%20navigate. Accessed 15 Feb 2024.
28. Metz, C. (2018). Mark Zuckerberg, Elon musk and the feud over killer robots. *The New York Times*. Available at https://www.nytimes.com/2018/06/09/technology/elon-musk-mark-zuckerberg-artificial-intelligence.html. Accessed 15 Feb 2024.
29. Miklosik, A., Evans, N., & Qureshi, A. M. A. (2021). The use of Chatbots in digital business transformation: A systematic literature review. *IEEE Access, 9*. https://doi.org/10.1109/ACCESS.2021.3100885
30. Murgai, A. (2018). Transforming digital marketing with artificial intelligence. *International Journal of Latest Technology in Engineering, Management & Applied Science (IJLTEMAS), VII*(IV), 260–262. ISSN:2278-2540.
31. Niininen, O. (2021). *Contemporary issues in digital marketing* (1st ed.). Routledge. https://doi.org/10.4324/9781003093909
32. Official Legal Text. (n.d.-a). Art. 22 GDPR – Automated individual decision-making, including profiling. Available at: https://gdpr-info.eu/art-22-gdpr/. Accessed 18 Dec 2023.
33. Official Legal Text. (n.d.-b). Art. 4 GDPR – Definitions. Available at: https://gdpr-info.eu/art-4-gdpr/. Accessed 18 Dec 2023.
34. Official Legal Text. (n.d.-c). Art. 6 GDPR – Lawfulness of processing. Available at: https://gdpr-info.eu/art-6-gdpr/. Accessed 18 Dec 2023.
35. Official Legal Text. (n.d.-d). Art. 7 GDPRConditions for consent. Available at: https://gdpr-info.eu/art-6-gdpr/. Accessed 18 Dec 2023.
36. Official Legal Text. (n.d.-e). Art. 83 General conditions for imposing administrative fines. Available at: Art. 83 GDPR – General conditions for imposing administrative fines – General Data Protection Regulation (GDPR) (gdpr-info.eu). Accesses 13 Mar 2024.
37. Official Legal Text. (n.d.-f). Chapter 1 – General provisions. Available at: https://gdpr-info.eu/chapter-1/. Accessed 15 Feb 2024.
38. Official Legal Text. (n.d.-g). General Data Protection Regulation (GDPR). Available at: https://gdpr-info.eu/. Accessed 18 Dec 2023.
39. Ogundele, T. (2022). A beginner's guide to privacy and security issues in Chatbots. *LinkedIn*. Available at: https://www.linkedin.com/pulse/beginners-guide-privacy-security-issues-chatbots-tolulope-ogundele. Accessed 14 Dec 2023.

40. Porutiu, T. (2023). Surfer SEO REview 2024: Features, pricing and alternatives. *AuthorityHacker*. Available at: https://www.authorityhacker.com/surfer-seo-review/. Accessed 11 Feb 2024.
41. Prasad, A., & Perez, D. (2020). The effects of GDPR on the digital economy: Evidence from the literature. *Information Policy, 27*(3), 3–18.
42. Priyanha, G. (2023). The effects of artificial intelligence on digital marketing. *Journal of Visual and Performing Arts, 4*(1), 158–167.
43. Rana, A., & Kaithwas, A. (2023). The shifting landscape of digital marketing: Trends and predictions for the future. *Phalanx: A Quarterly Review for Continuing Debate, 18*(2), 35–42.
44. Ransbotham, P., et al. (2018). Artificial intelligence in business gets real. *MIT Sloan Management Review and The Boston Consulting Group.*. Available at: https://sloanreview.mit.edu/projects/artificial-intelligence-in-business-gets-real/. Accessed 11 Feb 2024
45. Remsen, A. (2016). A lawyer Digs into Instragrams Terms of Use. *PetaPixel*. Available at: https://petapixel.com/2016/12/07/lawyer-digs-instagrams-terms-use/. Accessed 11 Feb 2024.
46. Ribeiro, T., & Reis, J. (2020). Artificial intelligence applied to digital marketing. In *Trends and innovations in information systems and technologies* (pp. 158–169). https://doi.org/10.1007/978-3-030-45691-7_15
47. Rijmenam, M. (2023). Privacy in the age of AI: Risks, challenges and solutions. *The Digital Speaker*. Available at: https://www.thedigitalspeaker.com/privacy-age-ai-risks-challenges-solutions/. Accessed 16 Feb 2024
48. Robinson, E., & Zhu, Y. (2020). Beyond "I Agree": Users' understanding of web site terms of service. *Social Media + Society, 6*(1), 1–13.
49. Sharma, N. (2021). 7 emerging digital marketing trends you need to know about. *Entrepreneur*. Available at: https://www.entrepreneur.com/article/369976. Accessed 11 Feb 2024.
50. Siegel, A., & Etzkorn, I. (2013). When simplicity is the solution. *The Wall Street Journal*. Available at: https://www.wsj.com/articles/SB10001424127887324000704578386652879032748. Accessed 11 Feb 2024.
51. Sipkus, R. (2023). The ethics of using AI in marketing automation. *LinkedIn*. Available at: https://www.linkedin.com/pulse/ethics-using-ai-marketing-automation-rimantas-sipkus. Accessed 18 Dec 2023.
52. Sterne, J. (2017). *Artificial intelligence for marketing: Practical applications*. Wiley.
53. Taylor, C. (2021). How online privacy issues will shape future use of artificial intelligence in advertising. *Forbes*. Available at: https://www.forbes.com/sites/charlesrtaylor/2021/03/01/how-online-privacy-issues-will-shape-future-use-of-artificial-intelligence-in-advertising/?sh=1fd4c1352262. Accessed 14 Feb 2024.
54. Theodorakis, N. (2018). The 2018 buzzword: "GDPR," and how it practically affects corporations in the EU and the US. *Transatlantic Technology Law Forum, 1*(32), 1–26.
55. Thomas, I. (2019). The impact of the General Data Protection Regulation on digital marketing and analytics. *Applied Marketing Analytics, 4*(3), 194–205.
56. Treku, D. (2020). Effect of penitence on social media trust and privacy concerns: The case of Facebook. *International Journal of Information Management, 50*(1), 171–181.
57. Valentino-DeVries, J., et al. (2018). Your apps know where you were last night, and they're not keeping it secret – The New York Times. *New York Times*. Available at: https://www.nytimes.com/interactive/2018/12/10/business/location-data-privacy-apps.html. Accessed 29 Feb 2024.
58. Van Dam, J. W., & Van De Velden, M. (2015). Online profiling and clustering of Facebook users. *Decision Support Systems, 70,* 60–72. https://doi.org/10.1016/j.dss.2014.12.001
59. Wilson, C., Hargreaves, T., & Hauxwell-Baldwin, R. (2017). Benefits and risks of smart home technologies. *Energy Policy, 103*, 72–83. https://doi.org/10.1016/j.enpol.2016.12.047
60. Yu, H., & He, W. (2019). The impact of GDPR on global technology development. *Journal of Global Information Technology Management, 22*(1), 1–6.
61. Zasloff, D. (2023). View larger image what marketers need to know about the death of third-party cookies. *Emodoinc*. Available at: https://www.emodoinc.com/blog/third-party-cookie-phase-out/#:~:text=Third-partycookiesallowbrands,todrivetheirmarketingstrategies. Accessed 11 Feb 2024.

Chapter 5
Assessing the Application of Artificial Intelligence and Machine Learning in Detecting Misinformation and Disinformation

Matthew Tomlinson, Joel Pinney, and Vibhushinie Bentotahewa

Abstract Digital platforms have become an increasingly important aspect for society as a source of information. As societies' reliance on digital information continues to increase, the proliferation of misinformation and disinformation presents new challenges that must be considered. These challenges, ranging from ethical dilemmas to political manipulation, will be analysed to assess the severity of misinformation and disinformation in a digital era. While exploring the formidable obstacles and challenges facing society, this chapter will examine the untapped potential of artificial intelligence (AI) and machine learning (ML) in combating the spread of false information. By drawing on the advancements in these fields, such as natural language processing, pattern recognition and anomaly detection, this chapter will discuss their viability to combat misinformation. Through the analysis of relevant literature, this chapter will provide a holistic overview of the challenges and opportunities in this domain while drawing on relevant case study examples. As digital landscapes evolve, the need for innovative detection techniques will shed light on the advancements that serve as the defence against the increasing threat of misinformation and disinformation. Content analysis methodologies will be examined to reveal how AI interprets content and verifies factual accuracy, providing insight into the evolving strategies designed to combat misinformation. An examination of algorithmic bias and fairness will highlight the ethical imperatives, ensuring that AI models contribute to the preservation of truth without inadvertently amplifying existing biases that many ML models are susceptible to. Furthermore, it will delve into the challenges of explainability and transparency in AI systems, emphasising the need for interpretable models to build user trust. This chapter will then proceed to analyse the future implications, exploring the potential consequences of AI-driven misinformation and disinformation detection systems on society, where ethical

M. Tomlinson (✉) · J. Pinney · V. Bentotahewa
Cybersecurity and Information Networks Centre (CINC), Cardiff School of Technologies), Cardiff Metropolitan University, Cardiff, UK
e-mail: mdtomlinson@cardiffmet.ac.uk; jpinney2@cardiffmet.ac.uk; vibentotahewa@cardiffmet.ac.uk

C. Hewage et al. (eds.), *Data Protection*,
https://doi.org/10.1007/978-3-031-76473-8_5

considerations must be prioritised, and the evolving legal frameworks necessary for effective data protection. By integrating insights from computer science, social sciences, law, and ethics, this chapter aims to reveal the path forward in safeguarding truth in the era of AI and ML.

5.1 Introduction

In an era dominated by information exchange, the prevalence of having more information has been seen as a common indication of trust in supporting the decision-making process of a source's credibility. Moreover, for many, the assumption is that information is a central part of the decision-making process and that having a higher quality will result in a superior outcome [49]. Unfortunately, when dealing with information, people must now consider the trustworthiness of the information presented. As information continues to play a prominent role in society, the danger of misinformation and disinformation threatens the credibility of information and poses a significant risk to the virtue of knowledge [42]. For instance, misinformation and disinformation disseminated through digital platforms can cause serious harm by undermining, for example, the integrity of election campaigns or public health messages [16]. Through mixing information that has been legitimately developed and distributed with intentionally (disinformation) or unintentionally (misinformation) inaccurate information, the credibility of the intended actual message may be lost in confusion.

In this chapter, we define misinformation as information that is false but has not been created with the direct intention of causing any harm, whilst disinformation is the creation of deliberately false information with the direct intention of causing harm to a person, organisation or social group [53]. As misinformation is often considered not to be purposely spread, it is most commonly disseminated through routes such as rumours and works of fiction [25] or parody and clickbait material [53]. Alternatively, with disinformation intentionally attempting to cause harm, it is often associated with the motivation to gain benefit, such as for personal or professional advantage (i.e. financial or political wealth) [16].

In recent years, it has been said that dissemination of misinformation and disinformation has increased rapidly due to the adoption of social media. As stated by Turcilo and Obrenovic [51], misinformation and disinformation are not new phenomena, but the widespread application and utilisation of social media have made this a more pressing issue. Another factor increasing the prevalence and awareness of misinformation and disinformation originated in response to the Russian disinformation campaigns. In response to these campaigns, the term 'fake news' was coined by the then-elected US president Donal Trump, who labelled disinformation as 'fake news' [16]. Since then, the term 'fake news' has seen widespread adoption, especially during the coronavirus disease of 2019 [40], where large quantities of inaccurate information were disseminated through social media sites, as demonstrated in Fig. 5.1, with over 62% of respondents of a global study conducted by

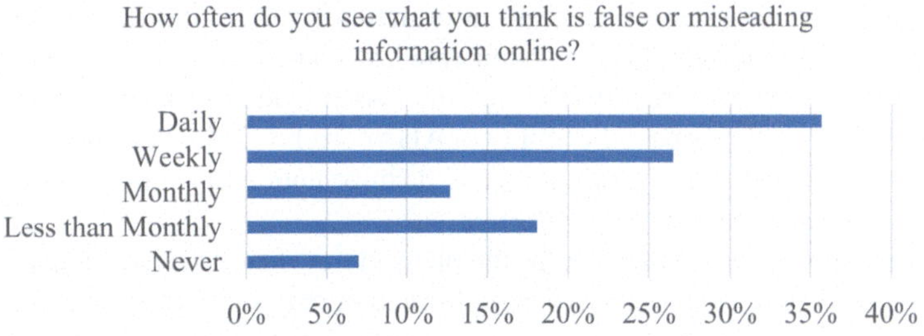

Fig. 5.1 Poynter [44]

Poynter [44] revealing that they believe to have encountered misinformation or disinformation online on a weekly basis.

As the widespread presence of misinformation and disinformation continues to contaminate credible and reliable sources of information, technologists must turn to new technologies to help limit the impact false information can cause. This chapter will explore how advancements in machine learning (ML) and artificial intelligence (AI) may offer a glimpse into restricting the dissemination of false information. Moreover, it will explore how cutting-edge detection techniques and content analysis tools can act as fact-checkers, ensuring the credibility of information.

5.2 The Rising Concerns of Misinformation and Disinformation

Documented in the official European Union report of 2018 [15], it was stated that misinformation and disinformation now influence at least two-thirds of citizens at least once a week. The report details how the widespread dissemination of false information has prompted users to think critically about the integrity of the information presented on their screens. However, some authors have claimed that the increase in exposure to misinformation and disinformation is predominantly down to social media. Due to its easy access, social media has become a platform allowing users to express ideas and publish news freely without any hindrance or control over the authentication of what is published [27]. However, due to this freedom and the vast number of users on the platforms, social media has become a hotspot for the spread of this false information.

The concerns surrounding misinformation and disinformation are highly disruptive and create a sense of uncertainty and doubt surrounding information, causing huge issues not only for society but also in business operations [2]. Interestingly, reported in the 2017 MIT Technology Review, Snow [48] stated that it was expected that by 2022, most individuals would be consuming more false information than valid and accurate information. As such, due to the rising concerns surrounding misinformation and disinformation, some authors have said AI and ML could help manage the dissemination of accurate information by detecting false information [14]. However, in the same report documented by Gartner, it was expected that the creation of a 'counterfeit reality' through the dissemination of misinformation and disinformation would surpass the ability of AI to detect it [38]. Therefore, the ability of AI systems to detect misinformation and disinformation may fall behind the production and dissemination of the information.

Additionally, with the increase in the adoption of data-driven systems, it is no surprise that a counterfeit reality of misinformation may surpass AI detection tools. Moreover, there is unparalleled access to information due to the introduction and rise of digital communication platforms such as social media, online news portals, and other digital channels. This has changed the traditional ways of sharing information. There is an information overload due to the vast amount of data accessible via digital platforms. Although having so much information empowers people to see things from different angles, distinguishing between false and accurate information is complex, with potentially disastrous consequences for individuals and communities. This has also caused profound disruptions to democracies, societies, and personal opinions.

The ease of sharing information globally has helped spread information virally through culturally diverse and geographically distant parts of the world instantaneously by exposing societies to diverse ideologies. This has raised concerns about cultural values at the individual level more than at the country level [22].

People in the twenty-first century gather information, especially news, from the internet. This accessibility has disadvantages in addition to many other benefits. The most significant disadvantage is the dissemination of misinformation and disinformation. Democracies are currently battling the impact of this false information. As we have seen in some elections worldwide, public opinion can sway based on online information. The danger of this is that it could result in people making illogical decisions, which may have dangerous consequences in the long term.

A survey indicates that a growing number of Americans get their news via social media [20]. Furthermore, according to a different study, younger individuals are more likely than older people to rely on Internet sources to obtain news [36]. In light of this, it is fair to say that if the accuracy is not checked, a considerable number of people will access misinformation and disinformation, making them more vulnerable. Although social media companies attempt to slow the spread of misinformation by implementing fact-checking programs, the volume and speed at which information spreads online are challenging. This makes programmes such as fact-checking programs struggle to keep pace with the flow of information on the Internet, resulting in them not acting as effectively as they could be.

On the other hand, finding the ideal balance between the necessity to restrict harmful content and the right to freedom of expression is one of the most significant issues. Although everyone has the freedom to voice their thoughts, this right is not absolute, particularly when it comes to spreading misinformation and disinformation. Therefore, social media businesses have also taken action to identify and ban the accounts that disseminate false information. Nevertheless, the question is whether those efforts are sufficient and adequate.

It is also the responsibility of social media users to identify and act against Fake News by reporting these items to the relevant parties and avoiding interaction with such information. In addition, promoting media literacy and critical thinking abilities among the public is essential to hold platforms accountable for spreading misinformation and disinformation. It is also imperative for policymakers to strengthen rules and enforcement protocols to prevent the dissemination of misinformation and disinformation on the internet.

Fundamentally, as information continues to firmly become a critical resource in modern technologically driven societies, the requirements for individuals and organisations to act upon misinformation and disinformation are essential. If fundamental steps are not taken, the integrity of digital information may be compromised beyond repair. It is, therefore, the technologist's role to incorporate future technologies to help battle and reduce the spread of misinformation and disinformation.

5.3 Application of AI and ML Detection Techniques

Manually detecting misinformation and disinformation is a time-consuming and labour-intensive task that is often completed on a case-by-case basis [23]. However, with the spread of false information (especially during the COVID-19 crisis), alternative methods of detection are required, explicitly applying artificial intelligence tools to detect false news automatically [47].

Artificial intelligence offers the ability to enable an AI-driven fact-checking system that can automatically compare information (such as contained within social media posts) against reliable sources to determine its accuracy [23]. The ability of these systems would enable a user to insert information that raises doubts and cross-reference it against information from sources that are deemed to be credible. According to certain authors, such as Nasir et al. [33], the application of AI techniques is particularly effective in exposing false information contained within social media platforms. However, other authors claim that the application of AI in the detection of disinformation requires a multifaceted approach that should also consider human verification [46]. Moreover, the use of AI is only one step of an intricate process in detecting misinformation, and human verification should also be applied to ensure the integrity of the information.

When detecting misinformation on social media platforms, it is important to remember that not all platforms are the same and that they all have unique features [11]. These features must be taken into account when developing detection algorithms.

5.3.1 AI and ML Misinformation and Disinformation Algorithms

Research on detection frameworks has seen considerable progress in recent years, with current ML and Deep Learning algorithms being adapted to tackle MI and DI online; the following is just a small number of algorithms aiming to improve detection.

5.3.1.1 Convolutional Neural Networks (CNN)

Yu et al. [55] propose the use of Convolutional Neural Networks (CNN). In addition to failing to establish meaningful interactions among features, the authors argue that traditional feature engineering techniques centred around Recurrent Neural Networks (RNN) are ill-suited to capturing the dynamic nature of social media and provide limited capability of early misinformation detection. By using CNN to extract important features and determine how they interact, the Convolutional Approach for Misinformation Identification (CAMI) overcomes these limitations and makes it possible to identify and detect misinformation more effectively and early on. CAMI's effectiveness in both tasks is validated by experimental results on large-scale datasets. Because of the quick development of social media, the paper emphasises the increasing importance of early misinformation detection. CAMI introduces paragraph vectors to represent microblog posts and utilises CNN to automatically extract key features from both misinformation and truthful information. CAMI's innovation lies in its capability to automatically extract significant local–global features, unveil high-level interactions, and flexibly identify key features dispersed within input sequences. Evaluation of large microblog datasets demonstrates CAMI's superiority over existing methods in both misinformation identification and early detection tasks.

5.3.1.2 Attention

Liu and Wu [26] presents research on the early detection of fake news on social media using convolutional and recurrent networks to classify the propagation path. The model combines recurrent and convolutional networks to capture user characteristics and focuses on classifying news propagation paths. According to experimental findings, the suggested model can accurately identify fake news within 5 minutes of its spread. The study emphasises the value of early detection as well as the shortcomings of current methods. When compared to baseline models, the suggested model detects fake news more effectively and efficiently. Future work surrounding this research will examine how user traits can be used to identify users who are likely to disseminate false information.

5.3.1.3 CNN + LSTM

Dhamani et al. [13] investigated the use of transfer learning and deep networks in combating misinformation. For text classification, character-level convolutional neural networks work well, enabling social media features and non-vernacular speech. For examining misinformation on social media, the CNN- Long Short-Term Memory (LSTM) model is a versatile tool. Transfer learning works well in natural language processing. The model can categorise emails and identify communications that are malicious. Moreover, the model can be used to identify and counteract misinformation and inauthentic behaviour. A semantic classifier with multiple labels and classes is the suggested approach. Bidirectional, dropout, max-pooling, and convolutional LSTM layers make up the model. Classifying content as intentionally neutral, intentionally beneficial, or intentionally harmful is crucial. Posts that are classified as harmful or damaging are accurately identified. Numerous domains can benefit from the CNN-LSTM classification approach. Large-scale disinformation campaigns can be understood by the model, and features for exploratory data analysis can be learned. Misspellings and unusual vocabulary usage are permitted by the character-level CNN. For sentence encoding and document classification, the model consists of two deep neural networks. After being trained on several datasets, the classifier obtains a test binary accuracy of 96.14 per cent.

5.3.1.4 BiLTSM

Zhang et al. [59] developed the FAEKDETECTOR model. When the researchers compared the model with comparable alternatives, they found that it has superior accuracy and credibility when deriving information from news articles. To identity fake news articles, creators, and topics, the model makes use of deep learning and heterogeneous network analysis techniques. The model takes into account information about the network structure and extracts explicit and latent features from text data. The model's exceptional ability to detect false news has been demonstrated through extensive testing on the PolitiFact dataset.

5.3.1.5 XFlag + LSTM

Chien, Yang, and Yu [10] propose an inventive explainable AI framework called XFlag, which makes use of the Situation Awareness-based agent Transparency (SAT) model to improve transparency in human–AI interaction, the Layer-wise Relevance Propagation (LRP) algorithm to explain the fake news detection model, and Long Short-Term Memory (LSTM) to identify fake news articles. The findings demonstrate how well the XFlag framework works to help users comprehend system operations and confirm the accuracy of the source. The study also looks at how user understandability, explainability, workload, and trust are affected by system transparency. The suggested XFlag framework raises transparency of AI models, helps users determine the legitimacy of sources, and enhances model accuracy while facilitating human–AI interaction. The research emphasises the significance of providing end users with clear explanations and transparent information. The model attains noteworthy outcomes in accuracy and F1 score, comparable to state-of-the-art models trained on the Weibo dataset, according to some aspects of the authors' conclusions, which claim to validate earlier research in this area. Chien acknowledges that the proposed XFlag framework has limitations in terms of the current dataset and the use of the aggregation approach in LRP. Future research will address these shortcomings by enhancing the framework's functionality and applying it to various datasets. Additionally, user input will be gathered to enhance the framework.

5.4 Content Analysis

Zhang et al. [57] reported on a computational approach to the detection of misinformation in which a framework processes data in real time, which was built by a research group at Auburn University. The approach involves utilising a web crawler to pull news events from legitimate sources to build a dataset of events. Traditional detection techniques have utilised the use of LDA to build topic models based on keywords. However, this approach utilises Stanford's CoreNLP to build a list of events from sentences within the news articles and categorise them into topics. It was found that this approach proved to have better accuracy and effectiveness when detecting fake news during testing compared to previous models such as FEND. The research group has expressed limitations in detecting fake news in real-time. The proposed model cannot handle evolving events or events where news outlets might disagree about the facts of an event, and in such cases, this might classify legitimate news as fake. Future work on this model would require the use of time-stamping events and purging outdated articles to improve accuracy.

5.5 Explainability and Interpretability

Explainable Artificial Intelligence (XAI) or Interpretability in the context of ML is an important consideration when designing an AI- or ML-based detection system, to ensure that the human user can trust the decision-making process of the models and bridge the gap between artificial and human intelligence.

The National Institute of Standards and Technology (NIST) [43] set out four main principles that XAI systems should address in order to foster confidence with the human user, no one of these principles can be used alone and for systems to truly gain user trust all must be used together and build upon the previous principles.

5.5.1 Explanation

Processes and/or outputs must be accompanied by evidence or reasoning that supports the decisions made by the system. Although it must be noted that NIST have not explicitly defined what types of evidence would be justifiable in this situation due to the large scope of use of AI and ML applications.

5.5.2 Meaningful

This is a measure of explanation; has the user understood the explanation that has been given? Miller [28] discusses how, as humans, we tailor our explanations of a topic to the individual receiving that information based on their knowledge of the topic, their cultural background, sex, and age. Building an XAI model based on existing human-generated explanations would improve the overall meaningfulness of explanations. Furthermore, Miller explains that research in social sciences has found that for an explanation to be meaningful, it must be built on simplicity, coherence, and generality.

5.5.3 Explanation Accuracy

Phillips et al. [43] explain that although the system's decision-making judgement might have been accurate, this is different from the accuracy of the explanation. Explanation accuracy expands on meaningfulness, and accuracy is dependent on whether the user has been able to understand the explanation given. An explanation for the decision of the system can be given accurately at a high level, but if the receiving user is not knowledgeable in the area, then the accuracy of the decisions made would be lost on the user. Papenmeier et al. [39] discovered that research suggests the only explanations that improved user trust and understanding were preceded by 'why' and 'why not', supporting the principle of Meaningful through keeping explanations simple to improve accuracy.

5.5.4 Knowledge Limits

Previous principles have assumed that a system has been able to accurately provide a response within its scope of application. An explainable application must also understand its own limitations and where an output does not fall within expected boundaries the system must be able to explain this to the user as to why a prediction was not achievable at this time, this could be down to an unexpected input or a lack of credible evidence to make an informed prediction [43].

5.6 Deepfakes

It must be remembered that misinformation and disinformation are not just a text-based problem, but with the advances being made in the creation of AI-generated images, the issue of Deepfake generation must also be considered. Deepfakes are AI-generated images or videos that, to the human eye, appear to be genuine, with the most common form of deepfakes created through the manipulation of genuine human imagery [29]. Research by Ng [35] has shown that although users are sceptical when it comes to the trustworthiness of some online content, when a user already has a connection with an individual, such as a well-known celebrity or politician if the content appears to be in line with what they would expect the individual to discuss, their likelihood to perceive a deepfake as trustworthy increases. Although current research shows that users possess the knowledge to identify a deepfake, the increasing capabilities of AI-generated imagery and videos mean that deepfake detection algorithms cannot be overlooked. Furthermore, as more users engage with visual content via social media platforms, the dangers that deepfakes hold may be significantly intensified in the coming years.

5.7 Algorithmic Bias and Fairness

Neumann et al. [34] address the lack of investigation into fairness in automated misinformation detection systems. The research emphasises how important it is to have justice frameworks that cover every facet of equity in information settings. Additionally, it discusses the possible drawbacks for various stakeholders such as information subjects and evidence providers. It is stressed how crucial it is to recognise algorithmic mistakes and the different ways in which they affect different groups. The study recommends for the creation of technical countermeasures to disinformation.

Algorithms in newsfeed could potentially encourage the spread of misinformation and disinformation while also undermining the authority of genuine sources [30]. Users become isolated and have less variety in their news sources due to echo chambers and filter bubbles based on their browsing history. Algorithms that recommend news items require better guidelines and standards. Paid content and algorithmic bias can also impact newsfeed performance. Discrimination and the deliberate dissemination of false information can result from an algorithm's lack of transparency.

Cheng et al. [9] introduce the concept of Socially Responsible AI Algorithms (SRAs), where the need to understand the connections among different dimensions of AI is highlighted. AI systems have obligations that fall within four categories: philanthropic, legal, ethical, and functional. The misuse of data and the absence of user consent are emphasised as problems within AI algorithms. SRAs should be centred around several important goals, including safety, fairness, transparency, and trust. However, fairness in AI is difficult to achieve because of its subjective and evolving nature. Within the context of SRAs, the research defines fairness within the decision-making process as something that is absent of prejudice or favouritism. Errors made by Stanford's COVID-19 vaccine allocation algorithm, which gave older employees priority over frontline workers, are used to highlight the need for transparency and fairness in AI algorithms.

5.8 Adversarial Attacks

Adversarial machine learning can be applied to amplify attacks and avoid detection [4]. Compared to its counterparts in computer vision, research on adversarial defence in natural language processing is still in its infancy [5].

Mosca et al. [31] propose an adversarial detector that uses logit variation interpretation. While this method has shown promise in identifying adversarial images, it has not yet been used in Natural Language Processing (NLP) adversarial defence. Current character-level defence techniques have proven successful in identifying syntax or spelling errors but have not proven effective against attacks that preserve language correctness. The technique uses machine learning models and feature attribution explanations to distinguish between adversarial and original samples and assign word relevance scores to an event. The adversarial detector receives the data from the pipeline, which gauges the model's response to inputs. Many detector architectures were examined, and word-level adversarial detection using the suggested approach performs better than the previous state of the art. Across various models, datasets, and attacks, the detector demonstrates generalisation capabilities.

Zhou et al. [60] propose that their Dirichlet Neighborhood Ensemble (DNE) defence method consistently outperforms existing methods. DNE uses a novel adversarial training algorithm to defend against strong attacks. Using the DNE method, words contained in input sentences are substituted with randomly selected points from a convex hull made up of the embeddings of the original word and its synonyms. Under the strongest attack, DNE achieves 77.2% accuracy, while other methods drop to 0.0%, 32%, 64.3%, and 8.1% accuracy. DNE substantially improves robust accuracy without sacrificing performance on clean data. The defence algorithm is broadly applicable and can be incorporated into any neural network. Adversarial-trained smooth classifiers outperform existing defences on various network architectures and datasets. Deep neural networks in NLP tasks are vulnerable to attacks, and word substitution-based attacks are discussed.

Yuan et al. [56] described bridging the gap between CV and NLP. They proposed a gradient-based textual adversarial attack framework. This paper introduces a textual adversarial attack framework for generating adversarial texts in Natural

Language Processing (NLP). The framework includes an effective attack method called Textual Projected Gradient Descent (T-PGD) that handles decision-based black-box attacks. The method transforms the problem of searching for adversarial samples in discrete text space into the continuous embedding space. Experimental results show that T-PGD outperforms other attack methods in terms of attack performance and adversarial sample quality. The method is evaluated on sentiment analysis, natural language inference, and news classification tasks. The success of the attack method is attributed to the more effective search process guided by gradient information. The quality of the adversarial samples generated by T-PGD increases with the text length. The method also yields higher Universal Sentence Encoder (USE) scores than baseline models, indicating more precise manipulation of adversarial samples. However, the grammatical performance of T-PGD is not the best on certain datasets. Yuan explains that the transferability of non-pretrained models to PLMs is still unknown, but this may not be a significant concern in practical scenarios when discussing possible improvements. The framework used to transfer adversarial attack methods from CV to NLP was instantiated with the PGD attack as an example, but it would be interesting to explore more attack methods and analyse their benefits for NLP. They propose that adversarial samples crafted by T-PGD have better quality than the baseline method in terms of grammar and fluency, according to human annotators. However, both BERT-Attack and T-PGD still have room for improvement in terms of grammatical correctness and fluency.

5.9 Case Studies

Misinformation and disinformation have grown to be serious issues not just in the United States but in other countries as well. After the Israeli government bombed Gaza and Hamas militants attacked Israel in 2023, verified users on Twitter published a vast amount of misinformation, recreating video game content, old videos, and altered photographs to promote partisan narratives [17].

People are already on edge after such incidents, and therefore, the dissemination of disinformation and misinformation has the potential to cause unnecessary stress, trauma, and psychological and emotional damage to people who have been directly impacted by the incident, as well as the other communities in the society. Additionally, if people panic, they may also act irrationally. Furthermore, it is typical for misinformation and disinformation to target specific racial, religious, or cultural groups, which can also lead to violence and increased isolation.

People occasionally use false and misleading information to fulfil their political agendas or incite social instability. If political and radical groups spread false information following an attack to pursue their personal agendas and goals, it will undoubtedly divide and polarise societies. On the other hand, the disinformation and misinformation may also cause law enforcement and investigators to incorrectly identify suspects, impeding efforts to stop similar attacks in the future and giving suspects the opportunity to flee the scene without being prosecuted. This could then weaken public trust in political institutions and law enforcement.

It is also vital to consider that when false news spreads faster than wildfire, it can divert resources away from where they are needed most. Emergency responders may have to respond to false alarms, and it may waste time and resources on a threat that does not exist. Because of that, the real victims may end up not getting the assistance they require. Therefore, it is crucial to filter information before it is made public for all of the previously listed reasons. Unfortunately, misinformation and disinformation have already influenced many aspects of society, including the following instances.

5.9.1 2016 Presidential Election

During the 2016 US presidential election, fake news predominated on social media more than real news (Center for Information Technology & Society, ND), which raises questions regarding the integrity of democratic processes. The average US adult encountered one or more fake news articles during the 2016 election period, with higher exposure to pro-Trump content [3]. Guess et al. [21] found that older Americans were more likely to spread fake news by up to four times as much as those from within the younger demographic and suggested that research indicates that memory decline has an impact on their susceptibility to sharing of misinformation, although acknowledges that further research is required in this field. Despite efforts at the time to detect false news, it remained in circulation since this more significant effort has been made to combat the spread of misinformation and disinformation.

Another prominent example of misinformation in presidential elections is that of Cambridge Analytica. In this case, Cambridge Analytica created psychological profiles of social media users globally using data from social media. They were able to predict people's political leaning by obtaining Facebook user data without authorisation (Center for Information Technology & Society, ND). Subsequently, a deliberate effort was initiated through the production of Facebook political advertisements and memes with the aim of undermining Hillary Clinton's presidential campaign and influencing Americans' views in support of Trump (Center for Information Technology & Society, ND). The content that was produced was inaccurate, aggressive, and provocative messages.

5.9.2 2013 Boston Marathon Bombing

Within 45 minutes of the Boston Marathon bombings, misinformation about the event started to spread [50]. Three prominent rumours were identified, including false claims about a girl dying in the marathon. A Tweet circulated of an 8-year-old dying in the bombing. Within 4 minutes of this Tweet, a false image of a girl had been added to aid in the spread of this false news. By only 33 retweets, the misinformation had started spreading on other platforms. The second speculation centred around the involvement of Navy Seals or Blackwater. Images of dark backpacks used in the bombing were released by the FBI. This resulted in a vigilante movement of one Reddit user to share images of men wearing similar backpacks in the area. One such image shared showed two men, one with a Navy Seals emblem, increasing speculation that the government were involved. The final rumour was the misidentification of Sunil Tripathi as a suspect: a university student who had gone missing some weeks before the bombing was falsely identified by a fellow student following the release of the police images of the suspects. This rumour gained less traction, though, following the release of the suspect's names by the FBI. The research found that due to the context of these rumours and the volume of misinformation, corrections were unable to slow the spread, with a misinformation-to-correction ratio of 44:1 for rumour one, 18:1 for rumour two, and 5:1 for rumour three.

5.9.3 Health Misinformation

The COVID-19 pandemic is another instance where, despite social media's potential benefits, fake news is mainly disseminated through it, making it harder for consumers to know what is correct and what is not. A significant amount of this false information went on to exacerbate phobias and anxiety. As a result of this, there were some odd patterns in the purchases of personal items such as toilet rolls [24] and the buying and using of drugs without authorised medical prescriptions [6]. These behavioural shifts may be detrimental and may have resulted in self-harm instances thus signifying the seriousness of disinformation. A number of reports went on to highlight real-world incidents, such as after hearing on the news that chloroquine could be able to treat COVID-19, a resident of the United States took it and passed away [6]. Another incident was identified in that the Iranian government disclosed that individuals had died from alcohol poisoning after drinking alcohol, believing that consuming illicit alcohol could protect them from the illness [1]. A study of more than 1700 US adults discovered that individuals, particularly those who rely more on intuition or have little scientific background, spread more false information about COVID-19 on social media due to a disregard for accuracy. A subtle reminder to verify information before reposting was found to make people less likely to spread misleading information [41].

Conspiracy theories are not new to a pandemic, and COVID-19 is not the first. Conspiracy theories about the Zika virus, Ebola, and possibly AIDS were also widely circulated [6]. The World Health Organisation alerted the world community to a developing 'infodemic' on social media in the early stages of the pandemic [54]. A report suggests that the WHO claims that this 'infodemic' is bringing about false medical advice, psychological panic, and economic disruption [32].

5.9.4 Mass Shootings

There was an incident where, after a major shooting in the USA, Conspiracy theorists were incentivised to post materials that were believed to have originated from Russian agents, political campaign operatives, and other fake news sources. These sources claimed that the shootings were staged to gain public support for gun control laws [8]. The story looked fake when they later examined the hashtags from tweets, web pages, and comments. Following the death of George Floyd in 2020, there was a national uproar in the United States, and racial and political divisions widened in the United States as a result of the widespread spread of fake news on social media, online rumours, and photoshop-manipulated photos [7]. In another horrific mass shooting that happened in a Maine restaurant and bowling alley towards the end of 2023 in the United States, false information about the perpetrator and his capture flooded social media platforms, including Facebook, Twitter, and TikTok [17].

The high levels of misinformation and false news pose a new major challenge for society. The extent of this challenge can have widespread effects, from shaping political procedures to instilling fear in the population. Events like the 2016 US presidential election, the Boston Marathon bombing, health misinformation during the COVID-19 pandemic, and mass shootings illustrate the gravity of this problem. Reducing the impact of false information and misleading content necessitates people to spend time verifying the reliability of sources. However, additional effort is needed to encourage responsible behaviour on social media platforms and improve media literacy.

5.10 Legal and Ethical Considerations

Due to the complex nature of detecting misinformation, ethical considerations need to be made surrounding the utilisation of AI for detection purposes. The use of AI for the detection of misinformation and disinformation raises issues relating to privacy, accountability, and transparency [46]. In order to successfully apply AI techniques to detect false information, Santos [46] states that we must be guided by ethical principles and fundamental human values.

Misinformation and disinformation are widespread problems in the current digital era, and they pose serious problems for people, governments, and societies everywhere. Therefore, investigating the legal and ethical challenges involved in addressing misinformation and disinformation is of paramount importance. However, regulating misinformation and disinformation is difficult, as a thin line separates upholding freedom of expression from reducing the damaging effects of misleading information. The ongoing dissemination of fake material online seems to have been facilitated by the absence of a thorough regulatory structure. It would take time to develop a robust structure, but it is time for the policymakers to formulate policies that work while upholding democratic values.

The anonymity of the source is another difficulty for policymakers and government officials dealing with misinformation and disinformation. This implies that people who disseminate false information occasionally attempt to conceal their identity by creating fake profiles. This damages the source's credibility and sense of trustworthiness. However, secretly disseminating false information does not exclude a person from governmental and judicial scrutiny. Regardless of the anonymity, rules and regulations are in place in many jurisdictions to handle the spread of fake news. If anonymous actors break any laws related to incitement, fraud, or defamation, they could be held legally liable. In addition, platforms may also use reporting tools and AI and ML algorithms to recognise and eliminate damage. This might also assist authorities and platform owners in combating misinformation and disinformation. In reality, even though anonymity provides a cloak of invisibility for spreading misinformation and disinformation online, anonymity does not shield content from removal or platform policies. On the other hand, if people can be extra cautious when accessing content from unidentified or anonymous sources, it would make it more difficult for anonymous sources to spread misleading information.

The capacity of criminals to intentionally fabricate and forge data in Big Data is also another concern [58]. This false data will unavoidably provide inaccurate outcomes. Sometimes, people may also fabricate data to create illusions that work in their favour and mislead others [58]. Some websites contain false comments and ratings, and users can easily be lured into buying goods and services based on those fake comments and ratings. On the other hand, it is also possible for people to fabricate datasets, which could cause doubt surrounding the authenticity and credibility of the open-source databases that researchers can access.

According to the International Human Rights Law, governments must guarantee adequate access to truthful information [51]. This type of access encompasses the freedom to look for and obtain information. In light of this, governments must support and safeguard unrestricted access to accurate and reliable information.

A number of nations have put laws and regulations in place to deal with misinformation and disinformation, especially in relation to social media and Internet platforms. These laws frequently imprison people and issue fines for those who disseminate false information. However, the effectiveness of these laws depends on the acceptance of existing legal frameworks. The online safety Bill in the United Kingdom addresses the issues regarding misinformation and disinformation [19]. The measure intends to hold digital companies accountable for the content posted on their platforms and has included new criminal offences related to online activities [45]. This will encourage authorities and organisations to keep an eye out for and halt the spread of false information. The president of Kenya signed the Computer and Cybercrimes Act into law in 2018 [37]. The act makes it illegal to broadcast fake news; if someone is found guilty, they will receive a two-year prison sentence, fines of up to nearly $50,000, or both. The law further states that it is illegal to publish false, misleading, or fraudulent material knowingly or to mislead people knowingly. The Penal Code of Sri Lanka also contains general provisions that address various types of false statements [18]. They are meant to enable action against those disseminating false information, particularly remarks that could jeopardise national security and encourage intergroup conflict.

Legislation against spreading false information through electronic devices was passed in Tunisia, although many people are against it because they believe it will limit freedom of expression. Article 24 in the legislation stipulates that anyone found to be disseminating false information and rumours online faces a maximum sentence of five years in prison and a fine of up to $15,000 [12]. Most importantly, that sentence doubles if the fake news is directed at a state official. Critics argue that the new legislation has given lawmakers an easy way to punish citizens as it does not specify what kind of information or rumours are prohibited [12]. Furthermore, critics claim that some clauses will let authorities check computers or telecommunication devices, seize equipment, and intercept data if authorities believe there is a violation of the new legislation [12].

5.11 Discussion

This chapter has explored how misinformation and disinformation on social media are pressing concerns today. Threat actors are using advanced technology to evade detection, making it harder to combat this problem with machine learning and artificial intelligence alone. Collaboration among various stakeholders is vital to address this challenge. Government institutions can enact legislation to regulate the spread of false information and penalise those who deliberately spread misinformation. They can also fund research into new technologies and methodologies to combat this issue.

Academia can contribute by studying the patterns and impacts of misinformation and disinformation, developing new strategies and tools to detect and counter them, and promoting educational initiatives to increase public awareness about the dangers of false information.

Social media platforms, as primary conduits of information in the digital age, have a significant role to play. They must promote awareness about the risks of misinformation and disinformation through user education programs and implement early warning systems that notify users of potential false information on their platforms.

Equipping users with knowledge to identify false information is critical. By leveraging AI and ML technologies to support users, they can become active participants in safeguarding themselves and society against these threats. This could involve developing user-friendly tools that help identify unverified information or suspicious sources.

The successful implementation of early warning systems and education initiatives has the potential to significantly reduce the impact of misinformation and disinformation on society, foster a more informed society, reduce polarisation caused by false narratives, and promote a healthier discourse in the digital public sphere.

5.12 Conclusion

In an information age, the threat of misinformation and disinformation poses significant risks to the credibility and integrity of information online. As this chapter has discussed, false information presented online can have significant implications that may result in mental or physical harm. The chapter has highlighted numerous case studies in which the dissemination of false information has resulted in varying degrees of impact. Primarily, this chapter has identified methods in which AI and ML may be utilised to identify and mitigate the risks posed by false information effectively. However, in identifying the potential of AI and ML, numerous technological restrictions and ethical dilemmas have been raised. However, the threat of misinformation and disinformation will continue to impair a reader's judgement on the integrity of information until companies (such as social media sites) adopt reforming changes to their operations. Moreover, implementing AI and ML directly into their operations to detect, confirm, and dispose of misinformation and disinformation in a legally abiding manner.

References

1. Aghababaeian, H., Hamdanieh, L., & Ostadtaghizadeh, A. (2020). Alcohol intake in an attempt to fight COVID-19: A medical myth in Iran. *Alcohol, 88*, 29–32. Available at:. https://doi.org/10.1016/j.alcohol.2020.07.006
2. Akhtar, P., Ghouri, A. M., Khan, H. U. R., Amin Ul Haq, M., Awan, U., Zahoor, N., Khan, Z., & Ashraf, A. (2022). Detecting fake news and disinformation using artificial intelligence and machine learning to avoid supply chain disruptions. *Annals of Operations Research, 327*. https://doi.org/10.1007/s10479-022-05015-5
3. Allcott, H., & Gentzkow, M. (2017). Social media and fake news in the 2016 election. *Journal of Economic Perspectives, 31*(2), 211–236.
4. Alsmadi, I., et al. (2021). Adversarial attacks and defenses for social network text processing applications: Techniques, challenges and future research directions'. *arXiv:2110.13980.*
5. Alsmadi, I., et al. (2023). Adversarial NLP for social network applications: Attacks, defenses, and research directions. *IEEE Transactions on Computational Social Systems, 10*(6), 3089–3108.
6. Barua, Z., et al. (2020). Effects of misinformation on COVID-19 individual responses and recommendations for resilience of disastrous consequences of misinformation. *Progress in Disaster Science, 8*(1), 100–119. https://doi.org/10.1016/j.pdisas.2020.100119
7. BBC News. (2020). George Floyd protests: Misleading footage and conspiracy theories spread online, *BBC News*, 2 June. Available at: https://www.bbc.co.uk/news/52877751. Accessed 16 February 2024.
8. Center for Information Technology & Society (ND) *The danger of fake news in inflaming or suppressing social conflict.* Available at: https://cits.ucsb.edu/fake-news/danger-social. Accessed 05 February 2024.
9. Cheng, L., Varshney, K. R., & Liu, H. (2021). Socially responsible AI algorithms: Issues, purposes, and challenges. *Journal of Artificial Intelligence Research, 71*(1), 1137–1181. Available at:. https://doi.org/10.1613/jair.1.12814
10. Chien, S.-Y., Yang, C.-J., & Yu, F. (2022). XFlag: Explainable Fake News Detection Model on Social Media. *International Journal of Human–Computer Interaction, 38(18–20)*, 1808–1827.

11. Cohen, R., et al. (2020). Addressing misinformation in online social networks: Diverse platforms and the potential of multiagent trust modeling. *Information, 11*(11), 539. Available at:. https://doi.org/10.3390/info11110539
12. Cordall, S. S. (2023). *Tunisia anti-fake news law criminalises free speech: Legal group*. https://www.aljazeera.com/news/2023/7/18/tunisia-anti-fake-news-law-criminalises-free-speech-legal-group#:~:text=The%20principal%20focus%20of%20the,made%20about%20a%20state%20official. Accessed: 07 February 2024.
13. Dhamani, N. et al. (2019). Using Deep Networks and Transfer Learning to Address Disinformation. *arXiv*. Available at: http://arxiv.org/abs/1905.10412. Accessed 14 February 2024.
14. Endsley, M. R. (2018). Combating information attacks in the age of the Internet: New challenges for cognitive engineering. *Human Factors, 60*(8), 1081–1094.
15. European Union. (2018). *Flash Eurobarometer 464: Fake news and disinformation online*. Available at: http://data.europa.eu/88u/dataset/S2183_464_ENG. Accessed: 6 February 2024.
16. Gibbons, A., & Carson, A. (2022). What is misinformation and disinformation? Understanding multi-stakeholders' perspectives in the Asia Pacific. *Australian Journal of Political Science, 57*(3), 231–247.
17. Gilbert, D. (2023). *Maine mass shooting disinformation floods social media as suspect remains at large, Wired*. Available at: https://www.wired.com/story/maine-mass-shooting-disinformation/. Accessed: 06 February 2024.
18. Government of Sri Lanka. (2023). *Penal Code- Volume VI*. Available at: https://www.srilankalaw.lk/revised-statutes/volume-vi/878-penal-code-ordinance.html. Accessed 07 February 2024.
19. Government of United Kingdom (ND). *Online Safety Bill*. Available at: https://publications.parliament.uk/pa/cm5803/cmpublic/OnlineSafetyBill/memo/OSB28.htm. Accessed 7 February 2024.
20. Grieco, E. (2017). *More Americans are turning to multiple social media sites for news*. Pew Research Center. Available at: https://www.pewresearch.org/short-reads/2017/11/02/more-americans-are-turning-to-multiple-social-media-sites-for-news/. Accessed: 06 February 2024.
21. Guess, A., Nagler, J., & Tucker, J. (2019). Less than you think: Prevalence and predictors of fake news dissemination on Facebook. *Science Advances, 5*(1) Available at:. https://doi.org/10.1126/sciadv.aau4586
22. Gupta, M., et al. (2023). Fake news believability: The effects of political beliefs and espoused cultural values. *Information & Management, 60*(2), 103745. https://doi.org/10.1016/j.im.2022.103745
23. Iqbal, A., et al. (2023). The relationship of Artificial Intelligence (AI) with fake news detection (FND): A systematic literature review. *Global Knowledge Memory and Communication, 67*(2), 1–22. https://doi.org/10.1108/GKMC-07-2023-0264
24. Leung, J., et al. (2021). Anxiety and panic buying behaviour during COVID-19 pandemic-a qualitative analysis of toilet paper hoarding contents on Twitter. *International Journal of Environmental Research and Public Health, 18*(3), 1127. https://doi.org/10.3390/ijerph18031127
25. Lewandowsky, S., et al. (2012). Misinformation and its correction : Continued influence and successful debiasing. *Psychological Science in the Public Interest, 13*(3), 106–131. https://doi.org/10.1177/1529100612451018
26. Liu, Y., & Wu, Y.-F. (2018). Early detection of fake news on social media through propagation path classification with recurrent and convolutional networks. *Proceedings of the AAAI Conference on Artificial Intelligence, 32*(1), 354–361. Available at: https://doi.org/10.1609/aaai.v32i1.11268
27. Madani, Y., Erritali, M., & Bouikhalene, B. (2021). Using artificial intelligence techniques for detecting Covid-19 epidemic fake news in Moroccan tweets. *Results in Physics, 25*(1), 1–10. https://doi.org/10.1016/j.rinp.2021.104266

28. Miller, T. (2019). Explanation in artificial intelligence: Insights from the social sciences. *Artificial Intelligence, 267*(1), 1–38. https://doi.org/10.1016/j.artint.2018.07.007
29. Mirsky, Y., & Lee, W. (2022). The creation and detection of Deepfakes: A survey. *ACM Computing Surveys, 54*(1), 1–41. Available at:. https://doi.org/10.1145/3425780
30. Mohseni, S., & Ragan, E. (2018). Combating Fake News with Interpretable News Feed Algorithms. *arXiv*. Available at: http://arxiv.org/abs/1811.12349. Accessed: 16 February 2024.
31. Mosca, E., et al. (2022). "That is a suspicious reaction!": Interpreting logits variation to detect NLP adversarial attacks. In *Proceedings of the 60th annual meeting of the Association for Computational Linguistics (volume 1: Long papers). Proceedings of the 60th annual meeting of the Association for Computational Linguistics (volume 1: Long papers)* (pp. 7806–7816). Association for Computational Linguistics. Available at: https://doi.org/10.18653/v1/2022.acl-long.538
32. Mourad, A., et al. (2020). Critical impact of social networks Infodemic on defeating coronavirus COVID-19 pandemic: Twitter-based study and research directions. *IEEE Transactions on Network and Service Management, 17*(4), 2145–2155. https://doi.org/10.1109/TNSM.2020.3031034
33. Nasir, J., Khan, O., & Varlamis, I. (2021). Fake news detection : A hybrid CNN-RNN based deep learning approach. *International Journal of Information Management Data Insights. Elsevier Ltd, 1*(1), 1–13. https://doi.org/10.1016/j.jjimei.2020.100007
34. Neumann, T., De-Arteaga, M., & Fazelpour, S. (2022). Justice in misinformation detection systems: An analysis of algorithms, stakeholders, and potential harms. In *2022 ACM conference on fairness, accountability, and transparency. FAccT '22: 2022 ACM conference on fairness, accountability, and transparency* (pp. 1504–1515). ACM. Available at: https://doi.org/10.1145/3531146.3533205
35. Ng, Y.-L. (2023). An error management approach to perceived fakeness of deepfakes: The moderating role of perceived deepfake targeted politicians' personality characteristics. *Current Psychology, 42*(29), 25658–25669. Available at:. https://doi.org/10.1007/s12144-022-03621-x
36. OfCom. (2023). Young adults turning to social media for news, finds Ofcom: News, Research Live. Available at: https://www.research-live.com/article/news/young-adults-turning-to-social-media-for-news-finds-ofcom/id/5115103. Accessed 06 February 2024.
37. Ombuor, R. (2018). Kenya approves controversial cybercrimes law. *Voice of America*, 17 May. https://www.voanews.com/a/critics-kenyan-cyber-crimes-law-to-crush-freedom-of-speech/4396882.html. Accessed: 06 February 2024.
38. Panetta, K. (2019). Gartner top strategic predictions for 2018 and beyond. *Gartner*. Available at: https://www.gartner.com/smarterwithgartner/gartner-top-strategic-predictions-for-2018-and-beyond. Accessed: 4 February 2024.
39. Papenmeier, A., et al. (2022). It's complicated: The relationship between user trust, model accuracy and explanations in AI. *ACM Transactions on Computer-Human Interaction, 29*(4), 1–33. Available at:. https://doi.org/10.1145/3495013
40. Pennycook, G., & Rand, D.G. (2021). The Psychology of Fake News. *Trends in cognitive sciences, 25(5)*, 388–402. https://doi.org/10.1016/j.tics.2021.02.007.
41. Pennycook, G. et al. (2020). Fighting COVID-19 Misinformation on Social Media: Experimental Evidence for a Scalable Accuracy-Nudge Intervention, *Psychological Science, 31(7)*, 770–780. https://doi.org/10.1177/0956797620939054
42. Perez-Escolar, M., Lilleker, D., & Tapia-Frade, A. (2023). A systematic literature review of the phenomenon of disinformation and misinformation. *Media and Communication, 11*(2), 76–87.
43. Phillips, P. J., Hahn, C. A., Fontana, P. C., Yates, A. N., Greene, K., Broniatowski, D. A., & Przybocki, M. A. (2021). *Four principles of explainable artificial intelligence*. National Institute of Standards and Technology (U.S.). https://doi.org/10.6028/nist.ir.8312
44. Poynter. (2022). *A Global Study on Information Literacy Understanding generational behaviors and concerns around false and misleading information online*. [online] Available at: https://www.poynter.org/wp-content/uploads/2022/08/A-Global-Study-on-Information-Literacy-1.pdf. Accessed: 7 May 2024.

45. Queen Mary University. (2022). Social media, misinformation and the law. Available at https://www.qmul.ac.uk/lac/our-legal-blog/items/social-media-misinformation-and-the-law.html. Accessed 7 February 2024.
46. Santos, C. C. (2023). Artificial intelligence in automated detection of disinformation : A thematic analysis. *Journalism and Media, 4*(1), 679–687.
47. Shim, J., Lee, Y., & Ahn, H. (2021). A link2vec-based fake news detection model using web search results. *Expert Systems with Applications, 184*(1), 1–12.
48. Snow, J. (2017). Can AI win the war against fake news?, *MIT Technology Review*. Available at: https://www.technologyreview.com/2017/12/13/147095/can-ai-win-the-war-against-fake-news/. Accessed: 4 February 2024.
49. Stahl, B. (2006). On the difference or equality of information, misinformation, and disinformation: A critical research perspective. *Informing Science Journal, 9*(1), 83–95.
50. Starbird, K., et al. (2014). Rumors, false flags, and digital vigilantes: Misinformation on Twitter after the 2013 Boston Marathon Bombing. In *iConference 2014 proceedings. iConference 2014 proceedings: Breaking down walls. Culture - context - computing, iSchools*. Available at: https://doi.org/10.9776/14308
51. Turcilo, L., & Obrenovic, M. (2020). *A companion to democracy #3*. Heinrich-Böll-Stiftung e. V., International Democracy Programme. Available at: https://www.boell.de/sites/default/files/2020-08/200825_E-Paper3_ENG.pdf
52. United Nations. (no date). Universal Declaration of Human Rights | United Nations. Available at: https://www.un.org/en/about-us/universal-declaration-of-human-rights. Accessed: 07 February 2024.
53. Wardle, C., & Derakhshan, H. (2017). *Information disorder: Towards an interdisciplinary framework for research and policy-making*. Council of Europe. Available at: https://rm.coe.int/information-disorder-toward-an-interdisciplinary-framework-for-research/168076277c. Accessed 4 February 2024.
54. World Health Organization: WHO. (2020). *Managing the COVID-19 infodemic: Promoting healthy behaviours and mitigating the harm from misinformation and disinformation*, World Health Organisation, Available at: https://www.who.int/news/item/23-09-2020-managing-the-covid-19-infodemic-promoting-healthy-behaviours-and-mitigating-the-harm-from-misinformation-and-disinformation. Accessed 07 February 2024.
55. Yu, F., et al. (2017). A convolutional approach for misinformation identification. In *Proceedings of the twenty-sixth international joint conference on artificial intelligence. Twenty-sixth international joint conference on artificial intelligence, Melbourne, Australia: International joint conferences on artificial intelligence organization* (pp. 3901–3907) Available at: https://doi.org/10.24963/ijcai.2017/545
56. Yuan, L. et al. (2023). Bridge the gap between CV and NLP! a gradient-based textual adversarial attack framework. *arXiv*. Available at: http://arxiv.org/abs/2110.15317. Accessed 16 February 2024.
57. Zhang, C., et al. (2023). A computational approach for real-time detection of fake news. *Expert Systems with Applications, 221*(1), 119656. Available at: https://doi.org/10.1016/j.eswa.2023.119656
58. Zhang, D. (2018). *Big Data security and privacy protection.* 8th international conference on management and computer science (ICMCS 2018). https://doi.org/10.2991/icmcs-18.2018.56.
59. Zhang, J., Dong, B., & Yu, P.S. (2019). FAKEDETECTOR: Effective fake news detection with deep diffusive neural network. *arXiv*. Available at: http://arxiv.org/abs/1805.08751. Accessed 14 February 2024.
60. Zhou, Y. et al. (2020). Defense against adversarial attacks in NLP via Dirichlet neighborhood ensemble. *arXiv*. Available at: http://arxiv.org/abs/2006.11627. Accessed 16 February 2024.

Chapter 6
Trust and Trustworthiness: Privacy Protection in the ChatGPT Era

Shasha Yu, Fiona Carroll, and Barry L. Bentley

Abstract The emergence of large language models (LLMs), exemplified by ChatGPT, has ushered in a significant transformation in our interactions with artificial intelligence (AI). However, the widespread integration of AI technologies has raised significant concerns about user privacy protection. To establish a robust relationship between users and AI while safeguarding user privacy, trust, and trustworthiness are fundamental factors in building user confidence and ensuring the responsible and ethical use of AI. Within this context, the authors reviewed previous research to examine the current global landscape of public trust in ChatGPT and its evolution over time. Furthermore, this chapter investigates the trustworthiness of AI by examining the risks and threats posed by ChatGPT from technical, legal, and ethical perspectives. The authors also explore the factors that influence user trust in AI, and propose strategies aimed at enhancing the trustworthiness of AI systems in safeguarding user privacy. In doing so, supporting people to be more aware and sensible of what they should and should not trust online with regard to AI. This chapter offers insights to a broad audience, including industry professionals, policymakers, educators, and the general public, all collaborating to achieve a harmonious equilibrium between user trust and AI trustworthiness. The goal is to enable the public to harness the benefits of advanced technology while safeguarding their privacy in the ChatGPT era.

S. Yu (✉)
School of Professional Studies, Clark University, Worcester, MA, USA
e-mail: ShaYu@clarku.edu

F. Carroll · B. L. Bentley
Cardiff School of Technologies, Cardiff Met University, Cardiff, UK
e-mail: fcarroll@cardiffmet.ac.uk; bbentley@cardiffmet.ac.uk

C. Hewage et al. (eds.), *Data Protection*,
https://doi.org/10.1007/978-3-031-76473-8_6

6.1 Introduction

The evolution of AI technologies, exemplified by the launch of OpenAI's ChatGPT in November 2022, has dramatically transformed our interactions with digital platforms, offering more personalised and intelligent responses and fundamentally changing how users interact with technology. As a cutting-edge development in large language model (LLM) technology, ChatGPT, originally built on the GPT-3.5 architecture, represents a significant technological leap. The subsequent upgrade to GPT-4 introduced multi-modal capabilities, allowing the model to engage beyond mere text-based interactions through enhancements like Reinforcement Learning from Human Feedback (RLHF) [1]. This process was specifically designed to minimise the generation of harmful or misleading content.

ChatGPT has been trained on an extensive dataset, enabling it to generate responses that closely mimic human conversational patterns across a wide array of topics. This advancement in natural language processing has paved the way for other AI models, such as Google's Gemini [2] and Microsoft's Copilot [3], further broadening the AI landscape. Other AI models, like Midjourney, Diffusion, and DALL-E [4], extend AI capabilities into creative domains such as text-to-image generation, while models like Sora advance into text-to-video creations [5]. These developments signify a substantial expansion in AI's role across various sectors, including virtual assistance, image recognition, and analytics [6].

The proliferation of such AI technologies has significantly impacted sectors like customer service, education, and content creation, where the demand for personalised and intelligent responses is high. However, the extensive data processing capabilities of these models also raise critical concerns about user privacy and data security [7]. The ability of AI to handle and analyse vast amounts of personal data without sufficient oversight can lead to privacy breaches and misuse of sensitive information.

In the context of this growing integration of AI into daily life and critical business functions, the issue of trust and trustworthiness becomes paramount. Ensuring robust privacy protections is crucial for maintaining user trust in AI technologies. As AI systems become more embedded in our personal and professional lives, the potential for privacy infringements increases unless stringent safeguards are put in place. Thus, ongoing research into effective privacy protection measures is essential to prevent misuse of data and to foster a trustworthy environment where AI's benefits can be fully realised without compromising individual privacy.

This chapter explores the intertwined concepts of trust and trustworthiness in the context of AI, with a particular emphasis on safeguarding privacy in the age of ChatGPT and analogous technologies. It examines the shifting global trust in ChatGPT, analysing how public trust has evolved. In addition, this chapter assesses the trustworthiness of AI by considering the technical, legal, and ethical hurdles it faces. The authors examine the factors that influence public trust in the ChatGPT era and suggest strategies to enhance the trustworthiness of AI. Furthermore, they underscore the need for a balance between user trust and AI trustworthiness.

Through this exploration, the study aims to contribute to the ongoing dialogue on creating a safe and trustworthy AI environment that prioritises and protects user privacy amidst the rapid expansion of the digital age.

6.2 Trust in ChatGPT Era

Trust is a complex notion that involves having confidence in the belief that another individual or entity will behave with benevolence, integrity, competence, and predictability [8]. When individuals place trust in someone, it suggests they are open to taking risks, including being emotionally exposed, within a connection or interaction [9]. Trust can also be described as an individual's ability to remain vulnerable and receptive to the actions or choices of another party without apprehension of adverse outcomes [10].

Given ChatGPT's transformative influence on public perception of AI [11], examining public trust in AI becomes crucial for its future adoption.

6.2.1 Public Trust in ChatGPT Worldwide

In an era where AI permeates nearly every aspect of our digital lives, understanding public trust in these systems is critical. As one of the most advanced AI language models, ChatGPT has become a touchstone for this inquiry. Its widespread adoption and diverse applications highlight the need to take the global pulse on its trustworthiness. Different cultural norms, regulatory environments, and societal expectations around the world influence the level of trust placed in such technologies. Examining public trust in ChatGPT on a global scale not only informs the use and development of AI but also sheds light on the broader societal relationship with technology.

Surveys conducted across various regions in 2023 have offered a comprehensive view of global public trust and perceptions regarding ChatGPT, highlighting the diverse responses and concerns among different demographics and professionals.

Southeast Asian In a February 2023 survey of 4000 participants from four Southeast Asian countries—Indonesia, Malaysia, Singapore, and Thailand—nearly half indicated apprehension about personal data collection in their use of chatbots like ChatGPT. Meanwhile, 42% expressed concerns over ethical matters, including data privacy and intellectual property rights. Furthermore, 53% of respondents feared that ChatGPT might disseminate machine-generated misinformation, lacking complete accuracy. Additionally, 49% were worried about the risk of ChatGPT being compromised or utilised for malicious purposes. Only a small fraction, 2%, reported having none of the listed or other concerns [12].

South Korea In a survey of 1016 South Korean participants aged 13 and older conducted in March 2023, approximately 26.5% found ChatGPT responses to be trustworthy, 62.1% remained neutral, and around 9.7% deemed them untrustworthy. Very few respondents considered it either very trustworthy or very untrustworthy, at 0.9% and 0.8%, respectively [13]. More than a third of respondents said they had tried ChatGPT at least once.

Japan A survey conducted in Japan in June 2023, involving 1056 respondents aged 15–69 years, revealed that 55.6% were somewhat interested in using the OpenAI chatbot ChatGPT in the future. Additionally, the survey indicated that 23.4% of participants were somewhat resistant to the idea of engaging with this artificial intelligence software, a figure slightly higher than the 20.6% who remained undecided on the matter [14].

China The use of ChatGPT faces unique challenges due to national restrictions, highlighting concerns around data privacy, content regulations, and its impact on public opinion, underscoring the complex regulatory landscape in China. Within the Chinese context, employing ChatGPT presents distinctive challenges and dynamics due to its national prohibition. In contrast to numerous other regions where ChatGPT enjoys widespread availability and usage, China has enforced a ban on its utilisation. This prohibition raises inquiries regarding data privacy, content regulations, and apprehensions regarding the technology's impact on public opinion and information dissemination. Furthermore, the ban underscores the unique regulatory framework in China, characterised by rigorous scrutiny and adherence to national policies and guidelines concerning the development and implementation of AI technologies [15].

The United States According to a survey conducted by the Pew Research Center from July 17–23, 2023, the majority of US adults have not utilised ChatGPT, with only 18% reporting usage. Among those familiar with ChatGPT, younger and college-educated adults are more likely to have engaged with it. While ChatGPT is commonly used for both entertainment and educational purposes, a limited number of Americans believe it will significantly influence their employment. Despite the potential implications of generative AI on professions requiring higher education, only 19% of employed adults anticipate a major impact on their roles. The survey underscores a prevalent scepticism regarding the practicality of chatbots in professional environments [16].

Global IT Professionals A June 2023 survey of 3240 IT professionals worldwide found that 81% used ChatGPT for learning new topics, and nearly 59% for code writing, indicating strong professional engagement with the tool for development purposes [17].

Medical Professionals Between April and May 2023, a survey of 456 urologists globally explored ChatGPT's use in academia and clinical practice. Nearly half utilised it for research, and about one-fifth in clinical settings, pointing out ethical dilemmas, such as plagiarism and the risk of generating misleading information, suggesting a need for regulatory guidelines [18].

These studies show that global perceptions of ChatGPT and AI vary significantly. Concerns primarily revolve around data privacy, ethical considerations, and the potential for spreading misinformation. Interest levels differ, with some regions showing keen engagement through rapid app adoption and use in professional contexts, while others display scepticism or regulatory restrictions. The global landscape of trust in ChatGPT is marked by cautious adoption amid optimism for its potential benefits.

6.2.2 Evolving Public Trust

Numerous studies, surveys, and government actions have tracked the evolving public trust in ChatGPT since its inception. This journey reflects an initial phase of curiosity and concern, typical for novel technologies, followed by cautious acceptance and ultimately, a somewhat naïve integration into various aspects of life.

The Google Trends chart (Fig. 6.1) depicting search trends for 'is ChatGPT safe?' reveals an initial peak around the end of March 2023, indicating a surge in public interest or concern. Subsequently, there is a gradual decline, suggesting a potential decrease in concern over time. However, minor fluctuations are observed throughout the timeline, with periodic rises and falls. The overall trend from the peak to the latest data point indicates that although public interest has diminished, there persists a consistent level of concern regarding the safety of ChatGPT.

Late 2022–Early 2023 The introduction of ChatGPT ignited significant public interest and sparked discussions across various platforms. ChatGPT reached one million users within 5 days of its launch and crossed 100 million users in under 3 months. In contrast, TikTok took 9 months, Instagram took 30 months, and Spotify took 55 months to hit the same milestone [19]. Data from Google Trends indicates a continuous rise in searches related to ChatGPT, suggesting growing curiosity and engagement with the topic (Fig. 6.1). ChatGPT-related videos on YouTube attracted more than 65 million views, significantly outperforming other content types in engagement and subscriber growth, regardless of the channel's existing subscriber

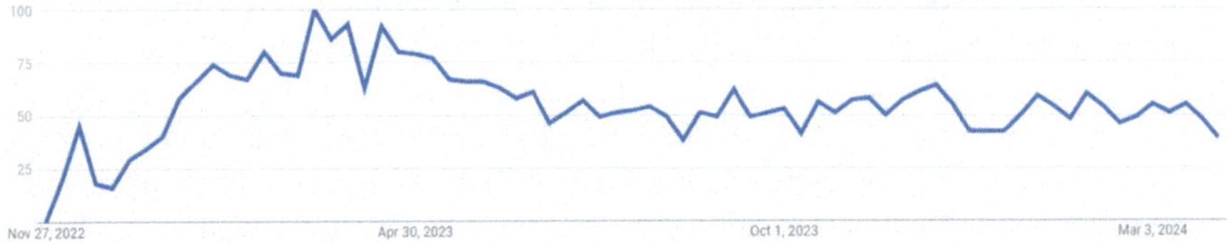

Fig. 6.1 Google Trends search for 'Is ChatGPT Safe?' since ChatGPT's launch to March 31, 2024

base [20]. This demonstrates strong public interest and underscores the growing curiosity and educational demand surrounding AI technologies. A Twitter sentiment analysis up to March 2023 revealed an overall positive attitude towards AI in e-learning, with dominant emotions of trust and joy [21]. Conversely, on Weibo—China's widely used social network—conversations focused on the technical capabilities and prospective uses of AI, reflecting a mix of hope for educational and professional progress and concerns over potential job losses and the spread of misinformation [15]. During the same period, the European Union initiated proactive measures within the AI Act to tackle the unexpected challenges posed by generative AI technologies such as ChatGPT [22].

March 2023 During this period, interest in ChatGPT and similar AI tools increased significantly. Nevertheless, the public adopted a cautious approach to these tools, primarily due to their novelty and the uncertainties they introduced. A Pew Research Center survey revealed that 58% of US adults were aware of ChatGPT, yet only a modest 14% had actually used it. Awareness levels vary by education and income, with higher levels correlating to greater familiarity. Asian Americans are the most likely ethnic group to have heard of ChatGPT. Users generally find ChatGPT somewhat useful, despite concerns over its accuracy and the potential for misinformation. Usage trends and perceptions suggest mixed feelings about the tool's reliability and broader implications for AI technology [23].

Public apprehension and scholarly attention towards AI's ethical and societal implications peaked. A notable incident was a collective call from technologists and scientists for a pause on AI advancements beyond GPT-4, highlighting the growing unease about AI's rapid development and its unforeseen consequences [24]. Some individuals advocated for an extended pause beyond 6 months [25], while others went as far as proposing the cessation of large-scale AI development activities worldwide, calling for an indefinite suspension [26]. This was also a time when Google Trends observed a surge in searches questioning ChatGPT's safety, marking a peak in public concern (Fig. 6.1).

April 2023 The sensitivity around AI's impact on data privacy and intellectual property was underscored by a notable increase in sensitive data leaks and source code exposures via ChatGPT. From April 9–April 15, 2023, there were 319 instances of sensitive data leaks per 100,000 employees globally via ChatGPT, marking a 60% increase from the period between February and March 2023 [27]. This reflects the security vulnerabilities introduced by AI technologies.

In response to the growing influence of AI in academic circles, the scholarly community did take steps to clarify the role of AI in research and publication. Notably, many academic journals have begun establishing policies regarding AI authorship. As reported, by April 1, 2023, over half of the top 300 academic journals had set guidelines for acknowledging AI contributions in research papers, with many implementing policies to prevent AI systems like ChatGPT from being listed as co-authors [28]. Moreover, several leading educational institutions around the world have set restrictions on the use of such AI tools [29].

Amid scepticism about ChatGPT, Italy temporarily restricted access to ChatGPT, citing non-compliance with data privacy regulations but later reversed the decision [30].

May–July 2023 Launched in May 2023, the ChatGPT mobile app swiftly attracted over a million users within 5 days, setting a new speed record for Internet service user growth. Following its launch, the app saw a consistent increase in monthly downloads, with February 2024 alone witnessing over 3.25 million downloads on the App Store and Google Play in the United States [31].

As ChatGPT's mobile app saw record user growth, governments worldwide started to delineate their stance on AI, with the United States establishing a working group on GenAI [32], and China introducing regulations for AI service management [33]. These actions represented a shift from observation to active management of AI's societal integration.

August–October 2023 The sentiment analysis of Reddit posts showed that although Reddit users' perceptions of ChatGPT were slightly negative shortly after its launch, these opinions have gradually become more positive over time [34]. This suggests an adaptation phase where the initial scepticism began to give way to recognition of ChatGPT's utility and potential [34]. Concurrently, a survey by the Pew Research Center conducted from September 26 to October 23, 2023, among US teens aged 13–17, found that about one-fifth of those aware of ChatGPT used it for schoolwork, which represents 13% of all US teens. This usage was more prevalent among older students, with varying levels of awareness and adoption based on race, ethnicity, and household income but no significant gender differences in usage. The survey also explored the acceptability of using ChatGPT for academic tasks, revealing mixed opinions among teens [35].

Meanwhile, the US executive order on AI indicated a strategic approach to harnessing AI's benefits while mitigating its risks [36]. The French government formed an interministerial commission dedicated to GenAI and initiated public calls for projects [37, 38]. The G7 nations introduced the Hiroshima AI Process, which aims to develop global recommendations for trustworthy AI, with a particular focus on what is commonly referred to as 'advanced' AI, including GenAI [39]. The UK government's approach stands out for its originality. Known for its pragmatic and liberal stance, it opted not to introduce any new specific regulations. Instead, it provided the private sector with a sandbox dataset to facilitate testing of AI models' regulatory compliance, GenAI included [40].

Late 2023–Early 2024 Surveys by the Pew Research Center revealed nuanced public engagement with ChatGPT in the United States, from its use among teens for schoolwork to adults' perspectives on its reliability and impact on employment [16]. Despite a slight increase in usage, trust issues persisted, particularly concerning

ChatGPT's reliability for election information, highlighting ongoing concerns about misinformation [41].

Throughout this period, the response to ChatGPT and generative AI has been characterised by a cycle of intrigue, concern, adaptation, and indulgence. There is a sense of AI indulgence, with many people exploring its capabilities excitedly without fully considering its implications, while others might become overly reliant on AI without critically evaluating its limitations. The global dialogue on ChatGPT's role in society underscores the need for a balanced approach that fosters innovation while addressing ethical, privacy, and security concerns.

To summarise, the public's trust in ChatGPT has experienced a nuanced evolution from its initial launch to widespread adoption. Early fascination gave way to security and ethical concerns, reflected in global search trends and scholarly debates. Despite varying levels of user engagement, with more awareness than actual usage, there has been a shift towards cautious acceptance, particularly in educational settings. This caution seems linked to the newness and unknown qualities of the product. Once users overcome initial hesitations, there appears to be a trend towards naively indulging in the tool. Concerns about AI's rapid growth spurred calls for regulatory pauses and resulted in policy adaptations and increased governance, aiming to balance innovation with ethical and privacy considerations. Over time, public sentiment has shown signs of positive adaptation, though concerns persist, particularly regarding misinformation and the impact on employment, underscoring the ongoing dialogue about the responsible integration of AI into society.

6.3 Trustworthiness in AI

The prevalence of ChatGPT and similar AI technologies has brought AI closer to the general public. With an increasing number of individuals using AI to process their personal data, the trustworthiness of these systems becomes crucial. Despite advancements in AI, challenges persist in ensuring its reliability, including technical limitations, legal concerns, and ethical dilemmas. Addressing these challenges is crucial for maintaining user trust and ensuring the responsible use of AI in handling sensitive information.

6.3.1 Technical Challenges

As AI systems like ChatGPT become more prevalent across various sectors, the issues of trust and reliability gain heightened importance. These systems, designed to mimic human-like interactions, face numerous technical challenges that can undermine their effectiveness and the trust users place in them. Ensuring the trustworthiness of AI involves addressing a complex mix of misinformation, security vulnerabilities, and ethical concerns, all of which are pivotal in determining how these technologies will be integrated into daily life and critical infrastructures.

Misinformation and Fake Information Despite extensive training, ChatGPT may inadvertently propagate false information, such as in areas like market trends or financial forecasts, leading to inaccuracies and a spread of misinformation, including hallucination [42]. 'Hallucination' refers to the phenomenon where AI models generate information that is not based on factual data or logical reasoning, potentially contributing to the dissemination of false or misleading content [43]. The dynamic nature of false information necessitates continuous updates and training for AI to effectively recognise and mitigate new forms of misinformation [44].

Limited Contextual Understanding and Causal Reasoning While ChatGPT has made strides in understanding context, its limitations in maintaining complex conversations and lacking true causal reasoning can lead to misleading outputs. Additionally, it may lack genuine comprehension, leading to potentially misleading interactions with users [45]. This is highlighted by variability in reliability across domains and difficulties in identifying unanswerable questions, which often result in irrelevant guesses [46].

Security Vulnerabilities AI models, including ChatGPT, are susceptible to various security threats throughout their lifecycle. From training to deployment, these models can be targeted by attacks designed to compromise their integrity, steal sensitive data, or manipulate outputs. Malicious users may exploit ChatGPT's capabilities to generate harmful content or engage in social engineering attacks. During the training phase, attackers can introduce poisoned or adversarial samples to undermine the performance of the model [47], or they can execute backdoor attacks to manipulate the results generated by the model [48]. Even after deployment, there is a risk of attackers stealing the AI model or its functionality through sophisticated methods such as intelligent model theft attacks [49]. As more complex strategies are adopted by large AI models such as ChatGPT, they become vulnerable to a wider range of security threats, including emerging ones such as jailbreak and prompt injection [50, 51]. Such vulnerabilities pose significant risks, especially as AI models become integrated into critical infrastructure and sensitive applications.

Dependency and Data Quality The performance and reliability of ChatGPT heavily depend on the quality and diversity of its training data. Over-reliance on flawed or biased data can lead to undesirable outputs. Similarly, the way users phrase prompts can significantly affect the AI's responses, underscoring the importance of careful interaction design [52].

Malicious Applications The potential for ChatGPT and similar AI tools to be used for harmful purposes, such as creating malicious code or facilitating cyberattacks, is a growing concern. Instances of cybercriminals exploiting ChatGPT's capabilities for malware creation or state-sponsored actors considering its use for geopolitical aims underscore the dual-use nature of AI technologies [53]. A study indicates that 51% of IT professionals believe we are close to witnessing a successful cyberattack attributed to ChatGPT. Additionally, 71% think that foreign states might already be using ChatGPT for malicious activities against other nations [54].

Transparency and Explainability The 'black box' nature of AI systems complicates efforts to build trust. Making AI decisions transparent and explainable is crucial for users to understand, trust, and effectively interact with AI technologies. Research into interpretability techniques aims to address this challenge, ensuring that AI's decision-making processes can be scrutinised and understood [52].

6.3.2 Legal Challenges

Beyond the technical hurdles, AI systems also face significant legal challenges. These include the risks of malicious use, the complexities of intellectual property rights, and the critical need for clear accountability. Addressing these challenges is critical to ensuring the responsible use and trustworthiness of AI technologies.

Disinformation and Malicious Use The rapid advancements in AI technology have led to an increase in disinformation and false evidence, including deep fake content and fake news [55]. These tools can be used to disseminate malware, ransomware, and phishing emails. ChatGPT's proficiency in generating convincing dialogues can facilitate the crafting of phishing emails that deceive recipients into clicking on harmful links, downloading malware, or divulging confidential information [56]. As a result, new forms of criminal activity, such as AI fraud, defamation, identity theft, and impersonation, have emerged [57]. For instance, ChatGPT's ability to produce deceitful and unethical responses can be misused by individuals with malicious intent to commit fraud, imitate speech patterns for impersonation, and create malicious code for hacking. This highlights the pressing need to establish mechanisms for tracing the origin and regulating materials generated by generative AI models to ensure accountability.

Intellectual Property (IP) and Ownership Intellectual Property (IP) and Ownership concerns significantly impact AI's trustworthiness and user trust. When AI generates content or innovations, determining the rightful owner of this IP can be complex, leading to legal ambiguities [58]. Users and creators may feel hesitant to fully engage with AI technologies due to fears of misappropriation of their ideas or creations. Ensuring clear guidelines and protections for IP rights in the context of AI is essential for fostering a trustful environment, encouraging innovation, and ensuring that creators feel secure in their contributions and usage of AI technologies.

Liability and Accountability Liability and accountability significantly influence AI's trustworthiness and users' trust. Determining who is responsible when AI systems make errors or produce unethical outcomes is challenging. This ambiguity can undermine users' trust, as they fear misuse or mistakes without adequate recourse. Establishing clear accountability and liability frameworks in AI development and deployment is vital for fostering trust. It reassures users that there are measures to responsibly and ethically address potential issues [59].

6.3.3 Ethical Challenges

As AI systems like ChatGPT become more integrated into everyday activities, it is crucial to address not only the risks related to privacy, bias, misuse, and intellectual integrity but also the broader implications for what it means to be human. AI could negatively influence how and what we trust. Exploring these ethical dimensions is essential not only to maintain public trust but also to protect the rights and privacy of users thereby upholding the standards of responsible AI use in various domains.

Privacy and Data Security Language models like ChatGPT often interact with users who may provide personal information to complete their queries, raising concerns about privacy and data security. Interactions with ChatGPT may result in the recording of sensitive user information, potentially creating privacy concerns if not properly managed. For example, ChatGPT has the ability to retain information related to conversations, user input, and data stored in cookies and logs during its interactions with users [60]. When ChatGPT communicates with patients and health care professionals, it might collect and keep confidential health details. This data could include medical records, outcomes of tests, medical conclusions, and more private information. Ensuring the security of this data is essential for preserving the privacy of patients and adhering to relevant privacy laws, like the Health Insurance Portability and Accountability Act (HIPAA) in the United States or equivalent regulations in other nations [61, 62].

Bias and Discrimination The challenge of bias and discrimination in AI is a critical issue influenced by the quality and diversity of their training data. AI derives its knowledge and intelligence from vast amounts of Internet data, which may contain biases inherent in the original material. These biases can manifest in the model's outputs, leading to unequal treatment or reinforcement of stereotypes [63]. Similar to other language models, ChatGPT may inadvertently perpetuate biases present in the training data, posing challenges in generating objective and impartial responses. This is particularly evident in how ChatGPT processes and generates content based on biases that may exist in the sources it learns from.

For example, one study observed that ChatGPT showed a preference for Irish liberal politicians over conservative politicians when tasked with creating limericks [64]. Further investigation of ChatGPT's responses to political inquiries revealed consistent tendencies: An analysis of 630 political questions revealed that Germany and the Netherlands tend toward left-leaning and environmentally friendly policies [65]. Likewise, when asked repeatedly the same political questions, ChatGPT demonstrated a bias against left-leaning political positions in several countries, including Brazil, the United States, and the United Kingdom [66, 67].

These instances highlight a broader issue in which biases in online messages related to factors such as gender, religion, race, politics, or geography can be inadvertently reflected or even amplified in ChatGPT responses [68]. These examples highlight the importance of addressing these challenges to ensure that AI produce unbiased and equitable responses, increasing their reliability and trustworthiness.

Misuse and Abuse The flexibility of ChatGPT allows for creative but potentially harmful uses, including the generation of misleading information or the bypassing of safety mechanisms. ChatGPT allows users to customise interaction styles by defining roles or tasks, with the default recommended role being that of an assistant. With the increasing use of applications, more and more system roles are being utilised. Some users can also assign 'jailbreak roles', such as DAN (Do Anything Now), to get around ChatGPT's safeguard mechanisms and coerce ChatGPT into answering inappropriate questions [46].

In addition, utilising ChatGPT for deceptive intentions, such as fabricating false reviews or spreading misinformation, raises ethical issues. The potential misuse of intelligent content generated by ChatGPT to create fake news and mislead audiences has arisen in the media industry [69]. Addressing these concerns requires robust safeguards and continuous monitoring.

Intellectual Property and Academic Integrity The utilisation of AI in academic writing brings to the forefront issues of plagiarism, the undermining of human expertise, and the potential for AI to impact the quality and integrity of scholarly work. The debate extends to the ethical use of AI in manuscript preparation, with calls for clear guidelines on AI authorship and the responsible use of AI tools in academic settings [70–73]. Additionally, AI has the potential to enable academic dishonesty, including cheating in classes, artistic plagiarism, and fraudulent academic publications, making such misconduct easier to commit and more difficult to detect. Similarly, apprehension arose that ChatGPT could further reduce the intellectual space of scholars, impede the expression of diverse ideas, and compromise the integrity of academic environment-monitoring functions [74].

6.4 Discussion

Trust is a complex, multidimensional construct that is difficult to operationalise, measure, and interpret [75]. In fact, there is more than one way to define and describe trust. As Simpson [75] shows, 'trust can be construed in different ways, and it might have varying importance at different stages of relationship development'.

Understanding the factors that influence public trust in the ChatGPT era is crucial for developing strategies to enhance AI trustworthiness. Moreover, maintaining a balance between user trust and AI trustworthiness is essential to ensure healthy and responsible use of AI technologies. This balance supports not just user confidence but also the sustainable integration of AI into daily activities.

6.4.1 Factors Influencing Public Trust in ChatGPT Era

Trust in AI, particularly in technologies like ChatGPT, is recognised as a fundamental factor influencing their successful integration and user adoption. This multidimensional concept encompasses users' confidence in AI's performance, transparency, ethical conduct, and the fulfilment of user needs. User trust in AI-enabled systems is increasingly recognised and proven as a key element in driving adoption [76].

Comprehensive analyses have identified several key factors that significantly impact public trust in the ChatGPT era:

Perception of Risk and Trust in AI Research highlights perceived risk and trust as critical to AI acceptance. Users' apprehensions about AI's potential negative impacts can undermine trust, which is essential for technology adoption. Trust is primarily derived from confidence in AI's decision-making accuracy and transparency in operations [77]. A study investigates factors influencing Indonesian college students' use of ChatGPT, focusing on trust, performance expectancy, and effort expectancy. Trust, particularly in data security and information correctness, significantly affects ChatGPT usage [78].

The media serves as a crucial source of public awareness about AI, including technologies like ChatGPT, thereby significantly influencing public perceptions of AI's trustworthiness. The portrayal of AI in the media often oscillates between its societal benefits and potential dangers. A study analysing UK news headlines from January to May 2023 reveals a predominantly sensationalised narrative, emphasising the potential threats posed by AI. While this type of media coverage can shape public attitudes towards fear and mistrust [79], it may also build curiosity, as bad news often gains more traction than good news.

Anthropomorphism and Design Novelty The attribution of human-like qualities to AI systems enhances user trust through familiarity. Furthermore, the unique design of AI interfaces can attract users, although trust ultimately depends on the system's performance. Institutional policies play a vital role in fostering trust by ensuring responsible and ethical AI deployment, especially in sensitive environments like higher education [80].

Institutional Policies and Ethical AI Deployment Institutional policies are crucial in fostering trust by ensuring responsible and ethical AI deployment. This is especially important in sensitive environments like higher education, where trust can be significantly influenced by the perception of AI's ethical use and accountability [80].

Capabilities Versus Benevolence An online survey indicated that people tend to trust AI's capabilities more than its benevolence, with technology expertise, AI knowledge, and familiarity with AI significantly affecting trust levels. This suggests that increasing public awareness and understanding of AI can positively influence trust [81].

Personalisation and User Experience While personalisation in AI interactions reduces perceived creepiness and boosts trust, motivations for task efficiency and social interaction can increase feelings of unease. Creepiness negatively affects intentions to continue using AI, whereas trust promotes continued engagement [82].

Conversational Skills and Human-Like Traits The human-like conversational abilities of AI, such as speaking and listening, significantly enhance user trust. AI chatbots perceived as warm and competent are viewed as more trustworthy, underscoring the importance of developing AI with human-like characteristics [10, 83].

Social Endorsement and Community Validation A survey highlighted the importance of perceiving ChatGPT as novel, human-like, and socially endorsed for widespread user acceptance. Trust in AI is cultivated through a combination of innovation, relatability, and community validation [84]. This structured approach categorises the various factors influencing trust in the ChatGPT era, highlighting how they interact and impact user adoption and acceptance.

These findings collectively emphasise the complexity of building trust in AI technologies like ChatGPT. Addressing these multifaceted aspects—from enhancing AI's human-likeness to ensuring ethical deployment and fostering a nuanced public discourse—can significantly impact the degree of trust and acceptance AI technologies receive from the public.

6.4.2 Strategies for Enhancing Trustworthiness

In the ChatGPT era, enhancing AI's trustworthiness is a multifaceted challenge that integrates technical solutions, ethical considerations, legal frameworks, and industry practices to foster a safe and reliable environment for AI interactions. The strategies for enhancing trustworthiness are designed to ensure that AI technologies like ChatGPT operate within a robust framework that prioritises user safety, data privacy, and ethical standards.

Technical Solutions

- *Data Protection Measures*

Implementing Encryption, Anonymisation, and Secure Data Storage Practices: These critical measures protect user data and enhance the reliability of AI systems. Encryption safeguards data in transit and at rest, while anonymisation helps in masking user identities, minimising personal data exposure. Secure data storage practices ensure that sensitive information is protected against unauthorised access and data breaches thereby enhancing the overall security and resilience of AI technologies [85].

- *Human-like Interaction Enhancements*

Enhancing ChatGPT's Human-like Traits: By improving the AI's ability to mimic human interactions, user expectations of its performance are positively influenced, leading to increased engagement. Technical improvements such as refining natural language processing algorithms, integrating context-aware responses, and enhancing emotional intelligence capabilities make interactions with ChatGPT more natural and satisfying [84]. While these enhancements help build trust by providing users with more relatable and intuitive experiences, it is crucial to maintain transparency that this is a machine, not a human. This balance prevents overtrust and encourages users to remain critical thinkers, even as the interface becomes more intuitive.

- *Content Moderation and Safety Implementing*

Moderation and Content Filtering Mechanisms: To prevent the model from generating and disseminating false information or engaging in harmful behaviour, it's crucial to employ moderation and content filtering mechanisms. These measures can reduce the likelihood of inappropriate responses and protect users from potentially harmful content.

- *Bias Detection and Mitigation*

Developing Techniques to Detect, Identify, and Mitigate Biases: Research efforts should focus on creating advanced techniques to ensure ChatGPT delivers fair and impartial responses. This involves detecting and correcting biases in the AI's outputs, which is crucial for maintaining the model's integrity and ensuring it treats all users equitably.

- *Transparency and Explainability*

Enhancing Transparency and Explainability in AI Models: The lack of transparency in how AI models, like ChatGPT, formulate their responses can lead to significant ethical concerns. As AI capabilities advance, it becomes increasingly important for users to understand the reasoning behind the AI's conclusions. Developing methods to provide meaningful explanations for the model's responses can significantly enhance user trust in the AI system.

Legal Frameworks and Ethical Guidelines

- *Ethical Frameworks for Trustworthy AI*

Development of Ethical Guidelines: The creation and adherence to ethical guidelines are crucial in ensuring AI's trustworthiness. The European Commission's High-Level Expert Group on AI has set forth the Ethics Guidelines for Trustworthy Artificial Intelligence. These guidelines emphasise that trustworthy AI must be lawful, ethical, and robust, encompassing principles such as human agency and oversight, technical robustness and safety, and privacy and data governance [86]. Key Requirements for Trustworthy AI: The guidelines specify seven key requirements for AI systems: Human agency and oversight; Technical robustness and safety; Privacy and data governance; Transparency; Diversity, non-discrimination, and fairness; Societal and environmental well-being; Accountability [86].

- *Research on Human Factors and Compliance*

Compared to the fast-paced advancements in technology, research into human factors and compliance in AI is still lacking [87]. It is critical to focus on the ethical and legal aspects of privacy protection in AI settings. Research should be directed towards key areas like informed consent, algorithmic bias and fairness, legal frameworks and compliance, as well as transparency, explainability, accountability, liability, and global cooperation on legal standards. These factors are crucial in creating AI environments that protect privacy and give individuals more control over their personal data. Addressing these issues is vital for the responsible development and use of AI systems.

- *Media's Role in Shaping AI Perceptions*

Balanced Media Narrative: Policymakers should advocate for a balanced media narrative that equally highlights the benefits and challenges associated with AI technologies. By promoting ethical journalism that avoids sensationalism, the media can provide a more nuanced understanding of AI, fostering informed public discourse [79].

- *Public Education and Awareness Improving AI Literacy* Supporting initiatives that improve AI literacy among the public, such as educational programmes and transparent reporting on AI developments, is vital. This approach ensures that citizens are well-equipped to appreciate the positive impacts of AI while being aware of potential risks, leading to more informed opinions and decisions regarding AI technologies.
- *Consent and Transparency in AI Interactions*

Informed Consent and User Awareness: Users interacting with AI systems like ChatGPT should be clearly informed that they are communicating with an AI. Providing comprehensive disclosures about the use of AI and obtaining informed consent from users can promote transparency and the ethical use of AI.

- *Addressing Dual-Use and Ethical Dilemmas*

Mitigating Dual-Use Risks: AI technologies, including language models like ChatGPT, have potential for both beneficial and harmful uses, such as in cyberbullying or disinformation campaigns. It is essential to conduct research and develop methods to address these dual-use scenarios, ensuring measures are in place to mitigate the potential adverse impacts of AI-generated content [88].

Industry Practices

- *Regulatory Compliance and Data Protection*

Enhancements to Data Protection Laws: Strengthening existing laws on data protection is critical. These enhancements should aim at safeguarding user data and aligning AI operations with higher ethical and legal standards. Such measures not only enhance trust but also ensure that AI technologies like ChatGPT operate within a secure legal framework [89].

- *Self-Regulation and Industry Standards*

Setting Industry Standards for AI Data Usage: Encouraging the setting of industry standards plays a vital role in ensuring consistent and ethical data handling practices across different sectors. By adopting these standards, AI developers can avoid ethical pitfalls and enhance the credibility of their technologies. Clear Accountability Mechanisms and Liability Definitions: Establishing clear accountability mechanisms and defining liability for AI development and deployment are essential. Transparent policies should be created to specify who is responsible for various aspects of AI behaviour and its outcomes. Additionally, legal frameworks that clarify liability in the event of errors or misconduct should be established to foster user trust [89].

- *User-Oriented Measures and Advocacy*

AI Literacy and Intelligence Augmentation: Advocating for improved AI literacy among the general public and stakeholders can demystify AI technologies and foster an informed user base. Intelligence augmentation, which focuses on enhancing human decision-making with AI support, can also play a crucial role in building trust by demonstrating the utility and reliability of AI in augmenting human capabilities [90].

Robust Auditing Processes: Companies should implement robust auditing processes to regularly assess AI systems for ethical compliance and performance accuracy. Such practices ensure that AI operations are continually refined and kept in check, reinforcing trust among users.

By integrating these strategies, the goal is to create a trustworthy AI ecosystem where users can interact with technologies like ChatGPT confidently, knowing that their data are protected, their rights are respected, and the AI they are interacting with adheres to the highest ethical and legal standards.

6.4.3 Balancing User Trust and AI Trustworthiness

Trust in AI is crucial for its integration and adoption, yet it also introduces significant risks with broad societal and economic impacts. Users' willingness to adopt AI technologies is heavily influenced by the information they receive. While positive reports often increase acceptance, negative information can both diminish it and spark curiosity, potentially creating a misalignment between user trust and AI trustworthiness. This discrepancy can lead to over-trust or under-trust, each presenting unique challenges. Specifically, when trust exceeds trustworthiness, misuse may occur, and when trustworthiness surpasses trust, disuse is likely [91].

Figure 6.2 demonstrates how the relationship between user trust and AI trustworthiness affects user engagement with AI. Ideal interaction happens when trust is aligned with trustworthiness, marked by the green area in the figure. In cases where

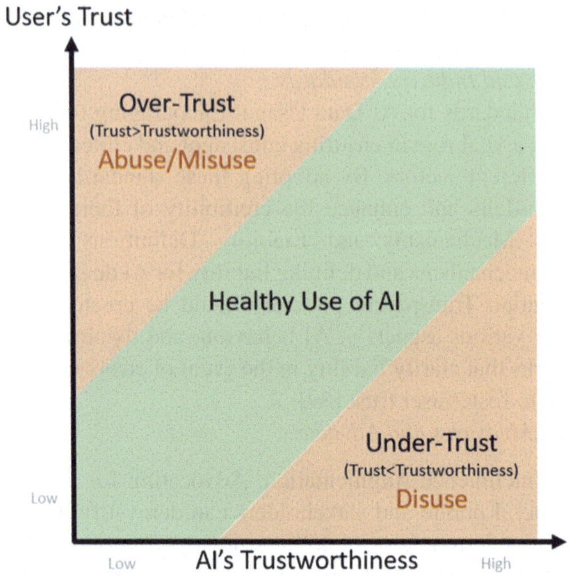

Fig. 6.2 Balancing user trust and AI trustworthiness [92]

AI shows low trustworthiness, perhaps due to unreliability or external manipulation, excessive trust can lead to misuse or abuse (indicated in the upper orange area). Conversely, a lack of trust compared to an AI's capabilities can result in its under-utilisation or complete disuse, shown in the lower orange area. These misalignments introduce risks of over-trust and under-trust.

Misuse in this context means overly relying on AI for tasks it may not be suited for, or without adequate oversight. Disuse refers to the insufficient application of AI capabilities, such as ignoring or turning off AI-driven functionalities that could enhance safety or efficiency. Abuse involves applying AI without careful consideration of its potential negative impacts on human systems [91]. These scenarios highlight the critical need for balancing trust with the proven reliability and ethical deployment of AI technologies to prevent potential negative consequences.

Balancing user trust with the trustworthiness of AI systems is a critical aspect of ensuring healthy interaction between humans and AI technologies. A misalignment between these two elements can lead to either misuse/abuse or disuse of AI, each carrying its own set of risks and consequences. Understanding this balance is essential for fostering a responsible and beneficial integration of AI into daily life and work processes.

When user trust exceeds the actual trustworthiness of an AI system, there is a risk of misuse or abuse. This situation arises when users place undue confidence in the AI's capabilities, assuming it to handle tasks or make decisions beyond its designed capacity. For instance, an excessively trusted AI system might be used to make critical medical or financial decisions without sufficient human oversight, leading to potential harms if the AI fails to consider all necessary factors or operates on biased data.

Misuse might also involve relying on AI for tasks it was not appropriately validated for, driven by the false belief that the AI understands more than it actually does. This can be particularly dangerous in fields like healthcare or law enforcement, where incorrect or inappropriate AI recommendations could lead to severe consequences.

Abuse, on the other hand, occurs when over-trust in AI capabilities leads to manipulative or unethical applications, such as deploying AI systems in ways that exploit user data without consent or in a manner that intentionally deceives users to benefit a particular agenda. The line between leveraging AI's capabilities for innovation and abusing its power can be thin, and it is often crossed when trust in the technology masks its underlying limitations or biases.

Disuse Due to Inadequate Trust

Conversely, when trust is significantly lower than AI's trustworthiness, it leads to disuse. This scenario occurs when AI systems are capable and reliable but are underutilised due to user scepticism or lack of understanding about what AI can safely and effectively do. An example can be found in older generations of users who may be more hesitant to rely on online banking AI systems that facilitate fraud detection and transaction processing, despite these systems having proven reliability.

Disuse results in missed opportunities for efficiency and enhancement of human capabilities. It can slow down the adoption of beneficial technologies, hindering societal and economic progress. Moreover, it can perpetuate a cycle of scepticism that prevents constructive feedback loops necessary for AI improvement and integration.

The key to a healthy use of AI is aligning user trust with the actual trustworthiness of AI systems. Users need clear, accessible information about what AI can and cannot do, how it makes decisions, and the safeguards in place to prevent errors or biases. By fostering an environment where AI's capabilities and limitations are transparently communicated, and its applications are ethically governed, both misuse due to over-trust and disuse due to under-trust can be mitigated. This balanced approach ensures that AI remains a powerful tool for innovation, aligned with human values and societal norms.

6.5 Conclusion

This chapter presents a thorough exploration of the evolving landscape of public trust towards AI technologies like ChatGPT, particularly focusing on the implications for privacy and data security. It begins by addressing the transformative impact of ChatGPT on human–AI interactions, emphasising the increasing concerns about privacy that arise as these technologies become integral to our daily lives. The fundamental thesis is that maintaining user trust is paramount, given the capacities of AI to handle vast amounts of personal information.

The authors dedicated to examining global attitudes towards ChatGPT, revealing a complex tapestry of trust and scepticism across different regions. For instance, surveys highlight varying levels of apprehension about data privacy, ethical considerations, and the potential misuse of AI. In Southeast Asia, a noticeable percentage of respondents express fears about personal data collection and the ethical implications of AI use. Similarly, the survey data from the United States and South Korea reflect a spectrum of trust and neutrality, suggesting that the cautious engagement with ChatGPT among the broader public may stem from its novelty and the unknown aspects it introduces.

The chapter also discusses the temporal dynamics of trust in AI, noting a pattern of initial curiosity and engagement, followed by a phase of scepticism due to concerns over data privacy and misinformation. This observation is supported by search trends and survey responses that indicate fluctuating levels of concern and acceptance over time. The authors argue that these shifts underscore the need for robust privacy protections, clear regulatory frameworks, and continuous improvements in AI's ethical standards to foster a stable trust relationship with users.

Furthermore, the authors propose several strategies to enhance the trustworthiness of AI systems. These include technical solutions like improving AI's human-like interaction capabilities and implementing robust data protection measures, alongside advocating for balanced media portrayals of AI. The discussion extends to the role of ethical guidelines and frameworks in establishing a foundation for trustworthy AI, emphasising the importance of transparency, accountability, and user-focused design in maintaining public trust. By examining the relationship between trust and trustworthiness and the risks posed by their imbalance, the authors advocate for a balanced approach to user trust and AI trustworthiness to promote healthier AI usage.

References

1. Wu, T., He, S., Liu, J., Sun, S., Liu, K., Han, Q. L., & Tang, Y. (2023). A brief overview of chatgpt: The history, status quo and potential future development. *IEEE/CAA Journal of Automatica Sinica, 10*(5), 1122–1136.
2. Google. (2023). Introducing Google Gemini AI. https://blog.google/technology/ai/google-gemini-ai/sundar-note. Accessed: 2024-04-14.
3. Microsoft. (2023, March). *Introducing Microsoft 365 copilot: Your copilot for work*, https://blogs.microsoft.com/blog/2023/03/16/introducing-microsoft-365-copilot-your-copilot-for-work/. Accessed: 2024-04-14.
4. Borji, A. (2022). Generated faces in the wild: Quantitative comparison of stable diffusion, midjourney and dall-e 2. *arXiv preprint arXiv:2210.00586*.
5. MIT Technology Review. (2024, February). *OpenAI's amazing new generative AI video model, Sora,* https://www.technologyreview.com/2024/02/15/1088401/openai-amazing-new-generative-ai-video-model-sora/. Accessed: 2024-04-14.
6. Yang, J., Jin, H., Tang, R., Han, X., Feng, Q., Jiang, H., Zhong, S., Yin, B., & Hu, X. (2023). Harnessing the power of LLMs in practice: A survey on chatgpt and beyond. *ACM Transactions on Knowledge Discovery from Data*.

7. Wu, X., Duan, R., & Ni, J. (2023). Unveiling security, privacy, and ethical concerns of chatgpt. *Journal of Information and Intelligence, 2*, 102.
8. Harrison McKnight, D., & Chervany, N. L. (2001). Trust and distrust definitions: One bite at a time. In R. Falcone, M. Singh, & Y. H. Tan (Eds.), *Trust in Cyber-societies* (pp. 27–54). Springer.
9. Schoorman, F. D., Mayer, R. C., & Davis, J. H. (2007). An integrative model of organizational trust: Past, present, and future.
10. Cheng, X., Zhang, X., Cohen, J., & Mou, J. (2022). Human vs. AI: Understanding the impact of anthropomorphism on consumer response to chatbots from the perspective of trust and relationship norms. *Information Processing & Management, 59*(3), 102940.
11. Adıgüzel, T., Kaya, M. H., & Cansu, F. K. (2023). Revolutionizing education with AI: Exploring the transformative potential of chatgpt. *Contemporary Educational Technology, 15*, ep429.
12. Leading user concerns held about chatgpt among respondents in southeast Asia as of February 2023. (2023). https://www.statista.com.goddard40.clarku.edu/statistics/1382944/sea-top-user-concerns-about-chat-gpt/. Accessed: 03-22-2024.
13. Perceived reliability of chatgpt in South Korea as of march 2023. (2023). https://www.statista.com/statistics/1381444/south-korea-chatgpt-result-reliability-perception/. Accessed: 03-22-2024.
14. Share of people who want to use chatgpt in japan as of June 2023. (2023). https://www.statista.com.goddard40.clarku.edu/statistics/137611 3/japan-share-of-people-who-want-to-use-chatgpt/. Accessed: 03-22-2024.
15. Lian, Y., Tang, H., Xiang, M., & Dong, X. (2024). Public attitudes and sentiments toward chatgpt in China: A text mining analysis based on social media. *Technology in Society, 76*, 102442.
16. Pew Research Center. (2023). *Most Americans haven't used chatgpt, few think it will have a major impact on their job*, https://www.pewresearch.org/short-reads/2023/08/28/most-americans-havent-used-chatgpt-few-think-it-will-have-a-major-impact-on-their-job/. Accessed: 2024-03-31.
17. Reasons for using chatgpt among technology professionals worldwide in 2023. (2023). https://www.statista.com.goddard40.clarku.edu/statistics/1450301/chatgpt-use-cases/. Accessed: 03-22-2024.
18. Eppler, M., Ganjavi, C., Ramacciotti, L. S., Piazza, P., Rodler, S., Checcucci, E., Rivas, J. G., Kowalewski, K. F., Belenchón, I. R., Puliatti, S., et al. (2024). Awareness and use of chatgpt and large language models: A prospective cross-sectional global survey in urology. *European Urology, 85*(2), 146–153.
19. Anonymous: Chatgpt on track to surpass 100 million users faster than Tiktok or Instagram, Ubs. (2023). https://finance.yahoo.com/news/chatgpt-on-track-to-surpass-100-million-users-faster-than-tiktok-or-instagram-ubs-214423357.html?guccounter=1. Accessed 15 April 2024.
20. Hussain, K., Khan, M. L., & Malik, A. (2024). Exploring audience engagement with chatgpt related content on youtube: Implications for content creators and AI tool developers. *Digital Business, 4*(1), 100071.
21. Stracqualursi, L., & Agati, P. (2024). Twitter users perceptions of AI-based e-learning technologies. *Scientific Reports, 14*(1), 1–14.
22. European Parliament. (2023). *AI act: A step closer to the first rules on artificial intelligence*, https://www.europarl.europa.eu/news/en/press-room/20230505IPR84904/ai-act-a-step-closer-to-the-first-rules-on-artificial-intelligence. Accessed: 2024-03-24.
23. Vogels, E. A. (2023). *A majority of Americans have heard of chatgpt, but few have tried it themselves*, https://www.pewresearch.org/short-reads/2023/05/24/a-majority-of-americans-have-heard-of-chatgpt-but-few-have-tried-it-themselves/. Accessed: 2024-03-31.
24. O'Brien, M. (2023). *Musk, scientists call for halt to AI race sparked by chatgpt (3 2023)*, https://apnews.com/article/artificial-intelligence-chatgpt-risks-petition-elon-musk-steve-wozniak-534f0298d6304687ed080a5119a69962
25. Noonan, P. (2023). *A six-month AI pause? no, longer is needed*, https://www.wsj.com/articles/a-six-month-ai-pause-no-longer-is-needed-civilization-danger-chat-gpt-chatbot-internet-big-tech-4b66da6e. Accessed on 11/11/2023.

26. Yudkowsky, E. (Mar, 2023). Pausing AI developments isn't enough. we need to shut it all down. *Time Magazine*. https://time.com/6266923/ai-eliezer-yudkowsky-open-letter-not-enough
27. Number of cases employees worldwide input sensitive data on chatgpt from February to April 2023. (2023). https://www.statista.com.goddard40.clarku.edu/statistics/1378692/corporate-sensitive-data-spill-chatgpt/. Accessed: 03-22-2024.
28. Lund, B. D., & Naheem, K. (2024). Can chatgpt be an author? A study of artificial intelligence authorship policies in top academic journals. *Learned Publishing, 37*(1), 13–21.
29. World Economic Forum. (2023). Why chatgpt raises issues of trust in AI and science. https://www.weforum.org/agenda/2023/02/why-chatgpt-raises-issues-of-trust-ai-science/. Accessed: 2024-03-31.
30. Supantha Mukherjee, G. V. (2023). Italy restores chatgpt after open AI responds to regulator, https://www.reuters.com/technology/chatgpt-is-available-again-users-italy-spokesperson-says-2023-04-28/. Accessed: 2024-03-24.
31. Number of monthly chatgpt and Gemini AI mobile app downloads in the united states from May 2023 to February 2024. (2023). https://www.statista.com.goddard40.clarku.edu/statistics/1454331/united-states-chatgpt-gemini-app-downloads/. Accessed: 03-22-2024.
32. The White House: Pcast working group on generative AI invites public input (2023, May). https://www.whitehouse.gov/pcast/briefing-room/2023/05/13/pcast-working-group-on-generative-ai-invites-public-input/. Accessed: 2024-03-24.
33. The State Council, The People's Republic of China: China moves to support generative AI, regulate applications (2023, Jul). https://english.www.gov.cn/news/202307/13/contentWS64aff5b3c6d0868f 4e8ddc01.html
34. Talafidaryani, M., & Moro, S. (2024). *Public perception of chatgpt on Reddit social media platform: Topic modeling and sentiment analysis study*. Available at SSRN 4716839.
35. Sidoti, O., & Gottfried, J. (2023). *About 1 in 5 u.s. teens who've heard of chatgpt have used it for schoolwork*. Pew Research Center. https://www.pewresearch.org/short-reads/2023/11/16/about-1-in-5-us-teens-whove-heard-of-chatgpt-have-used-it-for-schoolwork/. Accessed: 2024-03-31.
36. The White House. (2023). Executive order on the safe, secure, and trustworthy development and use of artificial intelligence, https://www.whitehouse.gov/briefing-room/presidential-actions/2023/10/30/executive-order-on-the-safe-secure-and-trustworthy-development-and-use-of-artificial-intelligence/. Accessed: 2024-03-24.
37. Gouvernement Français: Comité de l'intelligence artificielle générative. (2023). Available online at https://www.gouvernement.fr/communique/comite-de-lintelligence-artificielle. Accessed on 2024-03-24.
38. Bpifrance. (2023). Appel à projets : Communs numériques pour l'intelligence artificielle générative. https://www.bpifrance.fr/nos-appels-a-projets-concours/appel-a-projets-communs-numeriques-pour-lintelligence-artificielle-generative.
39. G7 Hiroshima AI process g7 digital tech ministers' statement. (2023). https://www.politico.eu/wp-content/uploads/2023/09/07/3e39b82d-464d-403a-b6cb-dc0e1bdec642-230906Ministerial-clean-Draft-Hiroshima-Ministers-Statement68.pdf. Accessed 2024-03-24.
40. Cantens, T. (2024). *How will the state think with chatgpt? The challenges of generative artificial intelligence for public administrations* (pp. 1–12). AI & Society.
41. McClain, C. (2024). Americans' use of chatgpt is ticking up, but few trust its election information, https://www.pewresearch.org/short-reads/2024/03/26/americans-use-of-chatgpt-is-ticking-up-but-fewtrust-its-election-information/. Accessed: 2024-03-31.
42. Jang, M. E., & Lukasiewicz, T. (2023). Consistency analysis of chatgpt. *arXiv preprint arXiv:2303.06273*.
43. Alkaissi, H., & McFarlane, S. I. (2023). Artificial hallucinations in chatgpt: Implications in scientific writing. *Cureus, 15*(2), e35179.
44. Khan, M. S., & Umer, H. (2024). Chatgpt in finance: Applications, challenges, and solutions. *Heliyon, 10*(2), e24890.
45. Page, D. (2023). *The AI arms race is changing everything*, https://time.com/6255952/ai-impact-chatgpt-microsoft-google/. Accessed: 2024-03-31.

46. Shen, X., Chen, Z., Backes, M., & Zhang, Y. (2023). In chatgpt we trust? Measuring and characterizing the reliability of chatgpt. *arXiv preprint arXiv:2304.08979*.
47. Tian, Z., Cui, L., Liang, J., & Yu, S. (2022). A comprehensive survey on poisoning attacks and countermeasures in machine learning. *ACM Computing Surveys, 55*(8), 1–35.
48. Chen, X., Liu, C., Li, B., Lu, K., & Song, D. (2017). Targeted backdoor attacks on deep learning systems using data poisoning. *arXiv preprint arXiv:1712.05526*.
49. Krishna, K., Tomar, G. S., Parikh, A. P., Papernot, N., & Iyyer, M. (2019). Thieves on sesame street! Model extraction of BERT-based APIs. *arXiv preprint arXiv:1910.12366*.
50. Li, H., Guo, D., Fan, W., Xu, M., & Song, Y. (2023). Multi-step jailbreaking privacy attacks on chatgpt. *arXiv preprint arXiv:2304.05197*.
51. Perez, F., & Ribeiro, I. (2022). Ignore previous prompt: Attack techniques for language models. *arXiv preprint arXiv:2211.09527*.
52. Reinhardt, K. (2023). Trust and trustworthiness in AI ethics. *AI and Ethics, 3*(3), 735–744.
53. Pa Pa, Y. M., Tanizaki, S., Kou, T., Van Eeten, M., Yoshioka, K., & Matsumoto, T. (2023). An attacker's dream? Exploring the capabilities of chatgpt for developing malware. In *Proceedings of the 16th cyber security experimentation and test workshop* (pp. 10–18).
54. BlackBerry.com. (2023). *Chatgpt may already be used in nation state cyberattacks, say it decision makers in blackberry global research*. https://www.blackberry.com/us/en/company/newsroom/press-releases/2023/chatgpt-may-already-be-used-in-nation-state-cyberattacks-say-it-decision-makers-in-blackberry-global-research. Online; accessed 3-January 2024.
55. Kharpal, A. (2023). Chinese police arrest man who allegedly used chatgpt to spread fake news in first case of its kind.
56. Wired. (2023). Large language models and phishing scams. https://www.wired.com/story/large-language-model-phishing-scams/. Accessed: 2024-04-01.
57. Weidinger, L., Mellor, J., Rauh, M., Griffin, C., Uesato, J., Huang, P. S., Cheng, M., Glaese, M., Balle, B., Kasirzadeh, A., et al. (2021). Ethical and social risks of harm from language models. *arXiv preprint arXiv:2112.04359*.
58. Lucchi, N. (2023). Chatgpt: A case study on copyright challenges for generative artificial intelligence systems. *European Journal of Risk Regulation*, 1–23. https://doi.org/10.1017/err.2023.59
59. Leboukh, F., Aduku, E. B., & Ali, O. (2023). Balancing chatgpt and data protection in Germany: Challenges and opportunities for policy makers. *Journal of Politics and Ethics in New Technologies and AI, 2*(1), e35166–e35166.
60. OpenAI. (2023). *March 20 chatgpt outage: Here's what happened*, https://openai.com/blog/march-20-chatgpt-outage
61. Su, Z., McDonnell, D., Bentley, B. L., He, J., Shi, F., Cheshmehzangi, A., Ahmad, J., & Jia, P. (2021). Addressing biodisaster x threats with artificial intelligence and 6g technologies: Literature review and critical insights. *Journal of Medical Internet Research, 23*(5), e26109.
62. U.S. Department of Health Human Services. (2023). *HIPAA for professionals: Security laws and regulations*. https://www.hhs.gov/hipaa/for-professionals/security/laws-regulations/index.html. Accessed: 2024-04-07.
63. Kocoń, J., Cichecki, I., Kaszyca, O., Kochanek, M., Szydło, D., Baran, J., Bielaniewicz, J., Gruza, M., Janz, A., Kanclerz, K., et al. (2023). Chatgpt: Jack of all trades, master of none. *Information Fusion, 99*, 101861.
64. McGee, R. W. (2023, February 15). Is chat gpt biased against conservatives? an empirical study. An Empirical Study.
65. Hartmann, J., Schwenzow, J., & Witte, M. (2023). The political ideology of conversational AI: Converging evidence on chatgpt's pro-environmental, left-libertarian orientation. *arXiv preprint arXiv:2301.01768*.
66. Motoki, F., Pinho Neto, V., & Rodrigues, V. (2024). More human than human: Measuring chatgpt political bias. *Public Choice, 198*(1), 3–23.
67. News, F. (2023). *Chatgpt faces mounting accusations of being 'woke,' having liberal bias*, https://www.foxnews.com/media/chatgpt-faces-mounting-accusations-woke-liberal-bias

68. Paul, J., Ueno, A., Dennis, C. (2023). Chatgpt and consumers: Benefits, pitfalls and future research agenda.
69. Pavlik, J. V. (2023). Collaborating with chatgpt: Considering the implications of generative artificial intelligence for journalism and media education. *Journalism & Mass Communication Educator, 78*(1), 84–93. https://doi.org/10.1177/10776958221149577
70. King, M. R. (2023). ChatGPT: A conversation on artificial intelligence, chatbots, and plagiarism in higher education. *Cellular and Molecular Bioengineering, 16*(1), 1–2.
71. Farhat, F., Sohail, S. S., & Madsen, D. Ø. (2023). How trustworthy is chatgpt? The case of bibliometric analyses. *Cogent Engineering, 10*(1), 2222988.
72. Salvagno, M., Taccone, F. S., & Gerli, A. G. (2023). Can artificial intelligence help for scientific writing? *Critical Care, 27*(1), 75.
73. Thorp, H. H. (2023). Chatgpt is fun, but not an author.
74. Hung, J., & Chen, J. (2023). The benefits, risks and regulation of using chatgpt in Chinese academia: A content analysis. *Social Sciences, 12*(7), 380.
75. Simpson, J. A. (2007). Psychological foundations of trust. *Current Directions in Psychological Science, 16.* https://doi.org/10.1111/j.1467-8721.2007.00517.x
76. Bach, T. A., Khan, A., Hallock, H., Beltrão, G., & Sousa, S. (2022). A systematic literature review of user trust in AI-enabled systems: An HCI perspective. *International Journal of Human-Computer Interaction.* https://doi.org/10.1080/10447318.2022.2138826
77. Yang, L., & Wang, J. (2023). Factors influencing initial public acceptance of integrating the chatgpt-type model with government services. *Kybernetes.*
78. Silvano, R., & Gui, A. (2024). Factors that influence college students in the use of chatgpt in Indonesia. In *2024 28th international conference on information technology (IT)* (pp. 1–4). IEEE.
79. Roe, J., & Perkins, M. (2023). 'What they're not telling you about chatgpt': Exploring the discourse of AI in UK news media headlines. *Humanities and social Sciences Communications, 10*(1), 1–9.
80. Polyportis, A., & Pahos, N. (2024). *Understanding students' adoption of the chatgpt chatbot in higher education: The role of anthropomorphism, trust, design novelty and institutional policy* (pp. 1–22). Behaviour & Information Technology.
81. Novozhilova, E., Mays, K., Paik, S., & Katz, J. E. (2024). More capable, less benevolent: Trust perceptions of AI systems across societal contexts. *Machine Learning and Knowledge Extraction, 6*(1), 342–366.
82. Baek, T. H., & Kim, M. (2023). Is chatgpt scary good? How user motivations affect creepiness and trust in generative artificial intelligence. *Telematics and Informatics, 83*, 102030.
83. Hu, P., Lu, Y., et al. (2021). Dual humanness and trust in conversational AI: A person-centered approach. *Computers in Human Behavior, 119*, 106727.
84. Ma, X., & Huo, Y. (2023). Are users willing to embrace chatgpt? Exploring the factors on the acceptance of chatbots from the perspective of AIDUA framework. *Technology in Society, 75*, 102362. https://www.sciencedirect.com/science/article/pii/S0160791X23001677. https://doi.org/10.1016/j.techsoc.2023.102362
85. National Institute of Standards and Technology (NIST). (2022). *Trustworthy AI: Managing risks in artificial intelligence.* https://www.nist.gov/speech-testimony/trustworthy-ai-managing-risks-artificial-intelligence
86. European Commission. (2019). *Ethics guidelines for trustworthy AI.* https://digital-strategy.ec.europa.eu/en/library/ethics-guidelines-trustworthy-ai
87. Yu, S., Carroll, F., & Bentley, B. L. (2024). Insights into privacy protection research in AI. *IEEE Access, 12*, 41704–41726. https://doi.org/10.1109/ACCESS.2024.3378126
88. Nazir, A., & Wang, Z. (2023). A comprehensive survey of chatgpt: Advancements, applications, prospects, and challenges. *Meta Radiology, 1*, 100022.
89. Gupta, M., Akiri, C., Aryal, K., Parker, E., & Praharaj, L. (2023). From chatgpt to threatgpt: Impact of generative AI in cybersecurity and privacy. *IEEE Access, 11*, 80218.

90. Fui-Hoon Nah, F., Zheng, R., Cai, J., Siau, K., & Chen, L. (2023). Generative AI and chatgpt: Applications, challenges, and AI-human collaboration, Journal of Information Technology Case and Application Research, *25*, 277.
91. Parasuraman, R., & Riley, V. (1997). Humans and automation: Use, misuse, disuse, abuse. *Human Factors, 39*(2), 230–253.
92. Yu, S., Carroll, F., & Bentley, B. L. (2024). Trust and risk: Psybersecurity in the AI era. In *Psybersecurity: Human factors of cyber defence*. Taylor and Francis. in press.

Chapter 7
Beyond the Black Box: XAI Strategies for Safeguarding Critical Infrastructure

Arjun Singh, Sarthak Kathait, Anshi Kothari, Shantanu Joshi, Yutika Agarwal, Saksham Badoni, Mohit Bhatt, and Preeti Mishra

Abstract Machine learning (ML) has become proficient in performing analysis and prediction of emerging threats more effectively and efficiently. It has been extensively used in protecting the critical computing infrastructure. However, the opaqueness with ML-based intrusion detection systems (IDSes) for protecting the data and applications of computing infrastructure raises a strong concern about the trustworthiness of such tools. In the realm of raising security incidents, transparency is crucial to trusting and confidently using ML models to develop robust security tools. Yet, the inherent opacity of black-box ML models and the increasing complexity of cumulative models pose potential security risks. In response, Explainable Artificial Intelligence (XAI) emerges as a crucial approach, focusing on the interpretability of AI systems that can clarify their decisions or predictions for users. XAI aims to enhance transparency, trustworthiness, and accountability, especially in high-stakes applications such as cybersecurity. This chapter offers a comprehensive understanding of the fundamental concepts of XAI and its applications for protecting critical computing infrastructure against emerging attacks. A comprehensive analysis of the taxonomy of ML models, XAI techniques, and libraries is provided to explain the importance of XAI in security. The future directions on XAI have also been discussed. A case study on the use of XAI is provided for protecting the data and applications from network malware attacks using open-source libraries and tools.

A. Singh · S. Kathait · A. Kothari · S. Joshi · S. Badoni · M. Bhatt · P. Mishra (✉)
School of Technology, Doon University, Dehradun, Uttarakhand, India
e-mail: mbhatt.cs@doonuniversity.ac.in; dr.preetimishra@doonuniversity.ac.in

Y. Agarwal
The NorthCap University, Gurgaon, Haryana, India

C. Hewage et al. (eds.), *Data Protection*,
https://doi.org/10.1007/978-3-031-76473-8_7

7.1 Introduction

The widespread adoption of cutting-edge technologies across multiple industries has highlighted the urgent need for comprehensive cybersecurity protocols across various countries [1]. The deep dependence on technology demands continual vigilance, resilient defenses, and proactive strategies to counter emerging cyber risks. Moreover, adversaries are refining their skills, harnessing advanced technologies like artificial intelligence (AI) to circumvent detection measures, and carry out meticulously targeted attacks with greater precision. A recent report by AV-TEST[1] documented 4618 cyberattacks across Europe, primarily comprised of Distributed Denial of Service (DDoS) and Ransomware incidents. These included 2525 DDoS attacks and 1066 Ransomware assaults, with an additional 1027 unspecified events encompassing various malicious activities like data theft for hacktivism or espionage. In 2024, AI models risk being targeted by malicious actors. Although generative AI (GenAI) and large language models (LLM) datasets pose challenges for threat actors to manipulate, cloud-based ML models with specialized applications are particularly vulnerable and appealing targets.[2]

To address evolving cyber threats, deploying effective Intrusion Detection System (IDS) is crucial in critical infrastructures like cloud computing [2, 3]. These systems encompass various techniques, including anomaly detection, misuse detection, introspection, and hybrid approaches. Misuse detection operates by aligning system behavior with predefined attack patterns, while anomaly-based detection identifies deviations without predefined patterns, effective against novel threats. Introspection techniques internally examine a system for potential threats. Hybrid approaches combine misuse, anomaly, and introspection for comprehensive intrusion detection in cloud environments [4].

Most of the existing IDS solutions are artificial intelligence (AI) based and have been used as black-box models [5, 6]. It is hard to trust such black-box AI models, wherein it is difficult to interpret their internal working mechanisms. Therefore, it is high time to go beyond black-box detection models and understand their internal workings. That's where Explainable Artificial Intelligence (XAI) comes into play.

XAI has shown their effectiveness by revealing the complex internal working of black-box models thereby making them more reliable and transparent in their decision-making process. It not only contributes to model debugging, data collection, and trust-building but also guides human decision-making. It employs two primary approaches such as intrinsic and post hoc. Intrinsic XAI focuses on elucidating the model's structure, with limitations on application to specific AI types. Conversely, post-hoc approaches analyze input and output, making them universally applicable. Additionally, XAI can be classified into global, explaining general model characteristics, and local, providing insights into individual decisions [7]. This adaptability enhances XAI's effectiveness in comprehending and optimizing

[1] https://www.av-test.org/en/news/cyber-incidents-in-numbers-year-2023/

[2] https://cyberriskleaders.com/proliferation-of-ai-driven-attacks-in-2024/

AI models across diverse scenarios. One notable area where this adaptability is crucial is XAI-IDS, which remains relatively unexplored in complex infrastructures, representing a critical frontier in the ongoing research and development of cybersecurity. The escalating demand for IDS necessitates accurate predictions coupled with transparent and interpretable explanations for their decisions. XAI-IDS seek to bridge the gap between the traditional AI's black-box nature and the need for insights understandable to humans into cybersecurity incidents [5]. Ultimately, by prioritizing transparency, trustworthiness, and robust security models, organizations can harness the power of trustworthy AI [8] while safeguarding sensitive data and mitigating cyber risks effectively.

Current research [9] emphasizes the implementation of XAI to enhance transparency and understandability, especially in domains such as healthcare and image analysis [10]. However, a notable literature gap exists in the form of a comprehensive survey paper [11] that assesses XAI-based solutions, beyond feature relevance parameters, across diverse cyber security applications. Although, there are existing papers on XAI [12] and cyber security [13] independently, none provide an exhaustive overview of XAI's applicability in terms of its parameters in diverse cybersecurity contexts. The need for comprehensive reviews and solutions in this intersection highlights a potential trajectory for further research and development in the field.

In this chapter, a comprehensive taxonomy of the machine learning (ML) models based on the extent of their interpretation has been discussed. The chapter provides an in-depth exploration of the core principles of XAI and its practical applications in safeguarding critical computing infrastructure from emerging threats. It delves into a thorough examination of XAI methodologies and their application in intrusion detection. Furthermore, available libraries are discussed, elucidating the significance of XAI in enhancing security measures. Furthermore, it outlines prospective avenues for the evolution and advancement of XAI in the context of cyber security. Figure 7.1 gives a basic outline of XAI. The major contributions of this chapter are as follows:

- The latest advancements in the XAI approaches are provided along with a taxonomy derived from an extensive literature review. This effort is aimed at establishing a foundation for formally defining explainability in the context of intrusion detection.
- A comparative discussion on various XAI techniques has been provided and their relevance across various models have also been elaborated.
- Future research directions are discussed to give research directions to readers.

The remainder of this chapter is organized as follows: Section 7.3 explores two types of attack families impacting virtualization. Section 7.4 presents a literature review on XAI, encompassing white-box and black-box models, along with our taxonomy for post hoc explainability. Section 7.5 delves into applications of XAI in various domains, and the chapter concludes in Sect. 7.6.

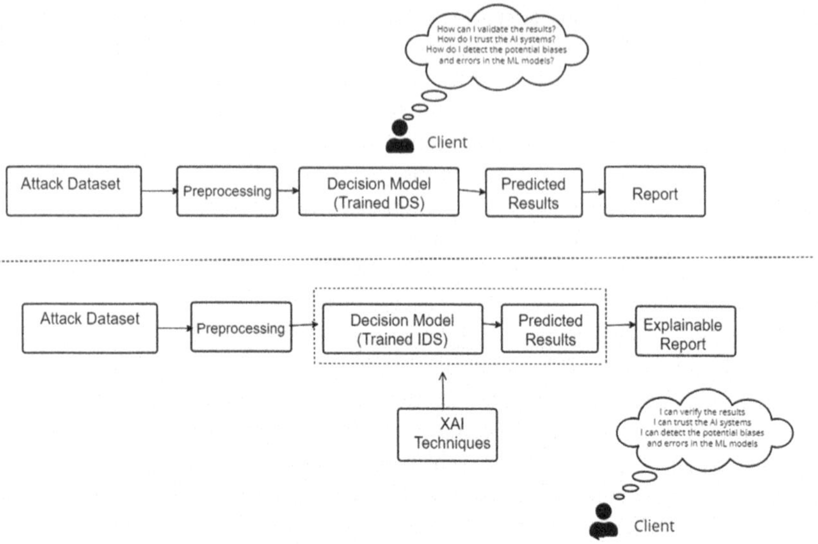

Fig. 7.1 Demonstration of XAI applicability in IDS

7.2 Related Works

The utilization of XAI in different domains of human life has been a popular topic in recent years. This section discusses the work that has been done.

Islam et al. [14] introduced a methodology to overcome challenges such as limited explainability and interpretability of successful AI models, hindering trust in predictive outcomes. This often necessitates human intervention, causing delays in responses. Addressing this issue, our study integrates well-established domain knowledge, specifically the CIA principles, into the AI model. Focused on network intrusion detection, our experimental findings demonstrate that this infusion of domain knowledge enhances both explainability and expedites decision-making. Moreover, the incorporation of domain knowledge fosters adaptability to unknown attacks and facilitates seamless integration with diverse IoT device-generated network traffic.

Security of AI models has always been a concern as it may provide confidential information, but confidentiality does not translate to security. Adadi et al. [15] reveal this property of AI models to trade secrets and pose threat to security. Oh et al. [16] provide an approach to make Deep Learning(DL) models more strudy toward exposing the confidential data. This indicates the necessity to develop such XAI tools that can explain ML models without compromising confidentiality.

Kuppa and Khac [17] carried out a deep security analysis of models that expose counterfactuals explanations. Four black-box attacks were designed that used XAI methods to attack confidentiality and privacy properties of the underlying classifiers. Through experiential and qualitative evaluation, the effectiveness of the attacks has been shown using different datasets, and attention was focused on the security threats of revealing to users and attackers alike. The aim was to explore the application of the counterfactual methods to shield from the attacks identical to the proposed in the work.

Zolanvari et al. [18] proposed XAI model named Transparency Relying Upon Statistical Theory (TRUST) that can utilized to defend against the attacks critical IOT systems. TRUST is a model agnostic, high-performing, and numerical compatible XAI model. TRUST XAI model gives explanations for new samples with a success rate of 98 percent. Factor analysis serves as a valuable methodology for the reconfiguration of input features into a novel collection of latent variables in academic and research contexts. TRUST uses multimodal Gaussian distribution to ascertain the probability of any new sample to each class. TRUST is superior to LIME in terms of performance, speed, and the method of explainability.

Siganos et al. [19] explained how the explosion of interconnected devices is the Internet of things (IOT) brings both convenience and heightened security risks. Traditional security measures struggle to keep pace with the evolving threats targeting resource-limited IoT devices. Machine learning offers a powerful tool for intrusion detection, but its "black box" nature makes it difficult to understand why an anomaly is flagged. This lack of explainability hinders trust and limits effective response. Explainable AI (XAI) emerges as a critical solution, using techniques like SHAP and LIME to make AI models interpretable. By providing insights into the data driving an anomaly, XAI empowers security analysts to not only identify threats but also understand their root cause. This newfound transparency allows for targeted responses, preventative measures, and ultimately, a more secure IoT ecosystem.

Sharma et al. [13] mentioned the rise of AI in cybersecurity offers powerful tools for threat detection and mitigation. However, a major hurdle remains—the "black box" nature of traditional AI models. Security professionals struggle to understand why these models flag certain events, hindering trust and effective response. Enter Explainable AI (XAI). By employing techniques like SHAP and LIME, XAI unlocks interpretability, allowing us to see "inside the box." This newfound transparency empowers security analysts. They can not only identify threats but also understand the reasoning behind the detection. This deeper knowledge enables more targeted responses, allowing them to focus on the most critical; vulnerabilities and ultimately strengthen overall cybersecurity posture.

7.3 Attack Families and Their Impact on Virtualization

Virtualization is a key component of cloud computing that creates virtual instances of cloud resources, networks, and servers, enabling them to run on multiple machines [20]. These machines play a crucial role in reducing workload and financial costs. Virtualization is implemented through the use of a hypervisor, which controls the functionality of the virtualized infrastructure. However, vulnerabilities exist in this layer, and cyber criminals continuously target both this layer and different virtualized instances. Many attacks impact virtualization, mainly categorized into two families such as network-based attack families and system-based attack families [21].

7.3.1 System-Based Attack Families

System-based attacks in virtual machines involve targeted actions aimed at exploiting vulnerabilities within the virtualized infrastructure and software stack that supports virtual machines [22]. These attacks focus on compromising the hypervisor, the core virtualization platform responsible for managing and allocating resources to virtual machines. The goal is to breach the security boundaries of individual virtual machines or compromise the overall integrity and performance of the virtualized environment.

7.3.1.1 Adload

Adload[3] is a kind of trojan downloader made using Visual Basic. Botnets generate various load variants. Malware is downloaded via an executable file that it downloads. Since executable files are saved in the cloud by Adload, the functionality of the file can change at any time. A virtual machine may become exposed if an ad load is put into it and the window proxy setup is deactivated.

7.3.1.2 Agent

A remote attacker might be able to access the victim's personal information if an agent[4] is injected into the virtual machine. It can also alter and remove the victim's files, as well as modify Windows settings. Other visitors could become infected by the agent malware strains.

7.3.1.3 Ceeinject

CeeInject[5] is of the VirTool kind, and it fools antivirus software by using obfuscation techniques. Malware that tries to evade detection by security software uses this technique. Bitcoin miners use a VM's power without permission if CeeInject infects

[3] https://www.malwarebytes.com/blog/detections/trojan-adload

[4] https://www.malwarebytes.com/blog/detections/trojan-agent

[5] https://www.microsoft.com/en-us/wdsi/threats/malware-encyclopedia-description?Name=VirTool

the VM. The firewall could be disabled by it self-mutating. Malware and unapproved websites are connected by it.

7.3.1.4 Fakerean

A family of rogue viruses is known as the Fakerean.[6] It poses as an anti-virus program. Malware, such as exploit kits, installs it. By showing randomly selected files, the malware seems to be a file scanner and raises false warnings about malware. To get rid of the purportedly scanned infection, it offers a chargeable service. The VM may modify the security settings and terminate the active program if it becomes infected with fakerean.

7.3.1.5 Renos

Renos[7] family is a common trojan downloader virus that can infect a computer and download anti-spyware without the user's permission. This software generates fake alerts, claims that spyware is present on the system, and demands payment to have the spyware removed. If the VM is infected with Renos malware, it will install a dynamic link library (DLL) in the Windows system folder. The virtual computer may become unstable due to the virus.

7.3.1.6 Vundo

Vundo[8] is a type of trojan downloader that displays the pop-up message. It can propagate through emails or be transmitted by other malware. Virtual machine performance may suffer if Windows and the task manager are disabled during safe boot. Vundo has the ability to cause a denial of service (DoS) or the "blue screen of death" to occur on websites such as Google and Facebook. It could spread to new virtual machines via network drives.

7.3.2 Network-Based Attack Families

These attacks are directed specifically at the networking components within virtualized environments, as the network infrastructure plays a pivotal role in connecting various virtual machines (VMs) and facilitating communication among them. The objective is to exploit vulnerabilities in the virtualized network [23–25], enabling

---

[6] https://www.microsoft.com/en-us/wdsi/threats/threat-search?query=win32/fakerean

[7] https://www.microsoft.com/en-us/wdsi/threats/malware-encyclopedia-description?Name=Win32%2FRenos

[8] https://www.f-secure.com/v-descs/trojan-w32-vundo.shtml

unauthorized access, data manipulation, or disruption of communication channels between VMs within the environment.

7.3.2.1 Emotet

It facilitates network infections and employs DLLs for continuous evolution and updates. Operating as self-propagating malware, it spreads through spam and phishing emails, establishing a foothold on a network. Subsequent malware may use this initial access to perform targeted attack [26].

7.3.2.2 Lokibot

Lokibot[9] is crafted to steal information from cryptocurrency wallets and execute network-related activities. Its primary mechanisms involve the use of a keylogger to capture credentials by monitoring browser and desktop activities. Furthermore, the malware establishes a backdoor in compromised systems, providing attackers with the ability to install additional payloads.

7.3.2.3 Lollipop

Lollipop[10] forces the user to act on some activities on an online advertisement and extract sensitive information. This program is commonly included as part of bundled packages with other free software that users download from the Internet.

7.3.2.4 Remcos

Remote access Trojans (RATs) are employed for exploiting systems remotely within a network. The Remcos malware [27] often employs methods such as process hollowing and process injection to avoid detection while functioning, though it may still leave traces within the operating system's registry.

7.3.2.5 Zeus

It transforms the computer into a remotely controlled part of a botnet and network of compromised machines [26].

---

[9] https://www.cisa.gov/news-events/cybersecurity-advisories/aa20-266a

[10] https://www.microsoft.com/en-us/wdsi/threats/malware-encyclopedia-description?Name=Adware:Win32/Lollipop

7.4 Explainable Technique for Attack Analysis

XAI plays a vital role in the analysis of attacks by offering transparency and interpretability into the decision-making processes of AI models employed for security purposes. In the domain of cybersecurity, where understanding the reasoning behind AI-driven decisions is crucial, XAI techniques provide insights into the factors influencing the conclusions reached. By shedding light on the features and parameters shaping the model's outputs, XAI assists in identifying potential vulnerabilities, recognizing attack patterns, and detecting adversarial manipulations [28]. This interpretability empowers security analysts to conduct comprehensive root cause analyses, uncovering the origins of security breaches or unusual behaviors.

Moreover, XAI promotes collaboration between human experts and AI systems, enabling analysts to make well-informed decisions and implement proactive measures in response to identified threats. Ultimately, the significance of XAI in attack analysis lies in enhancing transparency, facilitating comprehension, and strengthening the efficacy of defense mechanisms against cyber threats.

7.4.1 Key Terminologies

In this section, terminologies are explained which are useful to understand the XAI techniques.

Understandability
Understandability or intelligibility refers to the human capacity to comprehend the functionalities of the model without having any knowledge of its internal workings or structure [29].

Comprehensibility
Comprehensibility highlights the extent to which people, especially those who are non-experts, can comprehend the internal workings and decision-making processes of an AI model [30].

Interpretability
Interpretability is the capacity to communicate to a human being the outcomes of a model (its cause-and-effect relationship). It refers to understanding why the model has given a specific output. Interpretability is a passive or surface-level characteristic of a model at which it makes sense to a human observer [31].

Explainability
Explainability[11] refers to the ability to articulate the reasons behind a specific decision, recommendation, or prediction made by an AI system. Explainability is related

[11] https://www.algolia.com/blog/ai/what-is-explainable-ai-and-why-is-transparency-so-important-for-machine-learning-solutions/

to the idea of explanation as a human–machine interface that is both a human-understandable representation of the decision-maker and an accurate proxy for the decision-maker.

Transparency
A transparent model is the one which can be understood by itself [32]. Transparency ensures that the decision-making process of a model is clear and understandable to users, allowing them to easily comprehend why a particular prediction or decision was made. By addressing the "black box" nature of machine learning models, transparency enhances interpretability and accessibility for users thereby ensuring accountability, promoting ethical use of AI, and fostering user trust in AI systems.

Causality
Causality refers to the capacity of a model to explain the task of finding relationships between input and output data variables. It explains the output or the decision made in an informal way. Cause-and-effect relationships within a model analyse how different factors interact and influence each other to produce specific results [33].

Transferability
It involves the capability of an interpretable and explainable model to apply previously acquired knowledge to unfamiliar scenarios [34]. For example, if an interpretable model is trained to explain decisions in a medical diagnosis task, transferability would mean that the knowledge gained could be applied to another medical domain or a different diagnosis task.

Algorithmic Transparency
It refers to the clarity of the inner working of the model. It is concerned with the user's ability to understand, interpret, and explain the procedure followed by algorithmic models [35] and decisions to produce a particular outcome from the data it receives as input.

Correctability
It refers to the ability to identify and rectify errors or mistakes made by an AI model with the assistance of explanations provided by the system. It involves the capability to not only understand why a particular decision was made but also to correct it if it is deemed incorrect or undesirable [36].

7.4.2 Taxonomy of ML Models Based on Interpretability

In the ever-evolving field of ML, model interpretability has become a critical issue in many different fields. This need is especially evident in sectors where judgments affect lives of individuals, such as healthcare, IDS, and cybersecurity. Recognizing the various needs for interpretability, a structured method for categorizing ML models can be provided by a taxonomy based on interpretability. This taxonomy divides models into different groups, each characterized by the degree of explainability and transparency of the model. By systematically classifying ML models on the basis of their interpretability, this taxonomy provides researchers with an insightful structure to navigate the trade-offs between model complexity and comprehensibility. In the end, this encourages accountability, trust, and appropriate utilization of AI. Figure 7.2 represents the same. In the subsequent section, a brief overview of both transparent (white box) and opaque (black box) models has been provided. White-box models are readily interpretable and understandable for humans, on the other hand, black-box models are not readily interpretable due to their opaque and complex nature.

A. *White-Box/Transparent Models*

White-box models have inherent property to easily interpret how the model arrives at a particular decision. The internal workings of white-box models are known and are easy to understand, interpreted, accessible to the user. In this section, some white-box models like Decision Trees, GAMs, Linear models, and Rule-based models are discussed, including their strengths, limitations, and their inner workings showing how their transparency facilitates understanding, validation, and trust in the decision-making process.

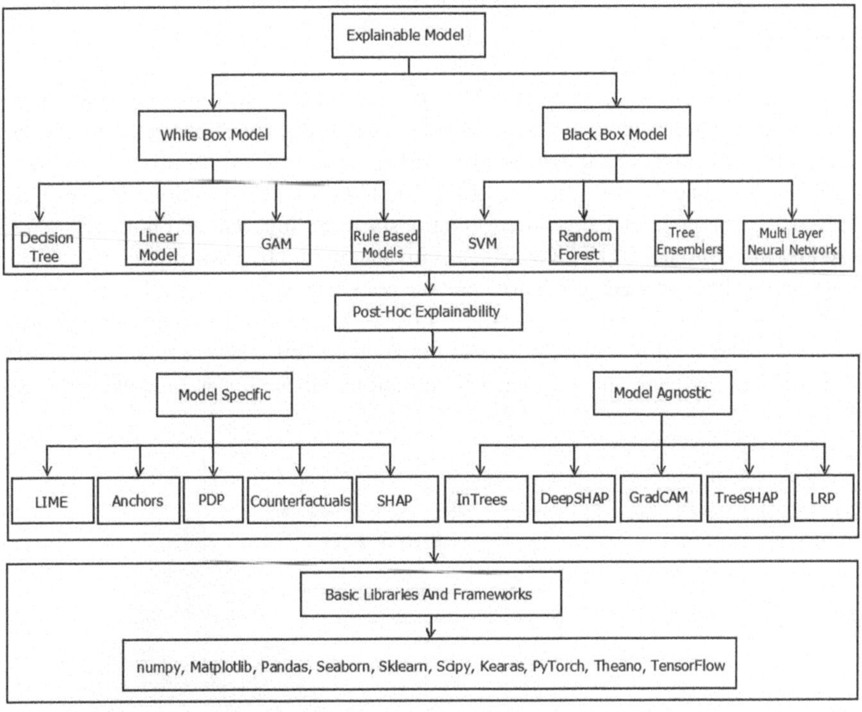

Fig. 7.2 Classification of ML models and the related XAI techniques based on interpretability

Decision Trees/Tree-Based Models A decision tree [37] is a widely used supervised machine learning algorithm effective for classification and regression tasks. Its structure resembles a tree, with nodes representing attribute tests and branches indicating outcomes. During training, it selects informative features and creates decision rules recursively, dividing the dataset until it achieves homogeneous segments or satisfies a stopping criterion. This process ensures accurate capture of underlying data patterns [38]. In regression tasks, decision trees predict target variable values by averaging instances within leaf nodes, enabling effective handling of categorical and continuous data.

Linear Models Linear classification models [39], like Logistic Regression (LR) and Support Vector Machines (SVMs) with linear kernels, predict discrete class labels by passing the output of the linear model through a threshold function [40]. These models offer simplicity, interpretability, and computational efficiency, making them advantageous for datasets with numerous features and linear or near-linear relationships with the target variable. However, they may struggle [41] with capturing complex, nonlinear patterns in the data. To enhance their performance, techniques like feature engineering, polynomial features, or kernel methods can be applied. Despite their limitations, linear classifiers remain valuable tools in machine learning, particularly when interpretability and computational efficiency are paramount.

Generalized Additive Models (GAMs) In traditional linear models, the interrelationship between predictors and response was supposed to be linear. In reality, it may be more complex and nonlinear. GAMs [42] address this limitation by allowing smooth and nonlinear functions of the predictors. This flexibility makes GAMs particularly valuable in fields where relationships are intricate and not easily captured by linear models. Thus, GAM is a statistical model by allowing for more flexible relationships between predictors and the response variable. By allowing for the exploration of nonlinear relationships, GAMs enhance our ability to uncover nuanced patterns [43] and improve predictive accuracy. They provide a valuable tool for data analysis and inference in situations where linear models fall short (Fig. 7.3).

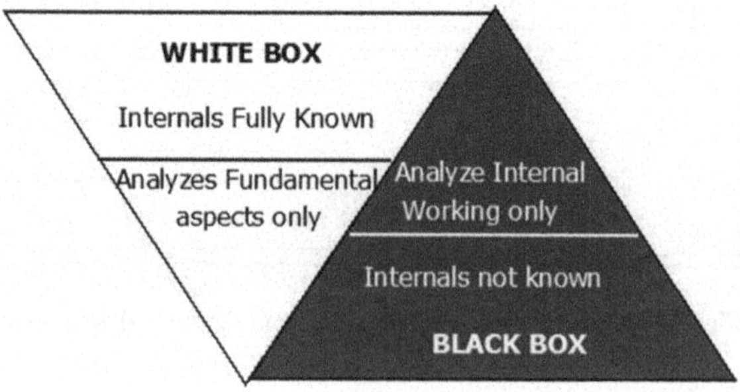

Fig. 7.3 Black box versus white box

Rule-Based Models Rule-based systems [44] convey knowledge through "if-then" rules, where conditions are stated, and decisions are made based on the fulfillment of these conditions. In the context of XAI, rules refer to the logic behind decisions. Thus, rule-based XAI involves characterizing the decision-making process [45] using explicit rules that are human-readable, transparent, and interpretable. This approach enhances understanding and trust in the system by making it easier for users to comprehend how decisions are reached. By providing clear explanations for each decision, rule-based XAI enables users to validate and potentially improve the model's performance and reliability.

B. *Black-Box Models*

Black-box models tend to have complex [46] internal workings which are opaque, with the model's decision-making process hidden behind layers of complex algorithms. Black-box models can still perform better when it comes to prediction, even when they are difficult to understand. This is especially true for jobs that include high-dimensional data or nonlinear relationships. Black-box models function more like "black boxes," where the input–output relationship is concealed. In this part, we'll discuss about a few black-box models, including SVM, Random Forest, Tree ensembles, and Multi-Layer Neural Network.

Support Vector Machine (SVM) It is an ML algorithm employed for classification purposes [47]. It operates by identifying extreme data points, known as support vectors, to establish a hyperplane that effectively divides the data into distinct classes within an n-dimensional space. Before employing SVM, specific data preprocessing steps are necessary. The algorithm seeks to reduce the gap between the support vectors and the hyperplane, effectively establishing the most suitable decision boundary for class differentiation. You can predict the output for the test set by changing the value of C, gamma, and kernel. The output of SVM is similar to the Logistic regression output.

Random-Forest (RF) It is a powerful ensemble learning algorithm used for classification and regression tasks in ML [48]. Each tree within the forest is constructed by utilizing a randomized portion of the training dataset and a random selection of features. This strategy fosters diversity among the trees and reduces overfitting. During inference, the predictions from all the trees are aggregated, resulting in a robust and accurate model. RF is renowned for its versatility, scalability, and proficiency in managing high-dimensional data effortlessly. It is widely used across various domains due to their effectiveness in producing reliable predictions while requiring minimal parameter tuning.

Tree-Ensemble Ensemble learning [49] is a machine learning method for supervised learning that combines weak or base learners, which are decision trees that have been trained separately and may not perform well on their own. However, when aggregated, these weak learners create a robust model, typically more accurate than its individual components. Ensemble learning includes bagging, boosting,

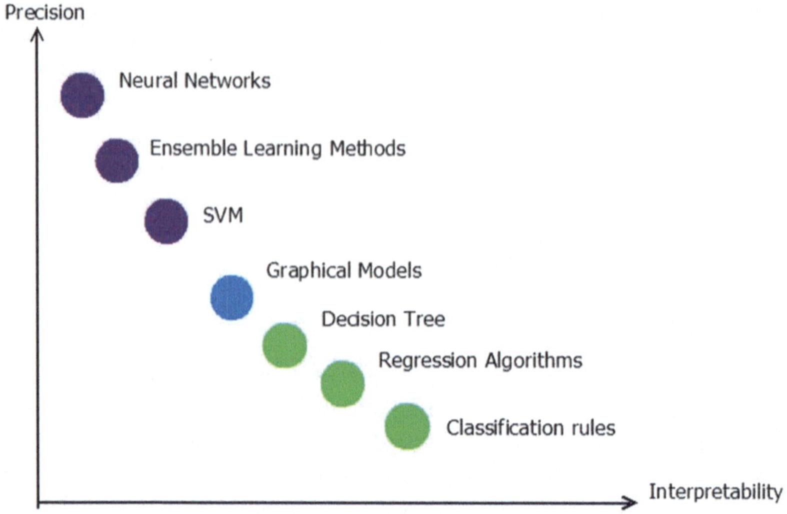

Fig. 7.4 Classification of machine learning algorithms depending on their predictive power (performance/accuracy) and interpretability adapted from

and gradient boosting as main methods, which can incorporate various weak learners, though this discussion focuses solely on trees.

Multi-Layer Neural Network Referred to as a deep neural network (DNN), a multi-layer neural network [50] is a robust tool comprising interconnected nodes structured hierarchically. Through training, the network can learn to represent complex patterns in data by adjusting its weights and biases. This makes it especially well suited for applications such as image recognition and natural language processing, as it excels at capturing complex nonlinear patterns within the data. If you want to tackle complex data analysis problems, a multi-layer neural network is an excellent choice (Fig. 7.4).

7.4.3 Post Hoc Explainable AI (XAI) Techniques for Model Interpretability

In this section, we'll explore post hoc explainability methods, categorized as either model-agnostic or model-specific approaches. Before delving into these methods, we'll briefly explain key parameters that give the basis of the output of ML algorithms. A post hoc explainable artificial intelligence (XAI) approach takes a trained or tested AI model as input and then generates understandable approximations of how the model works and makes decisions. This facilitates users in identifying essential features, evaluating their importance, replicating decisions from opaque models, and uncovering potential biases in both the model and the data [51]. There are various parameters to provide the explanation of ML results such as (i) Feature relevance, (ii) Local explanation, (iii) Global Explanation, (iv) Simplicity, (v) Visual Representation, (vi) Example-Based, (vii) Rule-Based, and (viii) Textual Representation.

Feature relevance refers to evaluating and determining the significance or contribution of input variables (features) to the decision-making process or the output produced by a machine learning model.

Local explanations describe the steps taken by an ML model to get to a particular prediction. It consists of a list of features and how they impact the prediction for a tabular data for that particular instance.

Global explanations refer to the features that are crucial to the model as a whole. This can be measured by looking at effect sizes or evaluating which features have the most impact on model accuracy.

Simplicity in AI models refers to how easily understandable, readable, and straightforward their explanations are.

Visual representation involves the use of visual representations to convey the inner workings of machine learning models. Visualizations can take various forms including graphs, charts, heatmaps, decision trees, and interactive interfaces.

Example-based terminology entails utilizing particular instances or scenarios to demonstrate and elucidate the behavior of machine learning models.

Rule-based involves illustrating how a machine learning model makes decisions using clear and interpretable rules.

Textual representation involves using natural language descriptions to articulate the decision-making process, model behavior, or the logic behind specific predictions made by machine learning models.

A. *Model-Agnostic*

Model-agnostic approaches refer to methods or tools that can be universally applied across diverse types of models or algorithms, without being tailored for a specific one. These techniques are often preferred in scenarios where you want flexibility and the ability to analyze or manipulate models without being constrained by the specifics of a particular algorithm. It includes LIME, SHAP, PDPs, and Counterfactuals Table 7.1 provides a comparative analysis of model-agnostic XAI techniques against various parameters as discussed in Sect. 7.4.3.

- *Local Interpretable Model-Agnostic Explanations (LIME)*

- It is an approach created to interpret the predictions of black-box machine learning models [52]. It works by capturing the local behavior of the complex model to make predictions for specific instances. Alodibat et al. [53] in their research work did a study on LIME to elaborate the predictions of a decision tree model trained on the Microsoft Malware Classification Challenge dataset. Specifically, nine samples were chosen from the test set, representing each of the 9 malware families. LIME provided prediction probabilities for each malware family class, demonstrating the model's confidence in its predictions. For example, one sample had a 97% probability of belonging to the "Ramnit" family, indicating high confidence in this prediction.

Table 7.1 Feature table for model-agnostic XAI techniques

XAI techniques	Feature relevance	Visual representation	Example based	Textual & numerical	Simplification	Input data
SHAP	✓	✓	X	✓	✓	Img, Tab, Num, Text, cls
LIME	✓	✓	X	✓	✓	Img, Tab, Num, Text, cls
PIMP	✓	X	X	✓	X	Img, Tab, Text, cls
CF	X	X	✓	✓	✓	Img, Tab, Num, Text, cls
Tree SHAP	✓	✓	✓	X	✓	Tree-based models, cls
Global Surrogate	✓	X	X	✓	✓	Img, cls
Anchors	✓	X	✓	✓	✓	Img, cls
CEM	✓	X	✓	✓	✓	Img, cls
Integrated Gradients	✓	✓	✓	✓	✓	cls
Proto Dash	X	✓	✓	X	✓	cls
PDP	✓	✓	X	X	X	any
ICE	✓	✓	X	X	X	any

- *Shapley Additive exPlanations (SHAP)*
- It is an approach used to interpret the predictions of machine learning models, and it uses a game theory approach to do so [54]. It is also used for model explainability. A certain Shapley value is assigned to each feature, and the model is trained and tested to find the impact of the feature on the predictions made by the model. Wang et al. [55] in their research work utilized SHAP alongside a surrogate model to assess the importance of feature values in the model's decision-making process for each sample. The study constructed feature dictionaries for the entire dataset based on SHAP values, leading to the identification of suspicious communities within the data. The application of SHAP and a surrogate model was demonstrated on the EMBER and Contagio datasets, containing 12,000 and 6000 samples, respectively. The results highlighted the effectiveness of the proposed suspicious samples filtering process in identifying backdoored samples with a low false positive rate across different attack strategies.
- *Anchors*
- Anchor explanation [56] explains the behavior of complicated model via the implementation of highly precise rules knows as anchors. It is used to describe a rule that, for a given instance, has a significant local influence and "anchors" a

prediction. The stability of this rule means that modifications to the instance's residual feature values have little effect on the predictions. Anchors are instrumental in elucidating how ML-based IDS make decisions, especially in the context of malware detection and network intrusion alerts. By applying anchors to the model, specific conditions or patterns leading to potential threats can be identified, providing security analysts with clear and concise explanations for flagged network traffic or identified malware files. This approach is particularly valuable in analyzing zero-day threats, where traditional signature-based detection methods may fall short, revealing specific characteristics or behaviors indicative of malicious intent.

- *Partial Dependence Plots (PDPs)*
- It is a global and model-agnostic method; it demonstrates how the target variable affects the overall prediction of the model [57]. Due to its model-agnostic nature, PDPs can be utilized across various models. It can be effectively employed in the realms of IDS and cybersecurity, visualizing diverse features that play a role in identifying security threats. PDPs can help analysts understand the impact of changes in network traffic volume or the frequency of system calls on the likelihood of intrusion. Analyzing PDP helps security experts pinpoint feature threshold values that notably raise the likelihood of an attack, improving alert accuracy in IDS. For instance, discovering that elevated CPU usage and specific network traffic patterns signal a Distributed Denial of Service (DDoS) attack guides the development of more effective detection algorithms, enhancing sensitivity to real-world attack scenarios.
- *Counterfactuals* (CF*)*
- Counterfactuals [58] are statements that assert a hypothetical scenario where the initial condition is false, describing how the world would be different if that condition were true—a kind of "what-if" question. In XAI, counterfactuals serve as interpretive tools to identify necessary changes required to achieve a desired prediction. Unlike many XAI methods that primarily focus on explaining why a specific outcome was predicted by a black-box model, counterfactuals take a different approach by guiding users to understand which features they should alter to attain a particular outcome. Naderi-Afooshteh et al. [59] introduced "MalMax," an analysis infrastructure utilizing counterfactual explanations and cooperative sandboxing. Testing on a 1 TB real-world website dataset, MalMax outperformed existing tools by identifying 3986 malware files compared to 238 found by maldet. In the benchmark dataset, MalMax detected 57/58 malware with no false positives, showcasing its superior performance in malware detection accuracy and lower false positives compared to existing tools.

B. *Model-Specific*

Model-specific refers to techniques, methods, or tools that are designed or tailored for a specific machine learning model or algorithm. Unlike model-agnostic approaches, which are general and can be applied to various types of models, model-specific methods are crafted with the characteristics and requirements of a

Table 7.2 Feature table for model-specific XAI techniques

XAI techniques	Feature relevance	Visual representation	Example based	Textual numerical	Simplification	Input data	Supported models
GRAD-CAM	✓	✓	X	✓	X	Img	CNN
LRP	✓	✓	X	✓	✓	Img	CNN
Pattern Net	✓	✓	X	X	X	Img	CNN
Pattern Attribution	✓	✓	X	X	X	Img	CNN
ProtoP Net	✓	✓	X	✓	✓	Img	CNN
GAM	✓	X	X	✓	✓	Num, Cat, Txt	LR
Ne Dissect	✓	X	X	✓	✓	Img	CNN
Interp Net	✓	X	X	✓	✓	Num, Img, Txt	CNN, RNN
GBP	✓	✓	✓	X	✓	Img	CNN

specific model in mind. It includes InTrees, DeepSHAP, TreeSHAP, GradCAM, and LRP. Table 7.2 provides a comparative analysis of model-specific XAI techniques against various parameters as discussed in Sect. 7.4.3.

- *Interpretable Trees (inTrees)*
- It is a global method that provides explanations by extracting rules and maintaining model accuracy from tree ensembles. This approach focuses on interpretability and aims to produce explanations for models that use decision trees, such as random forests, regularized random forests, and boosted trees. It can also be used to extract frequent patterns. InTrees framework can also be employed in intrusion detection, malware detection, network anomaly detection, phishing detection, user behavior analytics, incident response, rule pruning and selection, and forensic analysis. It helps extract clear rules from decision trees, making it easier to understand the factors contributing to the identification of security threats or incidents. Deng [49] also in his work proposed to extract interpretable information using InTrees framework to extract, measure and process rules from a tree ensemble.

- *Deep Explainer (DeepSHAP)*
- DeepSHAP[12] is a model-specific explainability technique that is mostly used with deep neural networks. It is based on the use of SHAP to elaborate the predictions of a deep learning (DL) model by the DeepLIFT (Deep Learning Important FeaTures) algorithm. DeepLIFT is a method where each activation

[12] https://arize.com/glossary/deep-explainer-deep-shap/

neuron is compared with reference neuron by backpropogation allowing for the traceable connection and even visualization of interdependence between them. Alenezi and Ludwig [60] in their work proposed the use of three SHAP methods, one of them being DeepSHAP, to explain the feature contributions of the different features in the Malicious URLs dataset and Android Malware dataset both of which consisted of five classes. SHAP plots were utilized to determine that in the Malicious URLs dataset, the "Spam" category, and in the Android Malware dataset, the "SMS malware" category, were identified based on feature impacts. This finding was confirmed by examining the classification tables and confusion matrices across all machine learning models.

- *Tree Shapley Additive exPlanations (TreeSHAP)*
- TreeSHAP [61] is an algorithm which is used for analyzing tree-based machine learning models, particularly decision trees or ensembles of decision trees (e.g., random forests, gradient boosting machines) by feature attribution method. It is a method for explaining the output of a model by attributing the contribution of each feature to the final prediction for a specific instance. It is a fast and computationally efficient algorithm generally considered useful for XGBoost or LightGBM, leveraging the inherent structure of tree algorithms to extract and generate SHAP values efficiently. Wali and Khan [62] integrated XAI with conventional ML-based IDS by utilizing global explanation on TreeSHAP to deal with adversarial attacks.
- *Gradient-Weighted Class Activation Mapping (Grad-CAM)*
- Grad-CAM [63] is an approach that aims to improve the transparency of models based on Convolutional Neural Networks (CNNs) by providing visual explanations. In order to improve the interpretability of the decision-making process, the method attempts to generate visualizations that highlight the areas of an input image that significantly effects the model's decision-making process. Grad-CAM computes importance scores for individual neurons pertaining to a specific decision of interest, leveraging gradient information propagating back into the final convolutional layer of the CNN.
- *Layer-Wise Relevance Propagation (LRP)*

- LRP is an interpretable machine learning method designed for specific feature types, including text and saliency maps. Its primary goal is to explain the output of neural networks within their input domain. The method involves maintaining the back-propagation process unaltered from the output to the input layer. Brigugilio [64] in his thesis conducted an experiment using the iNNvestigate [65] python library's LRP implementation. The author puts forward a fully interpretable approach for malware detection which leverages convolutional neural networks (CNN).

Table 7.3 describes some of the basic libraries as well as frameworks that can be used to implement the different XAI techniques.

Table 7.3 Different libraries and frameworks

Library	numpy	matplotlib	pandas	scipy	keras	PyTorch	Theano	TensorFlow
LIME	✓	✓	✓	X	✓	✓	X	✓
SHAP	✓	✓	✓	✓	✓	✓	✓	✓
Anchors	✓	✓	✓	✓	✓	✓	✓	✓
PDP	✓	✓	✓	✓	✓	✓	✓	✓
Counterfactuals	✓	✓	✓	✓	✓	✓	✓	✓
InTrees	✓	✓	✓	✓	✓	✓	X	✓
TreeSHAP	✓	✓	✓	✓	✓	✓	✓	✓
DeepSHAP	✓	✓	✓	✓	✓	✓	✓	✓
GradCAM	X	✓	X	✓	✓	✓	✓	✓
LRP	✓	✓	✓	✓	✓	✓	✓	✓

7.5 Future Directions

With the development of XAI, its sphere of applications is also being explored and expanding exponentially. XAI is used in various aspects of human life. XAI is showing a lot of potential in the domain of cyber security. XAI is employed in intrusion detection and prevention of the same and is showing satisfactory results in the same direction. Explainability in cybersecurity empowers defenders by providing a comprehensive understanding of adversaries' tactics, facilitating the creation of effective countermeasures. In the realm of cyber security, XAI denotes AI-powered security solutions adept at detecting threats and offering clear explanations for their behaviors. This attribute is paramount, offering insights into the rationale behind decisions or actions, fostering trust, and enabling robust response strategies. This section discusses the use of XAI in cyber security and cyber security in different domains such as healthcare, banking, agriculture and industry.

7.5.1 XAI for Cybersecurity in Healthcare

The healthcare industry, embracing smart systems like wearable devices and the Internet of Medical Things (IoMT), faces increased vulnerability to data breaches motivated by the lucrative sale of sensitive healthcare data. Threats include SQL injection, Packet Sniffing, Privilege escalation, and system vulnerabilities, enabling unauthorized access to healthcare services or fraudulent claims [66]. Addressing this, AI systems play a vital role in enhancing healthcare reliability and security. These AI applications span clinical decision-making, image-based disease detection, and efficient management of high-volume healthcare data, especially in cloud-based systems. Efficient ML models are imperative for handling raw data at scale. However, the black-box nature of classic AI models necessitates the incorporation of XAI frameworks like SHAP and LIME. XAI ensures a white-box approach, explaining decision outcomes to healthcare workers, and integrating human intelligence contributes to accurate results.

7.5.2 XAI for Cybersecurity in Banking

Smart systems are transforming the financial industry with intuitive decisions, yet they are increasingly vulnerable to cyberattacks, often motivated by monetary gain. Attacks, including those through social engineering on social media and mobile banking, compromise consumer data [67]. Safeguarding against fraud and scams is crucial [68]. AI's characteristic properties make it adept at customer classification, credit assessment, auditing, and regulatory compliance, preserving the privacy of sensitive banking data. However, AI solutions in banking need enhancement, especially in decision processes and addressing data breaches. The role of XAI is pivotal, with frameworks like SHAP and LIME providing transparency and interpretability. This transparency builds trust by providing customers and regulatory bodies with insights into the decision-making processes involved in security-related banking matters.

7.5.3 XAI for Cybersecurity in Agriculture

In agriculture, smart devices like IoT, robots, and drones driven by cloud computing, big data, AI, and augmented reality face challenges such as data transparency and vulnerability to attacks [69]. The diverse and voluminous data received from smart devices in agriculture is effectively managed by AI techniques, enabling tasks like monitoring crop disease, assessing soil condition, and analyzing behavior patterns. While AI minimizes human intervention and optimizes decisions, it still confronts security challenges, including trustworthiness, data reliability, privacy concerns, and adversarial machine learning issues. To enhance resilience against security threats, the integration of AI solutions with XAI frameworks like SHAP, LIME, and EL15 is crucial. XAI not only fortifies against attacks but also enhances trust by providing interpretability and explainability for users.

7.5.4 XAI for Cybersecurity in Industry

The adoption of Industrial IoT (IIoT) has revolutionized industrial operations, offering safer, more flexible, and efficient manufacturing. However, securing these industries from both traditional and non-traditional threats poses a significant challenge [70]. AI technology, with its decision-making capabilities, is instrumental in manufacturing operations, supply chain tasks, and solving complex engineering problems. Despite its potential benefits, AI faces security concerns, particularly the "Black Box" nature hindering its standalone use in critical industries and a lack of transparency for industrial automation.

The role of XAI becomes crucial in addressing these challenges through its emphasis on explainability and transparency. XAI frameworks like SHAP and LIME play a pivotal role in countering various security threats. Integration of XAI with data analytics further enhances the security and robustness of industrial networks, providing a solution to the "Black Box" issue and improving transparency in industrial automation processes.

7.6 Conclusion

In this chapter, we provide a detailed discussion on explainable machine learning and its applications in the area of cyber security. The major focus is to highlight the significance of transparency and interpretability in ML models. We provide an in-depth exploration of the core principles of XAI and its usage in safeguarding critical computing infrastructure from emerging threats. Furthermore, we provide the comparative discussion on available libraries and their applications in both model-agnostic and model-specific XAI approaches. In addition, we outline prospective avenues for the evolution and advancement of XAI in the context of cyber security. We discuss future research challenges in refining XAI techniques for intrusion detection, integrating them with advanced ML models, deploying them in real-world scenarios, and addressing ethical and legal considerations. The article will help readers better understand the predictions of black-box models thereby enhancing trust in their results.

Acknowledgments The Science and Engineering Research Board, Department of Science and Technology (SERB-DST) provided funding for this research under Project File No. SPG/2021/002003. The authors would like to express their gratitude to SERB-DST for their intellectual generosity and research assistance.

References

1. Srivastava, G., Jhaveri, R. H., Bhattacharya, S., Pandya, S., Maddikunta, P. K. R., Yenduri, G., Hall, J. G., Alazab, M., Gadekallu, T. R., et al. (2022). Xai for cybersecurity: State of the art, challenges, open issues and future directions. *arXiv preprint arXiv:2206.03585*.
2. Mishra, P., Varadharajan, V., Tupakula, U., & Pilli, E. S. (2019). A detailed investigation and analysis of using machine learning techniques for intrusion detection. *IEEE Communications Surveys Tutorials, 21*(1), 686–728. https://doi.org/10.1109/COMST.2018.2847722
3. Zhang, Z., Hamadi, H. A., Damiani, E., Yeun, C. Y., & Taher, F. (2022). Explainable artificial intelligence applications in cyber security: State-of-the-art in research. *IEEE Access, 10*, 93104–93139. https://doi.org/10.1109/ACCESS.2022.3204051

4. Mishra, P., Pilli, E. S., Varadharajan, V., & Tupakula, U. (2017). Intrusion detection techniques in cloud environment: A survey. *Journal of Network and Computer Applications, 77*, 18–47.
5. Patil, S., Varadarajan, V., Mazhar, S. M., Sahibzada, A., Ahmed, N., Sinha, O., Kumar, S., Shaw, K., & Kotecha, K. (2022). Explainable artificial intelligence for intrusion detection system. *Electronics (Basel), 11*(19), 3079.
6. Alahmed, S., Alasad, Q., Hammood, M. M., Yuan, J.-S., & Alawad, M. (2022). Mitigation of black-box attacks on intrusion detection systems-based ml. *Computers, 11*(7). https://doi.org/10.3390/computers11070115
7. Chalmers, E., Gruber, A. J., & Luczak, A. (2023). Hippocluster: An efficient, hippocampus- inspired algorithm for graph clustering. *Information Sciences, 639*, 118999. https://doi.org/10.1016/j.ins.2023.118999
8. Vassilev, A. (2024). *Adversarial machine learning: Technical report*. Gaithersburg, MD.
9. Dwivedi, R., Dave, D., Naik, H., Singhal, S., Omer, R., Patel, P., Qian, B., Wen, Z., Shah, T., Morgan, G., & Ranjan, R. (2023). Explainable ai (xai): Core ideas, techniques, and solutions. *ACM Computing Surveys, 55*(9). https://doi.org/10.1145/3561048
10. van der Velden, B. H. M., Kuijf, H. J., Gilhuijs, K. G. A., & Viergever, M. A. (2022). Explainable artificial intelligence (xai) in deep learning-based medical image analysis. *Medical Image Analysis, 79*, 102470. https://doi.org/10.1016/j.media.2022.102470
11. A., S., R., S. (2023). A systematic review of explainable artificial intelligence models and applications: Recent developments and future trends. *Decision Analytics Journal, 7*, 100230. https://doi.org/10.1016/j.dajour.2023.100230
12. Barredo Arrieta, A., Dıaz-Rodrıguez, N., Del Ser, J., Bennetot, A., Tabik, S., Barbado, A., Garcia, S., Gil-Lopez, S., Molina, D., Benjamins, R., Chatila, R., & Herrera, F. (2020). Explainable artificial intelligence (xai): Concepts, taxonomies, opportunities and challenges toward responsible AI. *Information Fusion, 58*, 82–115. https://doi.org/10.1016/j.inffus.2019.12.012
13. Sharma, D. K., Mishra, J., Singh, A., Govil, R., Srivastava, G., & Lin, J. C.-W. (2022). Explainable artificial intelligence for cybersecurity. *Computers and Electrical Engineering, 103*, 108356. https://doi.org/10.1016/j.compeleceng.2022.108356
14. Islam, S. R., Eberle, W., Ghafoor, S. K., Siraj, A., & Rogers, M. (2019). Domain knowledge aided explainable artificial intelligence for intrusion detection and response. *arXiv preprint arXiv:1911.09853*.
15. Adadi, A., & Berrada, M. (2018). Peeking inside the black-box: A survey on explainable artificial intelligence (xai). *IEEE Access, 6*, 52138–52160.
16. Oh, S. J., Schiele, B., & Fritz, M. (2019). Towards reverse-engineering black-box neural net- works. In *Explainable AI: Interpreting, explaining and visualizing deep learning* (pp. 121–144).
17. Kuppa, A., & Le-Khac, N.-A. (2021). Adversarial xai methods in cybersecurity. *IEEE Transactions on Information Forensics and Security, 16*, 4924–4938.
18. Zolanvari, M., Yang, Z., Khan, K., Jain, R., & Meskin, N. (2021). Trust xai: Model-agnostic explanations for ai with a case study on iiot security. *IEEE Internet of Things Journal, 10*, 2967.
19. Siganos, M., Radoglou-Grammatikis, P., Kotsiuba, I., Markakis, E., Moscholios, I., Goudos, S., & Sarigiannidis, P. (2023). Explainable AI-based intrusion detection in the internet of things. In *Proceedings of the 18th international conference on availability, reliability and security*. ACM.
20. https://www.researchgate.net/publication/346816485. Virtualization in cloud computing a review. Accessed: 2024-2-28
21. Sierra-Arriaga, F., Branco, R., & Lee, B. (2021). Security issues and challenges for virtualization technologies. *ACM Computing Surveys, 53*(2), 1–37.
22. Bhatt, M., Gaur, A., Badoni, S., & Mishra, P. (2022). Advanced malware and their impact on virtualization: A case study on hybrid feature extraction using deep memory introspection. In *Proceedings of the 2022 fourteenth international conference on contemporary*

computing. IC3-2022 (pp. 74–80). Association for Computing Machinery. https://doi.org/10.1145/3549206.3549223
23. Hoque, N., Bhuyan, M. H., Baishya, R. C., Bhattacharyya, D. K., & Kalita, J. K. (2014). Network attacks: Taxonomy, tools and systems. *Journal of Network and Computer Applications, 40*, 307–324.
24. Lee, S. C., & Shields, C. (2001). Tracing the source of network attack: A technical, legal and societal problem. In *Proceedings of the 2001 IEEE workshop on information assurance and security* (Vol. 6). United States Military Academy, West Point.
25. Mishra, P., Varadharajan, V., Tupakula, U., & Pilli, E. S. (2018). A detailed investigation and analysis of using machine learning techniques for intrusion detection. *IEEE Communications Surveys & Tutorials, 21*(1), 686–728.
26. Grammatikakis, K. P., Koufos, I., Kolokotronis, N., Vassilakis, C., & Shiaeles, S. (2021). Understanding and mitigating banking trojans: From zeus to emotet. In *2021 IEEE international conference on cyber security and resilience (CSR)* (pp. 121–128). IEEE.
27. Ude, O., & Swar, B. (2021). Securing remote access networks using malware detection tools for industrial control systems. In *2021 4th IEEE international conference on industrial cyber-physical systems (ICPS)* (pp. 166–171). IEEE.
28. Capuano, N., Fenza, G., Loia, V., & Stanzione, C. (2022). Explainable artificial intelligence in cybersecurity: A survey. *IEEE Access, 10*, 93575–93600.
29. Montavon, G., Samek, W., & Muller, K. R. (2018). Methods for interpreting and under- standing deep neural networks. *Digital Signal Processing, 73*, 1–15. https://doi.org/10.1016/j.dsp.2017.10.011
30. Burkart, N., Brajovic, D., & Huber, M. F. (2022). Explainable AI: Introducing trust and comprehensibility to AI engineering. *Automatisierungstechnik, 70*(9), 787–792. https://doi.org/10.1515/auto-2022-0013
31. Ali, S., Abuhmed, T., El-Sappagh, S., Muhammad, K., Alonso-Moral, J. M., Confalonieri, R., Guidotti, R., Del Ser, J., Dıaz-Rodrıguez, N., & Herrera, F. (2023). Explainable artificial intelligence (xai): What we know and what is left to attain trustworthy artificial intelligence. *Information Fusion, 99*, 101805. https://doi.org/10.1016/j.inffus.2023.101805
32. Arrieta, A. B., Dıaz-Rodrıguez, N., Ser, J. D., Bennetot, A., Tabik, S., Barbado, A., Garcıa, S., Gil-Lopez, S., Molina, D., Benjamins, R., Chatila, R., & Herrera, F. (2019). Explainable Artificial Intelligence (XAI): Concepts, taxonomies, opportunities and challenges toward responsible AI.
33. Carloni, G., Berti, A., & Colantonio, S. (2023). The role of causality in explainable artificial intelligence.
34. Liu, Y., Chen, X., Liu, C., & Song, D. (2017). Delving into transferable adversarial examples and Black-box attacks.
35. Gyevnar, B., Ferguson, N., & Schafer, B. (2023). *Bridging the transparency gap: What can explainable AI learn from the AI act?* IOS Press. https://doi.org/10.3233/FAIA230367
36. Jung, J., Lee, H., Jung, H., & Kim, H. (2023). Essential properties and explanation effec- tiveness of explainable artificial intelligence in healthcare: A systematic review. *Heliyon, 9*(5), 16110. https://doi.org/10.1016/j.heliyon.2023.e16110
37. Ville, B. (2013). Decision trees. *WIREs Computational Statistics, 5*(6), 448–455. https://doi.org/10.1002/wics.1278, https://wires.onlinelibrary.wiley.com/doi/pdf/10.1002/wics.1278
38. Bhargava, N., Sharma, G., Bhargava, R., & Mathuria, M. (2013). Decision tree analysis on j48 algorithm for data mining. *Proceedings of International Journal of Advanced Research in Computer Science and Software Engineering, 3*(6), 1114–1119.
39. Maulud, D., & Abdulazeez, A. M. (2020). A review on linear regression comprehensive in machine learning. *Journal of Applied Science and Technology Trends, 1*(2), 140–147. https://doi.org/10.38094/jastt1457
40. Priya, K. S. (2021). Linear regression algorithm in machine learning through MATLAB. *International Journal for Research in Applied Science and Engineering Technology, 9*(12), 989–995.

41. Schisterman, E. F., Vexler, A., Whitcomb, B. W., & Liu, A. (2006). The limitations due to exposure detection limits for regression models. *American Journal of Epidemiology, 163*(4), 374–383. https://doi.org/10.1093/aje/kwj039. https://academic.oup.com/aje/article-pdf/163/4/374/242257/kwj039.pdf
42. Agarwal, R., Melnick, L., Frosst, N., Zhang, X., Lengerich, B., Caruana, R., & Hinton, G. (2021). Neural additive models: Interpretable machine learning with neural nets.
43. Bejarano, G., DeFazio, D., & Ramesh, A. (2019). Deep latent generative models for energy disaggregation. In *Proceedings of the thirty-third AAAI conference on artificial intelligence and thirty-first innovative applications of artificial intelligence conference and ninth AAAI symposium on educational advances in artificial intelligence. AAAI'19/IAAI'19/EAAI'19.* AAAI Press. https://doi.org/10.1609/aaai.v33i01.3301850
44. Nie, J. H., & Lee, T. H. (1996). Rule-based modeling: Fast construction and optimal manipulation. *IEEE Transactions on Systems, Man, and Cybernetics - Part A: Systems and Humans, 26*(6), 728–738. https://doi.org/10.1109/3468.541333
45. Huysmans, J., Dejaeger, K., Mues, C., Vanthienen, J., & Baesens, B. (2011). An empirical evaluation of the comprehensibility of decision table, tree and rule based predictive models. *Decision Support Systems, 51*(1), 141–154. https://doi.org/10.1016/j.dss.2010.12.003
46. Hassija, V., Chamola, V., Mahapatra, A., Singal, A., Goel, D., Huang, K., Scardapane, S., Spinelli, I., Mahmud, M., & Hussain, A. (2024). Interpreting black-box models: A review on explainable artificial intelligence. *Cognitive Computation, 16*(1), 45–74.
47. Hearst, M. A., Dumais, S. T., Osuna, E., Platt, J., & Scholkopf, B. (1998). Support vector machines. *IEEE Intelligent Systems and their Applications, 13*(4), 18–28. https://doi.org/10.1109/5254708428
48. Breiman, L. (2001). *Machine Learning, 45*(1), 5–32.
49. Deng, H. (2014). *Interpreting tree ensembles with in trees.*
50. Popescu, M.-C., Balas, V., Perescu-Popescu, L., & Mastorakis, N. (2009). Multilayer perceptron and neural networks. *WSEAS Transactions on Circuits and Systems, 8*, 579.
51. Kenny, E. M., Ford, C., Quinn, M., & Keane, M. T. (2021). Explaining black-box classifiers using post-hoc explanations-by-example: The effect of explanations and error- rates in xai user studies. *Artificial Intelligence, 294*, 103459. https://doi.org/10.1016/j.artint.2021.103459
52. Ribeiro, M. T., Singh, S., & Guestrin, C. (2016). *Why should i trust you?: Explaining the predictions of any classifier.*
53. Alodibat, S., Ahmad, A., & Azzeh, M. (2023). Explainable machine learning-based cybersecurity detection using lime and secml. In *2023 IEEE Jordan international joint conference on electrical engineering and information technology (JEEIT)* (pp. 235–242). https://doi.org/10.1109/JEEIT58638.2023.10185893
54. Lundberg, S., & Lee, S. -I. (2017). *A unified approach to interpreting model predictions.*
55. Wang, X., Liu, C., Hu, X., Wang, Z., Yin, J., & Cui, X. (2022). Make data reliable: An explanation-powered cleaning on malware dataset against backdoor poisoning attacks. In *Proceedings of the 38th annual computer security applications conference. ACSAC '22* (pp. 267–278). Association for Computing Machinery. https://doi.org/10.1145/3564625.3564661
56. Ribeiro, M. T., Singh, S., & Guestrin, C. (2018). Anchors: High-precision model-agnostic explanations. In *AAAI conference on artificial intelligence (AAAI).*
57. Molnar, C. (2022). *Interpretable machine learning* (2nd ed.). https://christophm.github.io/interpretable-ml-book
58. Chou, Y.-L., Moreira, C., Bruza, P., Ouyang, C., & Jorge, J. (2021). Counterfactuals and Causability in explainable artificial intelligence: Theory, algorithms, and applications. *Information Fusion, 81*, 59.
59. Naderi-Afooshteh, A., Kwon, Y., Nguyen-Tuong, A., Razmjoo-Qalaei, A., Zamiri-Gourabi, M.-R., & Davidson, J. W. (2019). Malmax: Multi-aspect execution for automated dynamic web server malware analysis. In *Proceedings of the 2019 ACM SIGSAC conference on computer and communications security. CCS '19* (pp. 1849–1866). Association for Computing Machinery. https://doi.org/10.1145/3319535.3363199

60. Khandpur, R. P., Ji, T., Jan, S., Wang, G., Lu, C.-T., & Ramakrishnan, N. (2017). Crowdsourcing cybersecurity. In *Proceedings of the 2017 ACM on conference on information and knowledge management*. ACM.
61. Yang, J. (2022). Fast TreeSHAP: Accelerating SHAP value computation for trees.
62. Wali, S., & Khan, I. (2021). Explainable AI and random forest based reliable intrusion detection system.
63. Selvaraju, R. R., Cogswell, M., Das, A., Vedantam, R., Parikh, D., & Batra, D. (2019). Grad-cam: Visual explanations from deep networks via gradient-based localization. *International Journal of Computer Vision, 128*(2), 336–359. https://doi.org/10.1007/s11263-019-01228-7
64. Brigugilio, W. R. (2020). *Machine learning interpretability in malware detection*. PhD thesis,. University of Windsor.
65. Aggarwal, C. C. (2018). *Neural networks and deep learning*. Springer.
66. Coventry, L., & Branley, D. (2018). Cybersecurity in healthcare: A narrative review of trends, threats and ways forward. *Maturitas, 113*, 48–52.
67. Panja, B., Fattaleh, D., Mercado, M., Robinson, A., & Meharia, P. (2013). Cybersecurity in banking and financial sector: Security analysis of a mobile banking application. In *2013 international conference on collaboration technologies and systems (CTS)* (pp. 397–403). IEEE.
68. Li, X., Yi, P., Wei, W., Jiang, Y., & Tian, L. (2021). Lnnls-kh: A feature selection method for network intrusion detection. *Security and Communication Networks, 2021*, 1–22.
69. Ferrag, M. A., Shu, L., Friha, O., & Yang, X. (2021). Cyber security intrusion detection for agriculture 4.0: Machine learning-based solutions, datasets, and future directions. *IEEE/CAA Journal of Automatica Sinica, 9*(3), 407–436.
70. Larriva-Novo, X., Villagra, V. A., Vega-Barbas, M., Rivera, D., & Sanz Rodrigo, M. (2021). An IOT-focused intrusion detection system approach based on preprocessing characterization for cybersecurity datasets. *Sensors, 21*(2), 656.

Chapter 8
Data Protection Challenges in the Processing of Sensitive Data

H. M. S. S. Herath, H. M. K. K. M. B. Herath, B. G. D. A. Madhusanka, and L. G. P. K. Guruge

Abstract Sensitive data proliferation and processing have become essential to many aspects of contemporary life, from social media to healthcare and banking, in an era of digital revolution. Safeguarding sensitive data to maintain privacy and compliance with changing data protection legislation continues to be the top priority as organizations leverage data for innovation and decision-making. The numerous difficulties that arise while handling sensitive data are examined in this study, focusing on data protection. By exploring the complex field of data management, this research seeks to pinpoint and discuss the challenges businesses face while handling sensitive data. The study takes a broad approach, considering the legislative and technical factors that add to the complexity of data security. The study investigates the technological risks related to the handling, transferring, and processing of private information. The advancement of artificial intelligence and machine learning technologies presents a growing danger of unauthorized access and exploitation of sensitive information. Therefore, it is imperative to establish a solid technical infrastructure and implement encryption measures to reduce these risks effectively. Another critical component of this research is organizations' regulatory obstacles in navigating the ever-changing world of data protection laws and regulations. The influence of rules on organizational practices, such as the General Data Protection Regulation (GDPR), is examined in this research. Additionally, the study assesses how legislative frameworks influence data protection techniques,

H. M. S. S. Herath (✉) · L. G. P. K. Guruge
Department of Industrial 4.0 Convergence Bionics, Pukyong National University, Busan, Republic of Korea

H. M. K. K. M. B. Herath
Faculty of Engineering, Sri Lanka Technological Campus, Padukka, Sri Lanka

Computational Intelligence and Robotics Research Lab, Sri Lanka Technological Campus, Padukka, Sri Lanka

B. G. D. A. Madhusanka
Computational Intelligence and Robotics Research Lab, Sri Lanka Technological Campus, Padukka, Sri Lanka

Faculty of Engineering Technology, The Open University of Sri Lanka, Nugegoda, Sri Lanka

C. Hewage et al. (eds.), *Data Protection*,
https://doi.org/10.1007/978-3-031-76473-8_8

highlighting the necessity of a proactive approach to compliance. By integrating technical and regulatory viewpoints, this study adds to a more comprehensive comprehension of the intricate network, including the handling of confidential information. To support the creation of practical methods to secure sensitive information in a world that is becoming more and more data-centric, the results are intended to educate organizations, legislators, and academics on the changing problems in data security.

8.1 Introduction

8.1.1 Overview of Sensitive Data

Sensitive data are any information that, if compromised, could cause financial, reputational, or personal harm. This includes personally identifiable information (PII) [1, 2] like names, addresses, and social security numbers, as well as financial data like credit card numbers and bank account information, private company information, trade secrets, and intellectual property. In today's interconnected world, sensitive data are critical to the operations of many industries, including banking, healthcare, technology, and government [3]. Personal data refer to identifying information about an individual, whereas sensitive data are personal information that requires specific protection, such as health or religious convictions. Private data include any confidential information, both personal and non-personal, that individuals or organizations want to keep hidden from the public. Figure 8.1 shows the

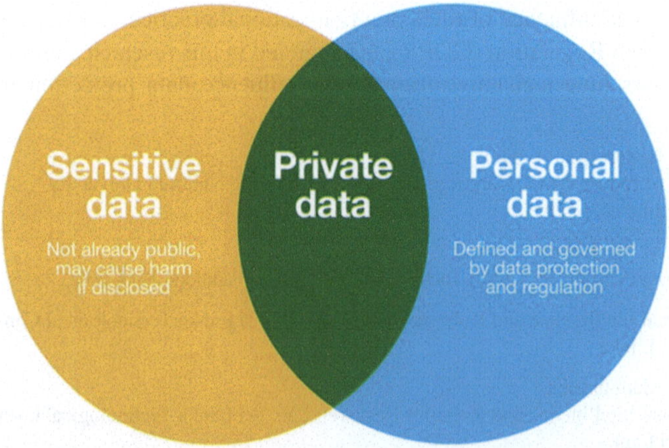

Fig. 8.1 Sensitive data, private data, and personal data. (*Adapted from Anonymisation and open data: an introduction to managing the risk of re-identification (report)*. https://theodi.org)

categorization of sensitive, personal, and private data according to the Anonymisation and Open Data report.

In finance, protecting sensitive data is critical for retaining client trust and meeting regulatory obligations. Financial institutions manage massive amounts of personal and financial information daily, making them perfect targets for hackers looking to exploit vulnerabilities for profit. A breach in this area can result in severe economic losses, regulatory fines, and irreparable reputational harm [4]. Similarly, in the healthcare industry, patient data security is critical not only for privacy and confidentiality but also for protecting the integrity of medical information and complying with rules such as the Health Insurance Portability and Accountability Act (HIPAA) [5].

The IT industry relies on sensitive data to drive innovation and gain a competitive advantage. Companies pay extensively in R&D to create cutting-edge products and services, so protecting intellectual property and private information is vital to maintaining growth and staying ahead of the competition. Furthermore, government organizations hold sensitive information, such as classified intelligence, national security data, and citizen records, making them perfect targets for state-sponsored cyberattacks and espionage [6, 7].

The digital age has offered unrivaled convenience and connectedness but has created unprecedented hurdles in protecting sensitive data. The growth of digital devices, cloud computing, and the Internet of Things (IoT) has significantly increased the volume and velocity of data generated, stored, and sent across massive networks [8–10]. While technology improvements have transformed company operations and increased efficiency, they have also broadened the attack surface for cyber-attacks, exposing organizations to new risks and weaknesses.

Cyberattacks have become more sophisticated and widespread, with threat actors infiltrating networks and stealing sensitive data using various strategies, including malware, phishing, ransomware, and insider threats [11, 12]. A data breach can have severe consequences, including financial losses, legal liability, brand reputation damage, and loss of customer trust. The notable incidents underscore the vulnerability of critical sectors to ransomware attacks. Change Healthcare, a prominent healthcare payment processor, fell victim to a significant ransomware assault in 2024, impacting a staggering one in three Americans and marking it as one of the most significant breaches in the healthcare domain. The repercussions reverberated throughout the sector, particularly affecting small and midsize healthcare providers, disrupting essential services such as electronic prescription filling and insurance reimbursement. The attack, executed using the Alphv ransomware variant by Russian-speaking cybercriminals, highlighted the sophisticated nature of modern cyber threats and the dire need for robust cybersecurity measures. This incident echoes similar assaults on critical infrastructure, such as the LockBit attack on the UK's Royal Mail in January 2023, which caused disruptions in international mail delivery, and the ESXiArgs attacks on VMware ESXi servers in February 2023, exploiting vulnerabilities and encrypting thousands of servers globally. These incidents serve as poignant reminders of the imperative for organizations to fortify their cyber defenses and remain vigilant against evolving threats to safeguard sensitive data and maintain operational resilience.

Furthermore, the growing regulatory landscape, characterized by stringent data protection laws such as the General Data Protection Regulation (GDPR) [13] and the California Consumer Privacy Act (CCPA) [14, 15], has increased the stakes for organizations that fail to adequately safeguard sensitive information, imposing hefty fines and penalties for noncompliance.

As organizations face an ever-increasing threat landscape, there is a rising realization of the importance of a proactive and multifaceted strategy for securing sensitive data. This chapter implements strong cybersecurity measures like encryption, access controls, intrusion detection systems, and security awareness training to reduce risks and strengthen defenses against evolving threats. Furthermore, developing a collaborative commitment to protecting sensitive information and maintaining data privacy standards requires adopting a security and accountability culture at all levels of the organization.

Furthermore, advances in technologies like artificial intelligence (AI) and machine learning (ML) are being used to strengthen cybersecurity defenses and improve threat detection capabilities [16, 17]. Using various strategies and tactics, AI algorithms significantly predict and mitigate cybersecurity breaches. By recognizing abnormalities that could be signs of a breach, artificial intelligence (AI) is excellent at detecting anomalies in network behavior. Anomalies can be classified and detected using methods like decision trees and neural networks in supervised learning and clustering and autoencoders in unsupervised learning [18, 19]. Furthermore, AI forecasts possible security problems based on past data trends using regression models and time-series analysis to predict analytics. Artificial intelligence (AI)-powered solutions improve breach mitigation by automating the scanning of massive databases to find suspicious activity and rank threats using graph analysis and natural language processing (NLP).AI helps real-time intrusion AI enhance detection and prevention systems by constantly updating threat signatures and behavioral analysis. AI-driven playbooks, such as those that automatically isolate and contain a compromised system and adaptive defense systems, which adjust security measures based on real-time threat intelligence, are examples of automated response mechanisms that speed up reaction times and minimize harm. AI also strengthens security posture by scoring risk, providing essential patch recommendations, and managing vulnerabilities [20, 21]. AI uses behavioral biometrics to track user behavior and transaction trends to detect fraud. Moreover, AI combines threat intelligence from several sources and uses correlation analysis to find coordinated assaults. Artificial intelligence (AI) dramatically improves cybersecurity by providing cutting-edge instruments for anticipating, identifying, and addressing threats, strengthening organizational security posture, and guaranteeing prompt and efficient mitigation of possible breaches. AI technology's ongoing development promises even more power in thwarting cyberattacks.

AI-powered systems can analyze massive volumes of data in real-time, detecting unusual behavior patterns and potential security concerns before they become full-fledged breaches. Furthermore, introducing blockchain technology offers the potential to revolutionize data security. Blockchain provides a decentralized, tamper-proof means for storing and verifying critical information. This decentralized nature eliminates the need for a central authority, reducing the risk of a single point of failure. The tamper-proof nature of blockchain ensures that once data are recorded, it cannot be altered without the network's consensus, making it highly secure. These features of blockchain technology make it a promising tool in the fight against cyberthreats.

Sensitive data are central to modern economies and society, supporting crucial activities across many industries. As the digital landscape evolves, protecting sensitive information becomes increasingly important. Organizations must stay alert and proactive in protecting data integrity, confidentiality, and availability to limit the ever-present threat of cyberattacks and maintain stakeholder trust and confidence. Businesses that adopt a holistic strategy for cybersecurity and leverage the power of emerging technologies can negotiate the complexity of the digital age and appear more robust and resilient in the face of evolving threats.

8.1.2 Types of Sensitive Data

Sensitive data exist in various forms, each with its level of relevance and security. Sensitive data encompass a broad array of information that necessitates heightened safeguarding measures owing to its potential to adversely affect an individual's privacy, security, or other fundamental rights. Personal data are at the forefront of these categories, containing information that directly identifies or can be used to identify a specific individual. Names, addresses, phone numbers, email addresses, social security numbers, and government-issued identity numbers are examples but are not exhaustive [22]. Personal data are at the heart of privacy concerns, as its disclosure can lead to identity theft, harassment, and other sorts of abuse.

Financial data are another critical category that includes information about an individual's monetary circumstances. This includes bank account numbers, credit card information, financial activities, and investment portfolios [23]. Disclosing such information can lead to financial loss, fraudulent activity, or unauthorized access to funds. Protecting financial data are critical for individuals, financial institutions, and enterprises that handle sensitive financial information.

Health-related data are critical in both personal and professional settings. It includes medical history, treatment records, prescriptions, diagnostic results, and other health-related data [24, 25]. Health data confidentiality is protected by ethical considerations and legislative requirements, such as the HIPAA [5] in the United States. Breaches in health data can have significant effects, such as medical identity theft, discrimination, or compromised healthcare delivery.

Beyond these core categories, there are numerous other sorts of sensitive data, each requiring its own level of security. Confidential company information may contain trade secrets, intellectual property, proprietary algorithms, and client lists. Disclosing such information can severely affect competitiveness, reputation, and market positioning. Intellectual property theft, industrial espionage, and corporate espionage are risks of disclosing secret business knowledge [26]. Another important category is government and classified information, including national security, defense, intelligence, and law enforcement data. Unauthorized access to such information can jeopardize the safety and security of a country, its residents, and its allies. As a result, rigorous security measures and strict access controls are required to protect government and secret information [27].

Educational information, such as transcripts, grades, and disciplinary records, are also considered sensitive data [28]. In the United States, the FERPA protects the privacy of educational information [29]. Unauthorized access to educational records can lead to identity theft, academic fraud, and lost educational prospects [30]. Furthermore, biometric data, such as fingerprints, iris scans, facial recognition data, and DNA profiles, have become crucial sensitive information. Biometric data are increasingly used for authentication and identification in various industries, including law enforcement, border security, healthcare, and finance. When biometric data are compromised, it can lead to identity theft, unauthorized access, or even bodily injury if used maliciously [31, 32]. Social media data, such as posts, messages, images, and location information, have become another sensitive area in the digital era. While people freely post vast amounts of personal information on social media networks, abuse or unauthorized access to this data can lead to privacy violations, reputational harm, or targeted advertising [33, 34].

Sensitive data fall into various categories, each requiring attentive protection to protect persons, organizations, and communities from the risks of misuse, exploitation, and harm. Personal, financial, health-related, secret commercial, government, educational, biometric, and social media data are among the critical categories that require strong security measures, ethical considerations, and legal safeguards to ensure privacy, integrity, and confidence in the digital age. Table 8.1 summarizes the types of sensitive data discussed above.

8.2 Legal and Regulatory Framework of Sensitive Data

In today's digital world, where data are seen as the new currency, safeguarding sensitive information has become critical. Governments and regulatory organizations worldwide have responded with many rules and regulations to protect individuals' privacy and properly manage sensitive data. Among the most renowned of these regulations are the European Union's General Data Protection Regulation (GDPR) [13, 35] and the United States Health Insurance Portability and Accountability Act (HIPAA) [5]. However, different industries have regulations particular to the data type they handle.

The GDPR, which went into effect in May 2018, is one of the world's most extensive data protection rules, affecting any organization that processes EU individuals' data [36, 37]. It specifies tight criteria for collecting, storing, and processing personal data, including consent requirements, data breach reporting, and the

Table 8.1 Types of sensitive data

Category	Description	Data types
Personal data	Information directly identifying individuals or can be used for identification.	Names, addresses, phone numbers, social security numbers, email addresses, government IDs
Financial data	Data related to an individual's monetary affairs, including bank accounts, credit cards, and transactions.	Bank account numbers, credit card details, financial transactions, investment portfolios
Health-related data	Information about medical history, treatments, prescriptions, and diagnoses.	Medical records, treatment history, prescriptions, diagnostic results
Confidential business	Sensitive business information includes trade secrets, intellectual property, and client lists.	Trade secrets, intellectual property, proprietary algorithms, client lists
Government & classified	Data concerning national security, defense, intelligence, and law enforcement.	Classified information, national security data, intelligence reports, law enforcement records
Educational records	Academic records, including transcripts, grades, and disciplinary history.	Transcripts, grades, disciplinary records
Biometric data	Unique biological characteristics are used for identification and authentication.	Fingerprints, iris scans, facial recognition data, DNA profiles
Social media data	Information shared on social media platforms, including posts, messages, and location data.	Social media posts, messages, photos, location information

appointment of Data Protection Officers (DPOs) [38]. The GDPR principles are based on transparency, accountability, and individuals' rights to control personal information. Noncompliance can result in hefty fines of up to €20 million or 4% of the company's global annual revenue, whichever is greater, making GDPR compliance a top priority for enterprises globally.

The healthcare industry in the United States is subject to strict HIPAA laws. HIPAA, enacted in 1996, aims to protect individuals' medical information while ensuring the confidentiality, integrity, and availability of electronic protected health information (ePHI) [39]. HIPAA requires covered entities, such as healthcare providers, health plans, and healthcare clearinghouses, to follow the privacy, security, and breach notification rule. These standards control a wide range of operations, from patient consent and data encryption to employee training and auditing, and noncompliance carries severe penalties, including fines and criminal charges.

On the other hand, a global biopharmaceutical company headquartered in California faced the daunting task of navigating the complex landscape of privacy regulations, including the General Data Protection Regulation (GDPR) and pertinent U.S. data privacy laws like HIPAA. Seeking clarity and assurance in their compliance efforts, the company engaged Kroll to craft a tailored solution. Kroll thoroughly assessed the company's data processing practices, meticulously examining their adherence to regulatory standards. Subsequently, a robust action plan was formulated, outlining specific measures to implement adequate controls and governance structures. The company achieved sustainable compliance and accountability through this collaborative effort, seamlessly aligning its data processing strategies with GDPR and HIPAA requirements. This concerted approach fortified privacy protection and bolstered the company's legal standing, ensuring resilience in an ever-evolving regulatory landscape.

Aside from these overall norms, different industries have developed guidelines to handle the specific issues given by their data environments. For example, the financial sector follows laws such as the Payment Card Industry Data Security Standard (PCI DSS) [40] and the Sarbanes-Oxley Act (SOX), which focus on financial transaction security and financial reporting accuracy, respectively. Similarly, the telecommunications industry adheres to legislation such as the Communications Assistance for Law Enforcement Act (CALEA),[1] which mandates service providers to support wiretapping and other surveillance actions authorized by law enforcement authorities. The technology sector presents unique issues in managing sensitive data, prompting the development of rules such as the California Consumer Privacy Act (CCPA)[2] and the Children's Online Privacy Protection Act (COPPA).[3]These rules seek to protect consumer privacy online, notably in the acquisition and sale of personal information by technology corporations. They also require rigorous parental authorization when collecting data from minors under thirteen.

In addition to industry-specific restrictions, many countries have passed additional legislation to improve data privacy within their boundaries. Canada's Personal Information Protection and Electronic Documents Act (PIPEDA)[4] and Brazil's Lei Geral de Proteção de Dados (LGPD)[5] follow international standards identical to the GDPR. These regulations reflect a global trend of tightening data protection safeguards and establishing consistency in managing sensitive information across borders [41, 42]. Despite the differences in legislation managing sensitive data, several fundamental ideas underlay their goals. These include safeguarding individuals' privacy rights, increasing openness and responsibility among data controllers and processors, and establishing redress mechanisms for data breaches or misuse. Furthermore, as technology advances and new data concerns arise, regulatory authorities must adapt to meet emerging threats and defend individuals' rights in an increasingly linked world.

The legal and regulatory framework governing sensitive data is complicated and multifaceted, with laws such as the GDPR and HIPAA establishing worldwide data protection standards and industry-specific regulations addressing unique difficulties within specific industries. Adherence to these standards is both a legal duty and a moral responsibility, as protecting sensitive information is critical to preserving

[1] https://ndcac.fbi.gov/calea

[2] https://theccpa.org/

[3] https://www.ftc.gov/legal-library/browse/rules/childrens-online-privacy- protection-rule-coppa

[4] https://laws-lois.justice.gc.ca/eng/acts/p-8.6/FullText.html

[5] https://www.gov.br/esporte/pt-br/acesso-a-informacao/lgpd

Table 8.2 Critical aspects of the data protection laws and industry-specific regulations

Regulation	Overview	Scope
GDPR	Comprehensive EU law governing personal data protection	Applies to organizations processing the personal data of EU citizens
HIPAA	US law protecting patients' medical information	Applies to healthcare providers, health plans, and clearinghouses
PCI DSS	Standards for securing financial transactions	Applies to entities processing payment card data
SOX	Regulates financial reporting and accounting accuracy	Applies to public companies
CCPA	California law protecting consumer privacy online	Applies to businesses collecting personal information about California
COPPA	Protects children's online privacy	Applies to operators of websites and online services directed at children
PIPEDA	Canadian law governing personal information handling	Applies to private-sector organizations collecting personal information.
LGPD	Brazilian law mirrors GDPR's principles	This applies to processing personal data in Brazil
CALEA	Requires telecommunications providers to facilitate law enforcement	Applies to telecommunications carriers

individuals' privacy rights and maintaining trust in the digital economy. As technology improves and new data concerns emerge, regulatory agencies must remain diligent in protecting these principles while adapting to an ever-changing reality of data privacy and security.

Table 8.2 summarizes the key aspects of the data protection law and industry-specific regulations discussed above.

8.3 Processing of Sensitive Data

8.3.1 Data Collection Challenges

Processing sensitive data necessitates extreme caution, rigorous attention to detail, and a firm commitment to ethical principles. Sensitive data, often known as personally identifiable information (PII) [1, 2], include a wide range of information that, if mismanaged, can result in significant implications ranging from privacy violations to identity theft and legal ramifications. Sensitive data include financial information, medical records, biometric data, and any other data that can directly or indirectly identify an individual. Given the ever-changing environment of data privacy legislation and the growing frequency of data breaches, organizations must take strong precautions to protect sensitive data throughout its lifespan.

Processing sensitive data begins with its acquisition, establishing the tone for future management. Transparency and informed permission are critical throughout this phase, as individuals must understand what data are being collected, why it is being collected, and how it will be utilized. Clear and straightforward privacy rules and procedures for getting an express agreement provide the foundation of ethical data-gathering activities. Furthermore, data reduction rules should lead organizations to collect only essential data for the intended purpose, lowering risk exposure [43, 44].

Once acquired, sensitive data must be maintained securely to prevent unauthorized access or disclosure. This means strengthening the digital fortress with powerful encryption techniques, access restrictions, and authentication systems. Encryption, in particular, acts as a powerful barrier against data breaches by making the information unreadable to unauthorized parties. Whether data are at rest or in transit, encryption techniques like AES (Advanced Encryption Standard) [45, 46] or RSA (Rivest-Shamir-Adleman) [47] are critical in maintaining its confidentiality. Furthermore, access controls should be fine-tuned to limit data access to those persons whose positions need contact with sensitive data. Role-based access control (RBAC) frameworks provide a structured approach to delineating access privileges based on job responsibilities thereby minimizing the risk of insider threats [48, 49]. In addition to technology precautions, physical security measures are critical in securing sensitive data and technology precautions. Data centers and server rooms should include biometric authentication systems, security cameras, and strict access restrictions to prevent unauthorized physical entry. Fire suppression systems and environmental controls make data storage facilities robust in the face of unanticipated events like fires or natural catastrophes. Regular audits and compliance assessments act as litmus tests for evaluating the effectiveness of security measures and identifying possible vulnerabilities that must be addressed. However, security measures are insufficient without a comprehensive data governance and compliance framework. Organizations must comply with various legal standards, including but not limited to the GDPR, the HIPAA, and the PCI DSS. These rules put tight requirements on organizations for collecting, storing, processing, and transferring sensitive data, with severe fines for noncompliance. As a result, it is critical to develop a robust data governance architecture that includes rules, processes, and controls that are compliant with legal requirements. This includes hiring a data protection officer (DPO) to manage compliance efforts, performing frequent privacy impact assessments (PIAs) to analyze data processing operations, and instilling a data privacy culture through training and awareness programs [50, 51].

Equally important is the feature of data integrity, which guarantees that sensitive data is accurate, dependable, and consistent throughout its lifespan. Data validation procedures should be deployed at several touchpoints to discover and remediate mistakes or inconsistencies that may jeopardize data integrity. Checksums, data validation criteria, and redundancy checks act as barriers to data corruption or manipulation, establishing confidence in the integrity of sensitive data. Furthermore, audit trails and logging methods aid forensic analysis and accountability by documenting all interactions with sensitive data, such as updates, accesses, and transfers. The final disposition of sensitive data signifies the end of its lifespan and requires careful attention to avoid unintended disclosure or unauthorized keeping. Data retention plans should provide clear rules for the retention periods of various types of sensitive data depending on legal, regulatory, and commercial considerations. Redundant or outdated data should be safely erased using industry-standard data sanitization procedures to avoid leftover data remnants getting into the wrong hands. Furthermore, data deletion should be thoroughly recorded to establish compliance with data protection requirements and reduce the risk of responsibility [52, 53].

Processing sensitive data require a comprehensive strategy incorporating technological, organizational, and procedural safeguards. Organizations can mitigate the inherent risks of handling sensitive data by following principles such as transparency, data minimization (the practice of collecting, storing, and processing only the minimum amount of personal data necessary for a specific purpose), encryption, access control, regulatory compliance, data integrity, and secure disposal. In an era where data privacy is more valued and scrutinized, responsible data processing is more than a legal responsibility; it is a moral necessity that highlights the essence of ethical stewardship in the digital age.

8.3.2 Data Storage and Security Concerns

Data storage and security considerations are crucial in today's digital environment, especially when dealing with sensitive information. Storing sensitive data offers a variety of security threats, including unauthorized access, data breaches, and compliance violations. Encryption is one of the most effective strategies for mitigating these dangers. Encryption is the process of encrypting data so authorized people can only access it with the necessary decryption key. Encrypting sensitive data in transit and at rest can significantly minimize the possibility of data breaches and unauthorized access. Advanced encryption algorithms like AES [45, 46] and RSA [47] offer robust security features to protect sensitive data from hostile attackers.

Access restrictions are an important part of data storage security. Access controls specify who may access specific data and what actions they can take with it. Implementing granular access controls ensures that only authorized persons or entities can access sensitive data, reducing the risk of insider threats and unauthorized disclosure. RBAC frameworks enable organizations to allocate access rights based on job responsibilities, ensuring that employees only have access to data required for their employment [54]. Furthermore, Multi-Factor Authentication (MFA) adds an extra layer of security by forcing users to give several verification forms before accessing sensitive data, strengthening access restrictions [55, 56].

Furthermore, organizations must have robust security procedures to safeguard sensitive data from external attacks. This involves installing firewalls, intrusion detection systems (IDS, security tools that monitor network traffic or system logs to detect suspicious or unauthorized activities.), and endpoint protection solutions to protect data from hostile assaults. Regular security audits and vulnerability assessments assist in discovering and addressing any flaws in data storage systems, offering ongoing protection against emerging security threats. Furthermore, organizations should prioritize employee training and awareness programs to educate staff on best practices for handling sensitive data and identifying possible security concerns.

8.3.3 Data Processing in the Cloud

Data processing in the cloud brings additional considerations and problems, particularly when handling sensitive data, as more and more businesses move their operations to cloud platforms. Data residency, or the legal and regulatory restrictions governing where data can be handled and stored, is one of the main issues. Data residency standards are subject to variation among jurisdictions. Hence, organizations must comply with them to avoid legal repercussions. Organizations can take advantage of the scalability and flexibility of cloud computing while adhering to data residency requirements by collaborating with cloud service providers who offer data centers in compliant regions.

Another crucial factor when processing sensitive data in cloud environments is compliance with industry-specific regulations like the PCI DSS, HIPAA, and GDPR. Although most cloud service providers give compliance certifications and assurances, it is still the organizations' responsibility to ensure that their cloud deployments comply with applicable regulations. This entails putting encryption, access controls, and audit trails into place to safeguard confidential information and prove adherence to pertinent laws. Furthermore, cloud computing's fundamental shared responsibility paradigm necessitates that organizations distinguish between the roles of the cloud service provider and the consumer. It is the customers' responsibility to secure their data within the cloud environment, while cloud providers are in charge of maintaining the security of the cloud infrastructure, including network and physical security. Organizations must take strong security measures to reduce the risk of unauthorized access and data breaches. These methods include encryption, access controls, and data loss prevention (DLP) policies [57–59].

Data storage and security issues are significant problems at a time when sensitive data are widely dispersed, and cybersecurity threats are constantly changing. The dangers of holding sensitive data are significantly reduced by encryption, access controls, and other security measures, and robust compliance programs guarantee that legal requirements are met. Organizations handling sensitive data in cloud settings must deal with issues including data residency, compliance, and the shared responsibility model to maintain data integrity and prevent unwanted access. Organizations may manage the risks of processing and storing sensitive data in both on-premises and cloud-based environments by implementing comprehensive security measures and staying alert to new threats. Transitioning from legal frameworks to data processing challenges, it's evident that a multifaceted approach, integrating regulatory compliance and robust security measures, is imperative in navigating the complexities of sensitive data management in cloud environments.

8.4 Data Protection Through Privacy by Design and Default

Privacy by Design (PbD) has emerged as a proactive method of protecting sensitive data in an age of data breaches and privacy concerns. PbD promotes treating privacy issues as an afterthought and instead incorporating them into the design process of systems, services, and products from the ground up. Organizations can enhance user data protection and reduce potential privacy issues by including privacy principles and defaults in the design phase.

Principles of Privacy by Design (PbD)

The concept of PbD centers on integrating privacy elements into systems' basic architecture and design instead of tackling them as afterthoughts. The intention is for data protection to be incorporated into the product or service by default, with privacy being a fundamental feature. Proactive rather than reactive action is one of the Privacy design's core tenets. PbD encourages organizations to foresee potential risks and implement privacy protections preemptively instead of waiting for privacy issues to develop and then fixing them. Doing this may increase user trust and transparency by stopping privacy abuses before they happen [60–62]. Seven foundational principles (see Fig. 8.2) form the basis of the PbD concept and are a great place to start when starting the PbD journey. These principles do not, however, provide precise instructions on how to carry out PbD or criteria for gauging the effectiveness of PbD procedures; instead, they only offer a broad framework.

The end-to-end security of PbD is another crucial feature. This principle highlights how vital it is to safeguard data at every stage of its lifespan, from gathering and storing it to processing and destroying it. Organizations may guarantee the security and confidentiality of sensitive information by putting strong security measures in place at every turn [60, 62]. PbD also encourages user- centricity and openness. Businesses should be open and honest with users about their data practices,

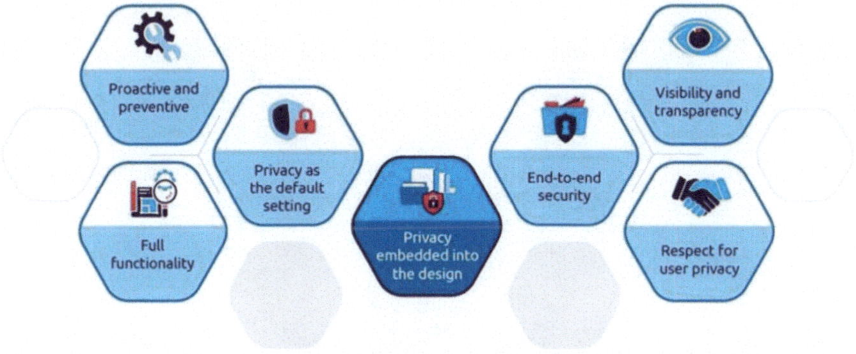

Fig. 8.2 PbD foundational principles. (Image from the article published on LinkedIn "An Engineer's Guide to Privacy by Design" https://www.linkedin.com)

including what information is gathered, how it is used, and who has access to it. Users should also be in charge of their data, allowing them to approve its acquisition and decide how it will be shared and used.

There are no single, universal strategies for putting PbD into practice. Several stakeholders must work together across the entire organization to implement it and it should be tailored to every business's particular needs and conditions. Six privacy engineering best practices have emerged to help engineers and product managers incorporate PbD into their systems and products, enabling privacy and innovation to live harmoniously (see Fig. 8.3).

Incorporating Privacy Considerations

Including privacy considerations in the development process can assist in reducing a number of data protection-related issues. Organizations can better handle privacy threats and lessen their impact on consumers by detecting them early on. Incorporating privacy elements into the design phase can simplify adherence to privacy standards and lower the risk of expensive penalties and legal consequences. For example, privacy should be considered from the beginning of the planning process while creating a new mobile application. This entails performing a privacy impact assessment and implementing privacy-enhancing measures such as data encryption, anonymization, and access controls to identify potential hazards. Developers can reduce privacy threats and improve the application's overall security by incorporating these characteristics into the design.

Additionally, companies can benefit from privacy by design frameworks, like the one created by Dr. Ann Cavoukian, which offers standards and best practices for integrating privacy into system and application design [63, 64]. By adhering to these

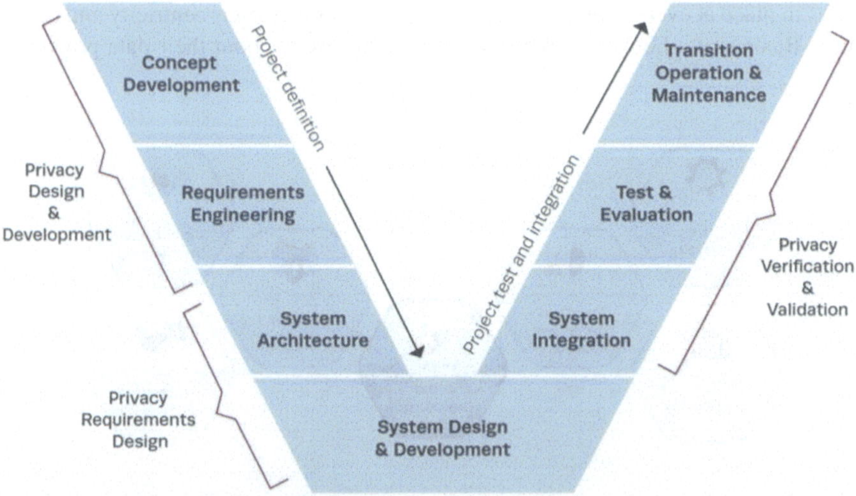

Fig. 8.3 Privacy engineering framework. (Image from the article published on LinkedIn "An Engineer's Guide to Privacy by Design" https://www.linkedin.com)

frameworks, organizations may guarantee that privacy is adequately considered during development, resulting in more reliable and privacy-friendly products and services.

Embedding Privacy Defaults

Privacy defaults are essential for safeguarding private information and maintaining user privacy. Users' level of privacy protection is influenced by default privacy settings, which dictate how data are treated by default. By using privacy-friendly settings, organizations can guarantee that consumers' privacy is protected from using a product or service [65].

Social media networks, for example, can choose privacy-friendly settings, such as restricting user-profiles and postings to friends only, instead of having them publicly available by default. Software programs can also be configured to automatically capture the minimal personal data required for their designed purposes instead of collecting a large amount of data that might be misused. Organizations can lower the risk of unintentional data disclosure while empowering users to make educated decisions about their privacy preferences by introducing privacy-friendly defaults [66, 67]. Furthermore, default privacy settings can act as a foundational layer of security, guaranteeing a respectable level of privacy to users even if they are unaware of potential privacy threats.

In the digital era, PbD and Default are fundamental concepts for guaranteeing data security and user privacy. Organizations can ensure compliance with privacy rules and improve user trust, transparency, and security by integrating privacy considerations into the design process and applying privacy-friendly defaults. Ultimately, putting privacy first by design and default benefits users' and businesses' long-term viability and standing in an increasingly data-driven society.

8.5 Data Breach History and Future Preventions

8.5.1 Incidents and Case Studies

Breach of sensitive data can have far-reaching consequences that impact the immediate victims, industry, organizations, and society. A person's name, financial information, medical history, or other identifying facts can be stolen, and they can be used fraudulently or for different types of abuse. Those who are impacted by this erosion of privacy may experience severe emotional pain as well as financial difficulty as they deal with the consequences of having their private information revealed. Moreover, companies that suffer from data breaches frequently incur significant monetary losses, legal ramifications, and harm to their reputation. The expenses incurred in looking into the breach, alerting the parties impacted, and putting security measures in place to stop similar events in the future can be extremely high. Furthermore, losing the trust of stakeholders and consumers can have long-term effects on the company's income streams and growth prospects. Many occurrences

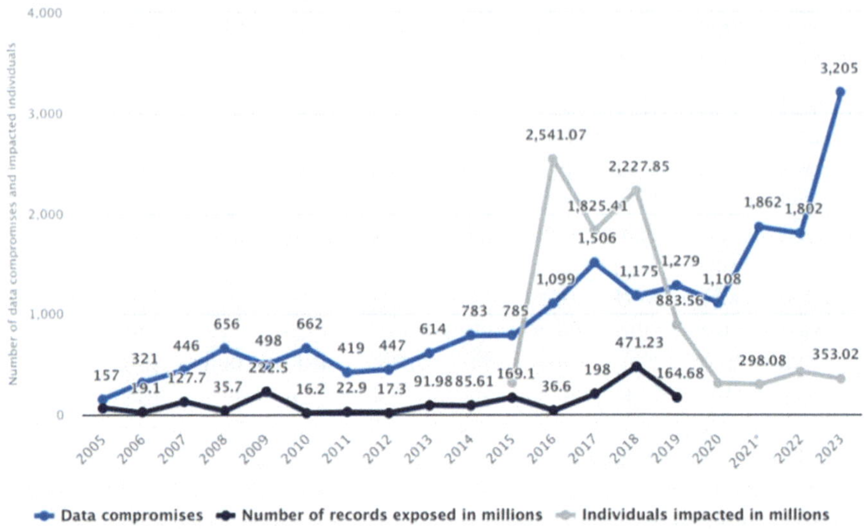

Fig. 8.4 The annual number of data compromises and individuals impacted in the United States from 2005 to 2023. (Image and data sources are adapted from https://www.statista.com)

have been reported in the United States in the last 10 years. The number of data breach events reported annually in the United States from 2005 to 2013 is depicted in Fig. 8.4. The graphic illustrates how the number of occurrences rises yearly, demonstrating the need for data security and protection.

Data breaches have a direct effect on specific companies. Still, they can also cause havoc across whole industries (see Fig. 8.5). A breach can potentially undermine customer confidence and trigger regulatory scrutiny in sectors like technology, healthcare, and finance, where protecting sensitive data are crucial. In reaction to high-profile breaches, stricter laws and compliance requirements could be implemented, adding to organizations' responsibilities and changing industry norms. Moreover, the consequences of data breaches on society are enormous. Personal data are becoming more valuable and in more significant quantities as digital technologies are incorporated into daily life. Given that data breaches disproportionately harm marginalized areas, the erosion of privacy threatens public confidence in digital systems and has the potential to worsen social inequality. Additionally, the frequency of data breaches undermines confidence in the digital economy and impedes technical innovation by contributing to a broader decline in trust in institutions.

Breach incidents involving sensitive data pose a severe risk to people, businesses, and society. Beyond the immediate victims, there are far-reaching financial, legal, reputational, and social ramifications. In light of the constantly changing digital world, addressing the underlying causes of data breaches, and improving cybersecurity procedures to reduce risks and preserve the integrity of sensitive data is critical.

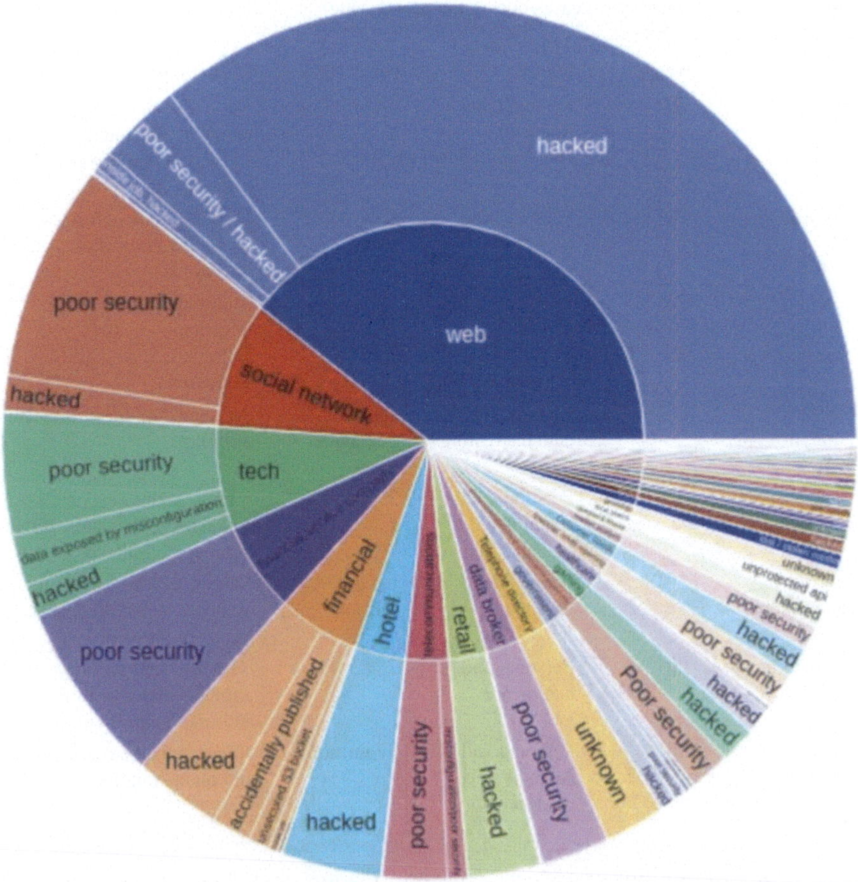

Fig. 8.5 Method of breaches in different sectors. (Adapted from Google Images) [85]

These are a few famous case studies, and instances of breaches involving sensitive data are shown in Table 8.3. Several high-profile data breaches have exposed millions of people's private information to unauthorized access in recent years. A hack at Equifax 2017 exposed 147 million people's financial and personal information, including driver's licenses and social security cards [68]. Yahoo's 2013–2014 security breaches impacted billions of users, which exposed hashed passwords, email addresses, and user identities [69]. The credit card and personal information of 110 million Target customers was compromised in a 2013 hack [70]. 2018 saw data on millions of Facebook users collected by Cambridge Analytica for political profiling [71]. The personal and account information of 77 million users was exposed in the 2011 breach on the Sony PlayStation Network [72]. The 2018 Marriott International hack may have exposed the names, addresses, and passport numbers of up to 500 million guests [73]. The personal and medical information of 78.8 million people was compromised by Anthem Inc. in 2015 [74]. Uber revealed

Table 8.3 Famous data breach cases during the past decade

Case	Year	Type of data breach	Affected population	Data compromised
Equifax [68]	2017	Personal and financial	147 million	Names, social security numbers, birth dates, addresses, and driver's licenses
Yahoo [69]	2013–2014	Personal and account credentials	Billions of users	Names, email addresses, telephone numbers, hashed passwords
Target [70]	2013	Payment card and personal	110 million	Credit and debit card information, names, addresses, and email addresses
Cambridge Analytica [71]	2018	Unauthorized data harvesting	Millions of Facebook users	Personal data for political profiling.
Sony PlayStation Network [72]	2011	Personal and account credentials	77 million	Names, addresses, email addresses, birth dates, PSN login credentials
Marriott International [73]	2018	Personal and payment	Up to 500 million	Names, addresses, phone numbers, email addresses, passport numbers
Anthem Inc. [74]	2015	Personal and healthcare	78.8 million	Names, social security numbers, dates of birth, addresses, employment
Uber [75]	2016	Personal and account credentials	57 million	Names, email addresses, and phone numbers

in 2016 that 57 million users' personal information, including names and email addresses, had been compromised [75]. These events highlight the importance of robust cybersecurity safeguards to protect sensitive data.

8.5.2 Emerging Technologies and Solutions for Data Breaches

In today's digital world, where the gathering and processing of sensitive data have become commonplace, privacy-preserving solutions have become essential. Homomorphic encryption is a cutting-edge technology that offers the best chance for safe data processing. Homomorphic encryption preserves privacy by enabling calculations to be done directly on encrypted data without decryption, unlike typical encryption techniques that require decryption to process data. This innovative technique protects sensitive data from possible breaches and unauthorized access by enabling the secure processing of such data even while it is encrypted [76, 77]. Homomorphic encryption satisfies the vital demand for privacy in data-driven applications, ranging from healthcare to finance, by preserving data secrecy during processing. As exploring the use cases, secure data monetization offers organizations an innovative pathway to explore new revenue streams by securely tapping into their existing data assets. Organizations can use homomorphic encryption to conduct machine learning computations on encrypted data, extracting valuable insights without sacrificing privacy. This technology extends its utility to cloud database queries, allowing for encrypted queries on databases stored in the cloud, ensuring data privacy while enabling essential operations. Moreover, homomorphic encryption facilitates the merging of sensitive datasets, deriving insights from multiple sources without compromising individual privacy. This capability proves instrumental for collaborative endeavors spanning organizational or jurisdictional boundaries, fostering a culture of data-driven decision-making while upholding stringent privacy standards.

Similarly, differential privacy has gained popularity as a way to preserve personal information while drawing insightful conclusions from datasets. Differential privacy ensures that the presence or absence of any individual's data cannot be identified with high confidence by adding noise to query replies. This statistical technique offers a strict framework for striking a balance between privacy and data utility, making it especially appropriate when protecting an individual's privacy, which is essential for medical research or the analysis of census data. Organizations can extract valuable information while protecting people's sensitive details by integrating differential privacy into data processing pipelines.

Blockchain technology's decentralized and unchangeable nature offers an additional means of improving the security and transparency of transactions involving sensitive data [78]. Blockchain guarantees that private information is secure from unauthorized access and tampering by distributing it throughout a network of nodes and encrypting transactions. Furthermore, because blockchain technology is transparent, stakeholders may monitor the flow of information and confirm its integrity without depending on a central authority. This provides for real-time visibility into data exchanges [79]. For applications like supply chain management or digital identity verification, where data integrity is critical, this transparency fosters trust in data exchanges.

Blockchain is used in various use cases for processing sensitive data, each of which takes advantage of its unique characteristics to solve a particular problem. One well-known application of blockchain technology is in healthcare, where it allows healthcare providers to securely share medical records while maintaining patient privacy and data integrity. Healthcare organizations may minimize the danger of data breaches, improve data interchange, and eliminate pointless procedures by storing patient data on a blockchain network. Additionally, blockchain gives patients more autonomy and privacy rights by letting them keep control over their medical records [79, 80]. Financial transactions have a strong use case for blockchain technology processing sensitive data, especially in digital asset management and cross-border payments. Because of characteristics like smart contracts and decentralized consensus procedures, blockchain-based solutions are more efficient and secure than traditional financial systems. By enabling safe and transparent financial transaction processing, these features reduce the risks related to centralization and intermediaries. Additionally, blockchain makes it easier for assets to be tokenized, which allows for fractional ownership and increased liquidity. This increases access to financial markets while maintaining regulatory compliance [80].

Blockchain can potentially protect sensitive data in industries like supply chain management, where traceability and transparency are critical, going beyond healthcare and finance. A blockchain ledger allows supply chain participants to follow product movement from the point of origin to the end of destination, guaranteeing authenticity and quality assurance [81–83]. Furthermore, smart contracts can be implemented using blockchain technology to automate agreements and expedite procedures, which lowers administrative overhead and improves operational efficiency. In addition to increasing participant trust in the supply chain, this automation and transparency help to reduce risks like supply chain disruptions and counterfeit goods [84].

Technologies that protect privacy, such as differential privacy and homomorphic encryption, provide creative ways to handle data securely while maintaining individual privacy and gaining insightful information. Similar to this, blockchain technology has the potential to completely transform a variety of industries, including supply chain management, healthcare, and finance, by improving the security and transparency of sensitive data transactions. Through these technologies, businesses can confidently handle the challenges of handling sensitive data and protecting data privacy, integrity, and trust in an increasingly digital environment.

8.6 Conclusion

In conclusion, the widespread distribution of sensitive data across various businesses within the quickly changing digital world highlights this data's critical function in modern society, affecting areas like social media, healthcare, and finance. Maintaining privacy standards and adhering to changing data protection laws require protecting sensitive information, mainly as businesses depend more and more on data for innovation and decision-making.

The different challenges organizations confront when handling sensitive data have been made clear by this study, highlighting the need for solid technical infrastructures and encryption techniques to reduce the possibility of personal data being misused or accessed without authorization. Proactive compliance strategies are essential given the complex regulatory environment—which is best illustrated by regulations such as the General Data Protection Regulation (GDPR). These strategies call for a thorough comprehension of the requirements of the rules and their incorporation into organizational data protection policies.

A proactive approach to protecting sensitive data in an age of data breaches and privacy concerns is Privacy by Design (PbD). Organizations can reduce privacy risks early on by incorporating privacy principles into the core architecture and design of systems, services, and products. PbD's emphasis on proactive action, end- to-end security, user-centricity, and transparency provides a robust framework for incorporating privacy considerations into regular operations. This increases user confidence and corporate credibility.

In the future, maintaining sensitive data protection and promoting long-term trust and sustainability for consumers and businesses in the digital era would require emphasizing privacy by design as default. Adopting a proactive strategy for data protection is becoming more and more necessary as technology develops to traverse the increasingly complex world of security and privacy successfully. Furthermore, utilizing cutting-edge technologies like blockchain, homomorphic encryption, and differential privacy provides creative ways to improve data security and privacy, lessen data breaches, and give businesses the resources they need to handle sensitive data in a constantly changing digital landscape properly.

Even though this study offers insightful information about protecting sensitive data, a few research gaps need to be filled. To advance our understanding and implementation of effective data privacy measures in the digital era, it is essential to pay attention to the following vital areas: assessing the practical effectiveness of emerging technologies, comprehending the long-term effects of data breaches, fostering cross-industry collaboration, investigating user perceptions of data privacy; addressing compliance challenges; and investigating the ethical implications of data protection practices.

Conflict of Interest The authors declare that they do not have any conflict of interest.

References

1. Schwartz, P. M., & Solove, D. J. (2011). The PII problem: Privacy and a new concept of personally identifiable information. *New York University Law Review, 86*, 1814.
2. McCallister, E., Grance, T., & Scarfone, K. (2010). Identifiable information (pii). *NIST Special Publication, 800*, 122.
3. Ervural, B. C., & Ervural, B. (2017). *Overview of cyber security in the industry 4.0 era* (Springer series in advanced manufacturing) (pp. 267–284). https://doi.org/10.1007/978-3-319-57870-5_16
4. Datoo, A. (2019). *Legal data for banking: Business optimisation and regulatory compliance*. Wiley.
5. Nosowsky, R., & Giordano, T. J. (2006). The health insurance portability and accountability act of 1996 (HIPAA) privacy rule: Implications for clinical research. *Annual Review of Medicine, 57*(1), 575–590. https://doi.org/10.1146/annurev.med.57.121304.131257
6. Herrmann, D. (2019). Cyber espionage and cyber defence. In C. Reuter (Ed.), *Information technology for peace and security*. Springer Vieweg. https://doi.org/10.1007/978-3-658-25652-4_5
7. Cyber-threats to critical national infrastructure: An intelligence challenge. (2024). *International Journal of Intelligence and CounterIntelligence*. https://www.tandfonline.com/doi/abs/10.1080/08850607.2013.780552
8. Harb, H., Idrees, A. K., Jaber, A., Makhoul, A., Zahwe, O., & Taam, M. A. (2018). Wireless sensor networks: A big data source in internet of things. *International Journal of Sensors, Wireless Communications and Control, 7*(2). https://doi.org/10.2174/2210327907666170906144926

9. Tyagi, A. K., & Nair, M. M. (2020). Internet of Everything (IoE) and Internet of Things (IoTs): Threat analyses, possible opportunities for future. *Journal of Information Assurance & Security, 15*(5), 153–177.
10. Ranjan, R., Rana, O., Nepal, S., Yousif, M., James, P., Wen, Z., Barr, S., Watson, P., Jayaraman, P. P., Georgakopoulos, D., Villari, M., Fazio, M., Garg, S., Buyya, R., Wang, L., Zomaya, A. Y., & Dustdar, S. (2018). The next grand challenges: Integrating the internet of things and data science. *IEEE Cloud Computing, 5*(3), 12–26. https://doi.org/10.1109/mcc.2018.032591612
11. Al-Turjman, F., & Salama, R. (2020). *An overview about the cyberattacks in grid and like systems* (pp. 233–247). CRC Press EBooks. https://doi.org/10.1201/9781003055235-11
12. Aslan, Ö., Aktuğ, S. S., Ozkan-Okay, M., Yilmaz, A. A., & Akin, E. (2023). A comprehensive review of cyber security vulnerabilities, threats, attacks, and solutions. *Electronics, 12*(6), 1333–1333. https://doi.org/10.3390/electronics12061333
13. The EU General Data Protection Regulation (GDPR). (n.d.). Retrieved April 21, 2024, from https://www.pagecorpgroup.com/wp- content/uploads/2019/04/p_14.pdf
14. Goldman, E. (2018). An introduction to the California Consumer Privacy Act (CCPA). *Social Science Research Network*. https://doi.org/10.2139/ssrn.3211013
15. Illman, E., & Temple, P. (2019). California consumer privacy act. *The Business Lawyer, 75*(1), 1637–1646.
16. Herath, H. M. K. K. M. B., & Mittal, M. (2022). Adoption of artificial intelligence in smart cities: A comprehensive review. *International Journal of Information Management Data Insights, 2*(1), 100076. https://doi.org/10.1016/j.jjimei.2022.100076
17. Herath, B., Mittal, M., & Kataria, A. (2024). Navigating the metaverse: A technical review of emerging virtual worlds. *Wiley Interdisciplinary Reviews: Data Mining and Knowledge Discovery*. https://doi.org/10.1002/widm.1538
18. Nassif, A. B., Talib, M. A., Nasir, Q., & Dakalbab, F. M. (2021). Machine learning for anomaly detection: A systematic review. *IEEE Access, 9*, 78658–78700. https://doi.org/10.1109/access.2021.3083060
19. Choi, H., Kim, M., Lee, G., & Kim, W. (2019). Unsupervised learning approach for network intrusion detection system using autoencoders. *Journal of Supercomputing, 75*(9), 5597–5621. https://doi.org/10.1007/s11227-019-02805-w
20. Kim, A. C., Park, M., & Lee, D. H. (2020). AI-IDS: Application of deep learning to Real-time web intrusion detection. *IEEE Access, 8*, 70245–70261. https://doi.org/10.1109/access.2020.2986882
21. Bouchama, F., & Kamal, M. (2021). Enhancing cyber threat detection through machine learning-based behavioral modeling of network traffic patterns. *International Journal of Business Intelligence and Big Data Analytics, 4*(9), 1–9.
22. Žliobaitė, I., & Custers, B. (2016). Using sensitive personal data may be necessary for avoiding discrimination in data-driven decision models. *Artificial Intelligence and Law, 24*(2), 183–201. https://doi.org/10.1007/s10506-016-9182-5
23. Shu, X., Yao, D., & Bertino, E. (2015). Privacy-preserving detection of sensitive data exposure. *IEEE Transactions on Information Forensics and Security, 10*(5), 1092–1103. https://doi.org/10.1109/tifs.2015.2398363
24. Christen, P., Ranbaduge, T., & Schnell, R. (2020). *Linking sensitive data*. SpringerLink. https://doi.org/10.1007/978-3-030-59706-1
25. Sensitive-by-distance: Quasi-health data in the algorithmic era. (2017). *Information & Communications Technology Law*. https://www.tandfonline.com/doi/abs/10.1080/13600834.2017.1335468
26. *Protecting intellectual property and sensitive information in academic campuses from trusted insiders | Proceedings of the 37th annual ACM SIGUCCS fall conference: communication and collaboration*. (2024). ACM Conferences. https://doi.org/10.1145/1629501.1629520
27. Sweeney, L., Crosas, M., & Bar-Sinai, M. (2015). Sharing sensitive data with confidence: The datatags system. *Technology Science*. https://techscience.org/a/2015101601/

28. Wang, X., & Wang, Y. (2018). Educational sensitive information retrieval: Analysis, application, and optimization. *IEEE Access, 6*, 30793–30803. https://doi.org/10.1109/access.2018.2844158
29. Young, E. (2014). Educational privacy in the online classroom: FERPA, MOOCs, and the big data conundrum. *Harvard Journal of Law & Technology, 28*, 549.
30. Abu Gosh, M. M., Atallah, R., & Samy, S. A.-N. (2016). Secure mobile cloud computing for sensitive data: Teacher Services for Palestinian Higher Education Institutions. *International Journal of Grid and Distributed Computing, 9*(2), 17–22.
31. Natgunanathan, I., Mehmood, A., Xiang, Y., Beliakov, G., & Yearwood, J. (2016). Protection of privacy in biometric data. *IEEE Access, 4*, 880–892. https://doi.org/10.1109/access.2016.2535120
32. Kindt, E. J. (2013). *Biometric data, data protection and the right to privacy* (Law, governance and technology series) (pp. 87–272). https://doi.org/10.1007/978-94-007-7522-0_3
33. Geetha, R., Karthika, S., & Kumaraguru, P. (2021). Tweet-scan-post: A system for analysis of sensitive private data disclosure in online social media. *Knowledge and Information Systems, 63*(9), 2365–2404. https://doi.org/10.1007/s10115-021-01592-2
34. Bioglio, L., & Pensa, R. G. (2022). Analysis and classification of privacy- sensitive content in social media posts. *EPJ Data Science, 11*(1), 12. https://doi.org/10.1140/epjds/s13688-022-00324-y
35. Voigt, P., & von. (2017). *The EU general data protection regulation (GDPR)*. SpringerLink. https://doi.org/10.1007/978-3-319-57959-7
36. Zaeem, R. N., & Barber, K. S. (2020). The effect of the GDPR on privacy policies: Recent Progress and future promise. *ACM Transactions on Management Information Systems, 12*(1) ACM Transactions on Management Information Systems (TMIS). https://dl.acm.org/doi/abs/10.1145/3389685
37. Mukiri-Smith, H., & Leenes, R. (2021). Beyond the'Brussels Effect'? Kenya's data protection act (DPA) 2019 and the European Union's general data protection regulation (GDPR) 2018. *European Data Protection Law Review, 7*, 502.
38. Recio, M. (2017). Data protection officer: The key figure to ensure data protection and accountability. *European Data Protection Law Review, 3*, 114.
39. Lawson, N. A., Orr, J. M., & Klar, D. S. (2003). The HIPAA privacy rule: An overview of compliance initiatives and regirements. *Defense Counsel Journal, 70*, 127.
40. DSS, W. I. P. (2010). *Payment Card Industry Data Security Standard (PCI DSS) 3.2*
41. Akanfe, O., Lawong, D., & Raghav Rao, H. (2024). Blockchain technology and privacy regulation: Reviewing frictions and synthesizing opportunities. *International Journal of Information Management, 76*, 102753–102753. https://doi.org/10.1016/j.ijinfomgt.2024.102753
42. Corte-Real, A., Nunes, T., & Paulo. (2024). Reflections about blockchain in health data sharing: Navigating a disruptive technology. *International Journal of Environmental Research and Public Health, 21*(2), 230–230. https://doi.org/10.3390/ijerph21020230
43. Dong, X., Li, R., He, H., Zhou, W., Xue, Z., & Wu, H. (2015). Secure sensitive data sharing on a big data platform. *Tsinghua Science and Technology, 20*(1), 72–80. https://doi.org/10.1109/tst.2015.7040516
44. Vashishth, T. K., Sharma, V., Sharma, K. K., Kumar, B., Chaudhary, S., & Panwar, R. (2024). *Security and privacy considerations in cloud-based data processing solutions for sensitive data* (Advances in systems analysis, software engineering, and high performance computing book series) (pp. 35–61). https://doi.org/10.4018/979-8-3693-5643-2.ch002
45. Heron, S. (2009). Advanced encryption standard (AES). *Network Security, 2009*(12), 8–12. https://doi.org/10.1016/s1353-4858(10)70006-4
46. Dworkin, M. J., Barker, E. B., Nechvatal, J. R., Foti, J., Bassham, L. E., Roback, E., & Dray, J. F. (2001, November 26). *Advanced Encryption Standard (AES)*. NIST.
47. Sihotang, H. T., Efendi, S., Zamzami, E. M., & Mawengkang, H. (2020). Design and implementation of Rivest Shamir Adleman's (RSA) cryptography algorithm in text file data security. *Journal of Physics. Conference Series, 1641*(1), 012042. https://doi.org/10.1088/1742-6596/1641/1/012042

48. Cruz, J. P., Kaji, Y., & Yanai, N. (2018). RBAC-SC: Role-based access control using smart contract. *IEEE Access, 6*, 12240–12251. https://doi.org/10.1109/access.2018.2812844
49. Bijon, K. Z., Krishnan, R., & Sandhu, R.. (2013). *A framework for risk-aware role based access control*. 2013 IEEE conference on communications and network security (CNS), National Harbor, MD, USA (pp. 462–469). https://doi.org/10.1109/CNS.2013.6682761.
50. Šidlauskas, A. (2021). The role and significance of the data protection officer in the organization. *Socialiniai tyrimai, 44*(1), 8–28.
51. Georgiou, D., & Lambrinoudakis, C. (2021). Data protection impact assessment (DPIA) for cloud-based health organizations. *Future Internet, 13*(3), 66–66. https://doi.org/10.3390/fi13030066
52. Pearson, S., & Benameur, A.. (2010). *Privacy, Security and Trust Issues Arising from Cloud Computing*. IEEE second international conference on cloud computing technology and science, Indianapolis, IN, USA (pp. 693–702). https://doi.org/10.1109/CloudCom.2010.66.
53. Lee-Gulley, P., & Drudi, J. (2022). Records management/records retention risk considerations. *Routledge EBooks*, 315–337. https://doi.org/10.4324/9781351107242-27
54. Ni, Q., Bertino, E., Lobo, J., Brodie, C., Karat, C. M., Karat, J., & Trombeta, A. (2010). Privacy-aware role-based access control. *ACM Transactions on Information and System Security (TISSEC), 13*(3), 1–31.
55. Taher, Kazi Abu, Nahar, T., & Syed Akhter Hossain. (2019). *Enhanced cryptocurrency security by time-based token multi-factor authentication algorithm*. 2019 International Conference on Robotics, Electrical and Signal Processing Techniques (ICREST). https://doi.org/10.1109/icrest.2019.8644084
56. Chaudhari, S., Tomar, S. S., & Rawat, A. (2011). *Design, implementation and analysis of multi layer, Multi Factor Authentication (MFA) setup for webmail access in multi trust networks*. https://doi.org/10.1109/etncc.2011.5958480
57. Charbonneau, S. (2011). The role of user-driven security in data loss prevention. *Computer Fraud & Security, 2011*(11), 5–8. https://doi.org/10.1016/s1361-3723(11)70112-9
58. Costante, E., Fauri, D., Etalle, S., Den Hartog, J., & Zannone, N. (2016). *A hybrid framework for data loss prevention and detection*. https://doi.org/10.1109/spw.2016.24
59. Lopez, G., Richardson, N., & Carvajal, J. (2015). Methodology for data loss prevention technology evaluation for protecting sensitive information. *Revista Politécnica, 36*(3), 69–69.
60. Schaar, P. (2010). Privacy by design. *Identity in the Information Society, 3*(2), 267–274. https://doi.org/10.1007/s12394-010-0055-x
61. Gürses, S., Troncoso, C., & Diaz, C. (2011). Engineering privacy by design. *Computers, Privacy & Data Protection, 14*(3), 25.
62. Rest, J., Boonstra, D., Everts, M., Rijn, M., & Paassen, R. (2014). Designing privacy-by-design. Lecture Notes in Computer Science, 55–72. https://doi.org/10.1007/978-3-642-54069-1_4
63. Cavoukian, A., Taylor, S., & Abrams, M. E. (2010). Privacy by design: Essential for organizational accountability and strong business practices. *Identity in the Information Society, 3*(2), 405–413. https://doi.org/10.1007/s12394-010-0053-z
64. Cavoukian, A., Polonetsky, J., & Wolf, C. (2010). SmartPrivacy for the smart grid: Embedding privacy into the design of electricity conservation. *Identity in the Information Society, 3*(2), 275–294. https://doi.org/10.1007/s12394-010-0046-y
65. Hansen, M., Schwartz, A., & Cooper, A. (2008). Privacy and identity management. *IEEE Security & Privacy, 6*(2), 38–45. https://doi.org/10.1109/msp.2008.41
66. Bieker, F., Meyer, J., Pape, S., Schiering, I., & Weich, A. (2023). *Privacy and identity management*. Springer Nature.
67. Johannes. (2013). *Options for achieving privacy in online social networks and E-commerce* (pp. 73–91). Springer EBooks. https://doi.org/10.1007/978-3-642-37913-0_6
68. Kenny, C. (2018). The Equifax data breach and the resulting legal recourse. *The Brooklyn Journal of Corporate, Financial and Commercial Law, 13*, 215.
69. Daswani, N., & Elbayadi, M. (2021). *The Yahoo Breaches of 2013 and 2014* (pp. 155–169). Apress EBooks. https://doi.org/10.1007/978-1-4842-6655-7_7

70. Manworren, N., Letwat, J., & Daily, O. (2016). Why you should care about the target data breach. *Business Horizons, 59*(3), 257–266. https://doi.org/10.1016/j.bushor.2016.01.002
71. Beyer, J. (2018, April 30). *Facebook and data privacy in the age of Cambridge Analytica – the Henry M. Jackson School of International Studies.* https://jsis.washington.edu/news/facebook-data-privacy-age-cambridge-analytica/
72. Bonner, L. (2012). Cyber risk: How the 2011 Sony data breach and the need for cyber risk insurance policies should direct the federal response to rising data breaches. *Washington University Journal of Law and Policy, 40*, 257.
73. Gangwar, Rashmi;Verma, Shilpi. (2014). Data breaches: A global exposure. *Journal of Library and Information Communication Technology, 11*(1), 1–8.
74. Shankar, N., & Mohammed, Z. (2020). Surviving data breaches: A multiple case study analysis. *Journal of Comparative International Management, 23*(1), 35–54. https://doi.org/10.7202/1071508ar
75. McGovern, V. (2024). *Uber: Cyber breaches.* SAGE Publications: SAGE Business Cases Originals eBooks. SAGE Publishing. https://doi.org/10.4135/9781071939994
76. Vithanwattana, N. (2022). *Securing mHealth – investigating the development of a novel information security framework.* [PhD thesis Middlesex University Computer Science].
77. Dixit, P. P. (2018). Conceptualising interaction between cryptography and law. *NUJS Law Review, 11*, 327.
78. Shahzad, A., Zhang, K., & Gherbi, A. (2021, March). Privacy-preserving smart grid traceability using blockchain over IoT connectivity. In *Proceedings of the 36th annual ACM symposium on applied computing* (pp. 699–706).
79. Siva Kumar, A., Godfrey Winster, S., & Ramesh, R. (2021). Efficient sensitivity orient blockchain encryption for improved data security in cloud. *Concurrent Engineering, 29*(3), 249–257.
80. Cha, S., Baek, S., & Kim, S. (2020). Blockchain based sensitive data management by using key escrow encryption system from the perspective of supply chain. *IEEE Access, 8*, 154269–154280. https://doi.org/10.1109/access.2020.3017871
81. Dutta, P., Choi, T.-M., Somani, S., & Butala, R. (2020). Blockchain technology in supply chain operations: Applications, challenges and research opportunities. *Transportation Research Part E, Logistics and Transportation Review, 142*, 102067. https://doi.org/10.1016/j.tre.2020.102067
82. Velmovitsky, P. E., Bublitz, F. M., Fadrique, L. X., & Morita, P. P. (2021). Blockchain applications in health care and public health: Increased transparency. *JMIR Medical Informatics, 9*(6), e20713–e20713. https://doi.org/10.2196/20713
83. Al-Farsi, S., Rathore, M. M., & Bakiras, S. (2021). Security of blockchain-based supply chain management systems: Challenges and opportunities. *Applied Sciences, 11*(12), 5585. https://doi.org/10.3390/app11125585
84. Dasaklis, T. K., Voutsinas, T. G., Tsoulfas, G. T., & Casino, F. (2022). A systematic literature review of blockchain-enabled supply chain traceability implementations. *Sustainability, 14*(4), 2439. https://doi.org/10.3390/su14042439
85. Liu, J., Huang, X., Zheng, J., Liu, B., Wang, J., Yoshie, O., Liu, Y., & Li, H. (2024). MM-Instruct: Generated Visual Instructions for Large Multimodal Model Alignment. ArXiv.org. https://arxiv.org/abs/2406.19736

Chapter 9
Metaverse Meets Robotics: Addressing Data Protection and Privacy in Robotic Environment

H. M. K. K. M. B. Herath, H. M. S. S. Herath, G. M. K. B. Karunasena, and S. L. P. Yasakethu

Abstract Within the quickly changing field of technological development, the intersection of robotics and the metaverse has become a focus point, creating new avenues for research and development. By exploring the complicated dynamics, this research sheds light on the complex problems of protecting user data in this complementary field. A paradigm change in the generation, transmission, and processing occurs as robotic systems, and the metaverse become more integrated. This study thoroughly investigates the various aspects of data protection, privacy, and security issues that emerge at the intersection of metaverse and robotics. We uncovered possible methods to protect data from real/physical robots to virtual/digital robots by handling security, privacy, and data protection. Moreover, we highlighted the urgent need for robust, flexible privacy protections to offset the changing threats in this mutually beneficial environment. The study proposed novel concept models for metaverse-based robots named MetRo and its security, privacy, and data protection approach called *MetRo-SP2*. Beyond theoretical frameworks, this study offers practical benefits for industry practitioners, technology developers, and policymakers. It adds to the current conversation about creating a safe, private ecology where robots and the metaverse coexist. A responsible approach to data protection ensures that

H. M. K. K. M. B. Herath (✉) · S. L. P. Yasakethu
Faculty of Engineering, Sri Lanka Technological Campus, Padukka, Sri Lanka

Computational Intelligence and Robotics Research Lab, Sri Lanka Technological Campus, Padukka, Sri Lanka
e-mail: kasunkh@sltc.ac.lk

H. M. S. S. Herath
Computational Intelligence and Robotics Research Lab, Sri Lanka Technological Campus, Padukka, Sri Lanka

Department of Industry 4.0 Convergence Bionics Engineering, Pukyong National University, Busan, Republic of Korea

G. M. K. B. Karunasena
Faculty of Engineering Technology, The Open University of Sri Lanka, Nugegoda, Sri Lanka

C. Hewage et al. (eds.), *Data Protection*,
https://doi.org/10.1007/978-3-031-76473-8_9

the promises of robots and the metaverse are fulfilled without jeopardizing people's security and privacy as society moves through this momentous transition.

9.1 Introduction

A new frontier that excites and worries technologists, legislators, and ethicists alike is emerging in the interaction between the metaverse and robots. In contexts where virtual reality (VR), augmented reality (AR), metaverse, and physical robots intersect, the merger of these two realms presents deep problems regarding data protection and privacy, even as it provides endless prospects for innovation and connectedness. This chapter explores the complexities of this convergence, highlighting the risks, ramifications, and possible ways forward for protecting private information and identity in the context of metaverse-driven robotics.

The idea of the metaverse, a communal virtual shared place formed by the fusion of physically persistent VR and virtually improved physical reality, is central to thistopic [1]. It's an environment where people may engage in real-time interactions with digital things and each other, blurring the boundaries between the actual and virtual worlds. Meanwhile, once restricted to manufacturing facilities and regulated environments, robots are finding their way into more and more aspects of daily life, including customer service, healthcare, and home duties. There is no denying the attraction of this confluence. Imagine a society in which, acting in place of their human counterparts, robotic avatars travel the metaverse, going to events, interacting with friends, and exploring virtual landscapes. Such a vision promises unprecedented degrees of ease, accessibility, and immersion. Beneath this technical marvel, however, are urgent worries about data security, privacy, and ethical issues. The sheer amount and variety of data produced in these hybrid situations is one of the main obstacles. Users who interact with virtual objects or assets, explore virtual environments, and connect with other users in the metaverse leave their digital footprints behind [2]. Meanwhile, a digital footprint refers to the traces or records left by an individual's online activities, encompassing data generated from interactions with digital platforms, websites, social media, and online services. Examples include browsing history (websites visited, pages viewed, and search queries), social media posts (Tweets, Facebook updates, and Instagram photos), cookies and tracking information (stored by websites to personalize content), and online purchases (records of items bought, payment methods used). In the metaverse, RoboTwin and AI avatars generate digital footprints as they engage with users, gather insights from data, and execute various tasks. These footprints contribute to refining user experiences and bolstering privacy management. Similarly, when robotic systems see and interact with the real world, they gather enormous volumes of data via sensors, cameras, and microphones. Concerns regarding unauthorized surveillance, data breaches, and the possibility of abuse by malicious actors are raised by these data flood.

In today's digital age, cyber threats pose significant risks to organizations and individuals. A prime example is the healthcare data breach, such as the 2015 Anthem breach [3], where inadequate security measures allowed an external attacker to access and compromise sensitive patient records, including medical histories, social security numbers, and billing information, leading to potential identity theft, financial losses, and reputational damage for the hospital. Similarly, IoT device vulnerabilities present serious risks, for example, a flaw in the firmware of a smart home security system could enable an attacker to remotely control cameras, as seen in the 2016 Mirai botnet attack that exploited vulnerable IoT devices to launch massive distributed denial-of-service (DDoS) attacks thus endangering residents' privacy and safety [4]. Social media platforms, which collect extensive user data for targeted advertising, are also vulnerable to privacy violations. The 2018 Cambridge Analytica Scandal exemplifies this, where Facebook mishandled user data, exposing private messages, location history, and personal preferences to third parties without consent [5]. Additionally, insider threats within organizations pose a significant risk; an employee with privileged access might intentionally leak sensitive customer data, as demonstrated by the 2014 JPMorgan Chase breach, where an insider stole data on 76 million households, resulting in legal consequences, loss of customer trust, and financial penalties [6]. These examples highlight the critical need for robust security measures to protect sensitive information from diverse cyber threats.

Combining robots and the metaverse challenges conventional ideas of autonomy and consent. As a virtual world filled with AI-driven bots and human users coexist, it gets more challenging to decide who owns the data and how it should be utilized [7]. Furthermore, concerns about people's ability to protect their data and uphold boundaries in these hybrid places emerge when robotic avatars erode the distinction between real and virtual presence. Furthermore, these technologies' overlapping nature creates new vulnerabilities and increases already-existing privacy problems. For instance, biometric information gathered by robots in the real world for identification might be used in the metaverse for illegal access or identity theft. Moreover, worries regarding algorithmic prejudice, discrimination, and the degradation of personal privacy rights are raised by the incorporation of AI algorithms into robotic systems [8, 9]. Addressing these issues requires a multifaceted strategy incorporating ethical principles, legal frameworks, and technical innovation. Developers must use robust encryption methods, data anonymization strategies, and privacy-by-design principles to reduce the dangers of data exposure and unauthorized access.

In the context of metaverse-based robotics, ensuring data confidentiality and security is critical. Two prominent encryption algorithms, Advanced Encryption Standard (AES) and Rivest–Shamir–Adleman (RSA), play essential roles in achieving these goals. AES is a symmetric encryption algorithm that operates on fixed-size blocks (128, 192, or 256 bits) using a secret key. It involves multiple rounds of substitution, permutation, and mixing operations, making it highly secure and efficient. AES is widely used in secure communications (e.g., HTTPS), file encryption, and disk encryption, ensuring the confidentiality of data transmitted within the metaverse. Conversely, RSA is an asymmetric encryption algorithm based on the mathematical properties of large prime numbers [10, 11]. It uses a public key for encryption and a private key for decryption. Although RSA is slower than symmetric algorithms, securing key exchange and digital signatures is crucial. RSA plays a vital role in TLS handshakes, certificate-based authentication, and ensuring data integrity and authenticity, which are essential for the secure operation of robotic systems in the metaverse. Both AES and RSA contribute significantly to privacy protections in metaverse-based robotics. AES secures data during transmission and ensures confidentiality over secure channels like HTTPS, while RSA facilitates secure key exchange and digital signatures. By combining AES for data encryption and RSA for key exchange in a hybrid encryption approach, robust privacy and security are achieved, ensuring the safe and reliable operation of robotic systems within the metaverse.

Furthermore, strategies like federated learning and differential privacy, as well as artificial intelligence (AI) and machine learning (ML) advances, may be used to improve data security and privacy [12–14]. ML models are essential for anomaly detection in the context of metaverse-based robotics. They identify typical patterns in data, like user interactions and activities in virtual environments, and raise alarms when deviations happen. Using algorithms such as Isolation Forests, One-Class SVM, and Autoencoders is essential in detecting anomalous activity or unapproved access in the metaverse. ML models for intrusion detection, such as Random Forests and Deep Learning (e.g., CNNs, RNNs), can detect harmful activity in real-time and effectively identify malware, phishing attempts, and virtual network intrusions. Predictive security in the metaverse leverages ML to forecast potential threats based on historical data using algorithms like Time Series Forecasting and Gradient Boosting, helping predict future attacks or vulnerabilities. Natural language processing (NLP) models analyze text for sentiment, intent, and context, which is particularly useful in detecting phishing attempts and social engineering within virtual communication channels. ML also automates vulnerability assessment in the metaverse through clustering and classification algorithms, identifying vulnerable software components. Additionally, behavioral biometrics, where ML learns user behaviors such as gesture patterns and navigation habits, is applied in continuous authentication and fraud detection, enhancing overall security measures in metaverse-based robotics.

Regulators must update current privacy laws to address the unique challenges metaverse-based robotics applications present. This involves establishing clear data ownership and consent rules, defining jurisdictional boundaries, and holding companies accountable for privacy breaches. Additionally, international cooperation is essential to ensure consistency and coherence in regulatory frameworks across different jurisdictions. The GDPR (General Data Protection Regulation) and CCPA (California Consumer Privacy Act) are vital in shaping data privacy standards, each with specific scopes and implications, particularly concerning robotics applications in the metaverse. The GDPR applies globally to any processing of personal data related to EU residents, covering both data controllers and processors. It broadly defines personal data, requiring companies to inform individuals about data processing through privacy policies and granting data subject rights to access, rectify, erase, and port their data. The GDPR ensures privacy by default for metaverse-based robotics applications, requiring robust consent mechanisms for user interactions. In contrast, the CCPA targets businesses operating in California that collect personal information from residents, exempting nonprofit organizations. It defines personal information as data that identifies, relates to, or describes consumers or households, requiring businesses to disclose such information at the point of collection. In the metaverse, the CCPA impacts virtual assistants, chatbots, and AI-driven avatars, emphasizing transparency and user control.

Robotic systems enabled by the metaverse are developed and implemented with ethical issues in mind. Developers and designers must prioritize transparency, accountability, and user empowerment to guarantee that people have absolute control over their digital identities and personal information. Moreover, in this quickly changing environment, stakeholders need to have an open discussion about new ethical problems and how to encourage responsible innovation [15]. Despite the enthusiasm and promise surrounding this development, it is essential to be aware of the underlying dangers and problems associated with the confluence of robots and the metaverse, especially about data protection and privacy. Robotic systems are becoming conduits for gathering, processing, and transferring enormous volumes of sensitive data as they increasingly connect with virtual worlds and digital ecosystems [16–18]. User privacy, security, and autonomy are significantly impacted by the data created in robotic settings inside the metaverse, ranging from biometric information and behavioral patterns to personal preferences and social interactions [19–21].

This study is driven by the necessity of taking proactive measures to address these risks and obstacles. The authors aim to shed light on the intricate dynamics and suggest solutions to reduce risks and preserve user rights by looking at the junction of robotics and the metaverse through data protection and privacy lenses. Furthermore, to make sure that innovation happens ethically and inclusively as society navigates the transformational implications of emerging technologies, it is critical to promote a conversation on ethical issues, legal frameworks, and best practices.

The rest of this chapter is organized as follows: Sect. 9.2 delves into the ever-expanding world of VR, investigating its possible applications and societal implications. Section 9.3 looks at the most recent advances in robotics, including the unique RoboTwin concept. Section 9.4 discusses the essential challenges of metaverse data security, privacy, and trust in robotics contexts, emphasizing the urgent need for solid frameworks. Section 9.5 proposes the innovative MetRo-SP2 concept model, which comprehensively answers these difficulties. Section 9.6 provides a full review of the consequences and possibilities given by the proposed paradigm. Finally, Section 9.7 concludes the study by discussing future directions.

9.2 Emerging Virtual World (Metaverse)

The advancement of technology has consistently altered our understanding of reality and broadened the scope of human experience. Every technological advancement, from the development of the Internet to the widespread use of smartphones, has pushed us closer to a digital future in which the virtual and real worlds coexist. The idea of the metaverse, a communal virtual environment that surpasses distinct platforms and experiences and provides a single world where users may explore, engage, and produce content, is at the vanguard of this new digital frontier. We examine how the diverse array of technologies forming the metaverse is changing our society, from blockchain technology and decentralized platforms to AR and VR [12, 19, 22, 23].

The word "*Metaverse*"was initially used by science fiction writer Neal Stephenson in his 1992 fictional novel "*Snow Crash*" to refer to a virtual reality- based Internet replacement [24]. Since its inception, the notion has captivated the interest of engineers, futurists, and entrepreneurs, igniting ideas of an entirely immersive virtual environment where people may reside, work, and have fun. Although the first forms of the metaverse were only found in science fiction, new technological developments have made this vision a reality. The development of social media, online gaming, virtual worlds, and AR/VR technologies has paved the way for creating networked digital environments, which are the foundation of the metaverse [25]. The metaverse is hailed by The Times Journal (Fig. 9.1 [24]) as the next big thing in

Fig. 9.1 Metaverse cover in Time Journal, Illustration by Micah Johnson for TIME. (Adapted from Refs. [24, 26])

digital technology that will completely transform our world. With an emphasis on addressing security and privacy problems from the onset, prudence is suggested, considering the nature of new technologies in cyberspace. To guarantee the secure assimilation of the metaverse into our everyday existence, it is critical to fully understand its nature and evaluate its present state of development.

From March 2019 to March 2024, the weekly interest levels in three digital reality technologies—metaverse, AR, and VR—are displayed in Figs. 9.2, 9.3, and 9.4. The interest levels show the relative popularity of each technology as search interest indexes. Interest rates show a cyclical trend from March 2019 to March 2024. At first, interest in VR was the strongest, followed by comparatively less interest in AR

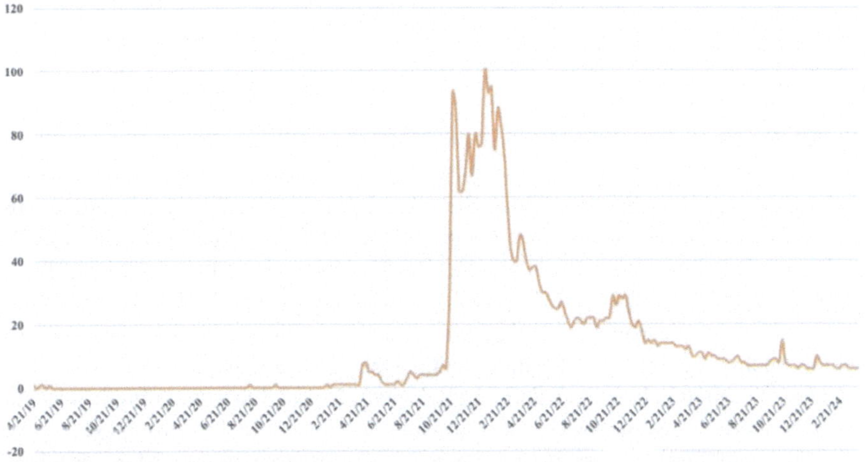

Fig. 9.2 Google search of the term "metaverse"since 2019. (Source Google Trend)

Fig. 9.3 Google search of "augmented reality" since 2019. (Source Google Trend)

Fig. 9.4 Google search for "virtual reality" since 2019. (Source Google Trend)

and metaverse. However, over time, there has been a noticeable rise in interest in metaverse, which had become more popular by the end of the era than AR and VR.

Figure 9.2 illustrates a notable increase in interest in the metaverse from September 2021 to April 2022, possibly attributed to the COVID-19 period. It appears that people were particularly interested in remote working, especially in sectors such as healthcare, education, and business operations.

There is much debate over defining the metaverse and incorporating it into the modern internet. The metaverse, in contrast to previous virtual environments, seeks to build a computerized version of our world with creative additions. Many developing technologies are shaping the immersive digital environment of the metaverse since it is still in its early stages of development. Because the phrase constantly evolves, accurately defining it academically is difficult. But for the metaverse to advance into the next stage of cyberspace, it has to have four essential characteristics [24]:

1. Socialization: The metaverse facilitates social interaction and collaboration among users, allowing them to engage with others in shared digital environments, much like in the real world.
2. Immersive interaction: Users can engage with the metaverse through immersive experiences, often using VR or AR technologies, which blur the lines between the digital and physical worlds.
3. Real world building: The metaverse aims to replicate aspects of the natural world, creating digital environments that mimic real-life spaces, activities, and interactions, concerns with imaginative extensions and possibilities.
4. Expandability: The metaverse is designed to be expansive and dynamic, capable of growing and evolving as new technologies, content, and user interactions are introduced, allowing for endless possibilities and experiences.

9.2.1 *Key Technologies Shaping the Metaverse*

Several crucial technologies that support the development and upkeep of the metaverse are at its core. Augmented reality, or AR, blurs the lines between the virtual and real worlds by superimposing digital material on the real environment. AR-enabled smart glasses and smartphone apps like Pokémon GO are just two examples of how technology enhances daily life and creates new communication and involvement opportunities. In contrast, VR takes users to fully digital settings, submerging them in realistic 3D scenes where they may explore, interact, and create. VR headsets, such as the Oculus Rift and HTC Vive, have transformed gaming, entertainment, and training through simulation by providing users with never-before-seen levels of presence and immersion [19, 27].

The metaverse is significantly shaped by blockchain technology, which offers the foundation for decentralized governance, safe transactions, and digital asset ownership [23]. The metaverse's thriving economy is supported by non-fungible tokens (NFTs), which facilitate the creation and trading of one-of-a-kind digital assets, including virtual land, artwork, and in-game goods. Blockchain-based decentralized networks provide users more freedom and control over their digital identities and belongings, promoting an atmosphere that is more welcoming and interactive [28]. Users may purchase, trade, and construct virtual land parcels through projects like Decentraland, CryptoVoxels, and Somnium Space, forming a decentralized network of linked virtual worlds [29].

9.2.2 *Current Development Discussion*

According to the [30], the metaverse, dependent on AI, VR/AR, and IoT technologies, combines the virtual and physical worlds to allow real-time interactions between people, computers, and robots. It promotes varied interactions via avatars and smart devices across languages and cultures, spanning education, business, health, and entertainment, and it produces massive data streams in the process. Advanced AI analysis improves user experiences and the sustainability of ecosystems, while robotics makes integrating the virtual and physical worlds easier. However, given the existing hardware limits, realizing a fully functional metaverse depends on technological and infrastructure developments. Tech companies are making significant investments in the metaverse's development because they see it as the Internet of the future, giving users authority over digital content production, communication, and avatars and possibly transforming economies. The growth of the metaverse and the internet will shape future digital landscapes.

The meta-universe expands the actual world through immersive simulations, thoroughly mapping reality. It is essential for optimizing VR as it reflects various real-world situations. Its ability to create virtual objects and pictures in its scenarios and drive and interact with them intelligently is extremely promising. Its scope includes many situations, each needing a different approach to execution. A methodology for mapping the metaworld to actual sceneries was presented by Wang et al. [31], with an emphasis on industrial settings. It presents industrial robots as basic virtual entities, using reinforcement learning to optimize their motion paths. This idea of a digital twin is essential for optimizing robot paths worldwide. The framework may be extended to other field scenarios, which makes it easier to design the virtual environment of the metaworld.

The difficulties in attaining quick switchover for various types of workpieces in vision-guided robotic picking systems were discussed by Tsai et al. [32]. Automation is hampered by laborious dataset collecting and manual redefining of vision software. A new method that used the RoboTwin and FOVision devices was demonstrated. To fine-tune robot motions and produce annotated virtual pictures for each workpiece's pickable location, RoboTwin provides a metaverse scenario. Using these photos, an AI model for FOVision was trained, combining edge computing and vision for the robotic manipulator. In practical situations, the system demonstrated high instance segmentation precision and choosing success rates. It sped up dataset creation by 80 times compared to human annotation, lowering simulation-to-real gap errors and facilitating quick line changeover in flexible production systems.

9.3 Physical Robots and RoboTwin Systems

Combining concepts from computer science, AI, data science, mechanical, electrical, and electronics engineering, robotics is an interdisciplinary domain. Fundamentally, robotics is the study and application of robot design, building, programming, and operation. These machines may operate autonomously or semi-autonomously, thanks to their sensors, actuators, and control systems; pre- programmed instructions or algorithms frequently lead them. Robotics integration across industries is revolutionizing them by improving safety, accuracy, and efficiency. Robots in manufacturing reduce workplace injuries and human mistakes by handling dangerous tasks and streamlining assembly lines. Surgical robots in healthcare make minimally invasive operations possible, which enhances patient outcomes. Agricultural robots increase production and sustainability by helping with planting, harvesting, and crop health monitoring. Automated transportation and sorting systems help logistics and warehousing by streamlining supply chain processes. Service robots in the hospitality and retail industries enhance customer experiences through individualized interactions. Educational robots improve learning opportunities by offering interactive, hands-on instruction in STEM fields and encouraging creativity and critical thinking. Robotics technology is becoming increasingly valuable, spurring innovation and economic growth in various industries.

The introduction of robot twin systems is one of the recent advances in robotics. A real robot and its digital twin, a virtual image of its physical counterpart, make up a robot twin system [33, 34]. The RoboTwin System is a versatile robotics platform that can perform various activities flexibly and efficiently [32]. It builds a flexible and agile robotic system by fusing cutting-edge AI, complex sensors, and durable mechanical parts. This creative platform, set to transform many sectors and change our connection with automated systems, results from the fusion of human intellect with state-of-the-art technology. Robotics is currently using the idea of digital twins, which was first made famous in sectors like aerospace and automotive. Figure 9.5 shows the diverse applications and uses of RoboTwin systems in the metaverse domain.

Essential components of the RoboTwin systems included physical robots and digital twins.

1. Physical Robot: The physical robot is a tangible entity designed to carry out specific tasks in the real world. It is equipped with sensors, actuators, and processing units to interact with its environment and execute programmed actions.
2. Digital Twin: The digital twin is a virtual replica of the physical robot, created using advanced modeling and simulation techniques. It mimics the physical counterpart's behavior, structure, and functionalities, providing a high-fidelity representation in a digital environment.

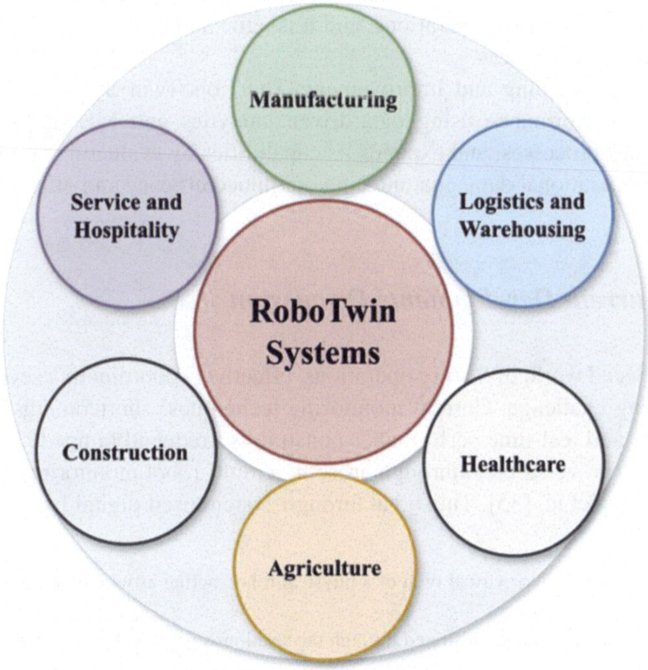

Fig. 9.5 Diverse applications and uses of RoboTwin Systems

Key features and capabilities of the RoboTwin systems included (1) modularity and scalability, (2) human–robot collaboration, (3) advanced sensing and perception, (4) decision-making and navigation, and (5) continued learning and improvements.

1. Modularity and Scalability: One of the RoboTwin System's best characteristics is its modular architecture, which enables customers to grow and modify the platform according to their unique requirements. The technology may be customized to handle various applications in manufacturing, logistics, healthcare, or another industry.
2. Human–Robot Collaboration: Unlike conventional robotic systems that function independently, the RoboTwin[1] systems are designed for smooth cooperation with human counterparts. It promotes a Symbiotic Interaction[2] between people and robots through user-friendly interfaces and sophisticated safety measures, improving productivity and safety in various work contexts.
3. Advanced Sensing and Perception: The RoboTwin System is outfitted with cutting-edge sensors and perception algorithms, which enable it to have an acute sense of its environment. Its sensory skills allow accurate and effective functioning in dynamic conditions, from navigating complicated areas to instantly detecting objects and barriers.
4. Decision-Making and Navigation in Autonomy: Thanks to AI-driven navigation algorithms, the RoboTwin System can plan and carry out operations independently with minimum human assistance. It can navigate congested warehouses and optimize logistical operations, and it is agile and proactive in responding to changing circumstances.
5. Continuous Learning and Improvement: The RoboTwin System continuously improves performance using data-driven analytics and ML. It finds trends, streamlines processes, and expands its capabilities by evaluating enormous volumes of operational data, guaranteeing continued efficacy and efficiency.

9.3.1 Current Developments Discussion

In the fast-paced world of factory operations, effectively coordinating several robots is an ongoing challenge. Current monitoring techniques' shortcomings, including convenience and real-time performance constraints, frequently impede smooth collaboration. However, a breakthrough arose in a multi-robot monitoring system disclosed by Zong et al. [35]. This breakthrough system used digital twin technology

[1] A RoboTwin is a digital or virtual twin of a physical robot, acting as its mirror image in a computer simulation.

[2] Symbiotic interaction can be illustrated through the relationship between a physical robot and its digital counterpart, known as a RoboTwin.

to revolutionize collision detection and simulation, making manufacturing operations more efficient. With great attention to detail, the system's architects created attractive user interfaces and optimized workflows, revealing previously untapped production monitoring capabilities. This technique aimed to maximize robot collaboration in real-world settings by improving condition monitoring efficacy and reducing disagreements.

Meanwhile, in industrial Internet-based collaborative production, teleoperation and multi-robot coordination dominated [36] introduced a cutting-edge multi-robot collaborative manufacturing system that used digital twin and AR techniques. Motion planning became more effective because of a sophisticated reinforcement learning approach, which flawlessly synchronized robot states. This innovative technology, including interactive AR-assisted DT modes, provided planned and real-time motion control, paving the way for increased efficiency and flexibility across various production situations.

The need for adaptable robotic applications increased as the industry adopted mass personalization trends [37] responded to this demand by developing a generalized technique based on the digital twin concept, enabling customizable robotic work cells to be built. By leveraging iterative refinement in digital and physical realms, our strategy accelerated commissioning and reconfiguration procedures while lowering manual labor requirements. The result is an increased productivity and streamlined processes in the age of flexible automation, as demonstrated in laboratory settings.

Beyond the limits of factory walls, the metaverse beckoned, promising immersive experiences and connection. However, the COVID-19 pandemic highlighted the value of physical contact, motivating a quest for remedies. Faisal [38] investigated the potential of digital twin technology through the concept of Robo Twins. Faisal's work revitalized the concept of tangible contact by creating underactuated robotic arms for handshake meetings between physically distant humans, providing a ray of hope in a socially distant society.

Faisal et al. [39] invented underactuated digital twin arms to perform physical handshake encounters across various locales to span physical distances and strengthen interpersonal bonds. The experimental findings demonstrated participants' interest in using the system for remote connections. They revealed intriguing links between handshake qualities and personality factors, giving insight into the intricacies of human contact in the digital age.

As the world dealt with the aftermath of the pandemic [40], investigated the possibilities of digital twin technology to ease the lack of physical contact in virtual interactions. This effort intended to guide future ways to meet the acute demand for physical connection in a pandemic-shaped virtual environment by systematically evaluating existing literature and identifying research gaps.

9.3.2 *Development of RoboTwin Systems*

RoboTwin systems are based on building robotic avatars or counterparts that emulate the movements and mannerisms of their human operators. Numerous industries, including manufacturing, healthcare, space exploration, and entertainment, have favored this concept [40]. RoboTwin systems significantly advance human–robot interaction, providing a safe and practical alternative to human presence in specific scenarios. To enable real-time remote control and cooperation across several domains, engineers and academics are creating robotic avatars that mimic the movements and behaviors of their human operators. To guarantee smooth communication between people and robots, these systems incorporate cutting-edge sensor technologies, accurate actuation mechanisms, complex control algorithms, and a robust communication infrastructure [41]. The creation of RoboTwin systems highlights a growing trend toward augmenting human capabilities through robotic counterparts, ultimately revolutionizing how we work, explore, and interact with our surroundings. Figure 9.6 shows the development steps of the RoboTwin systems for the metaverse.

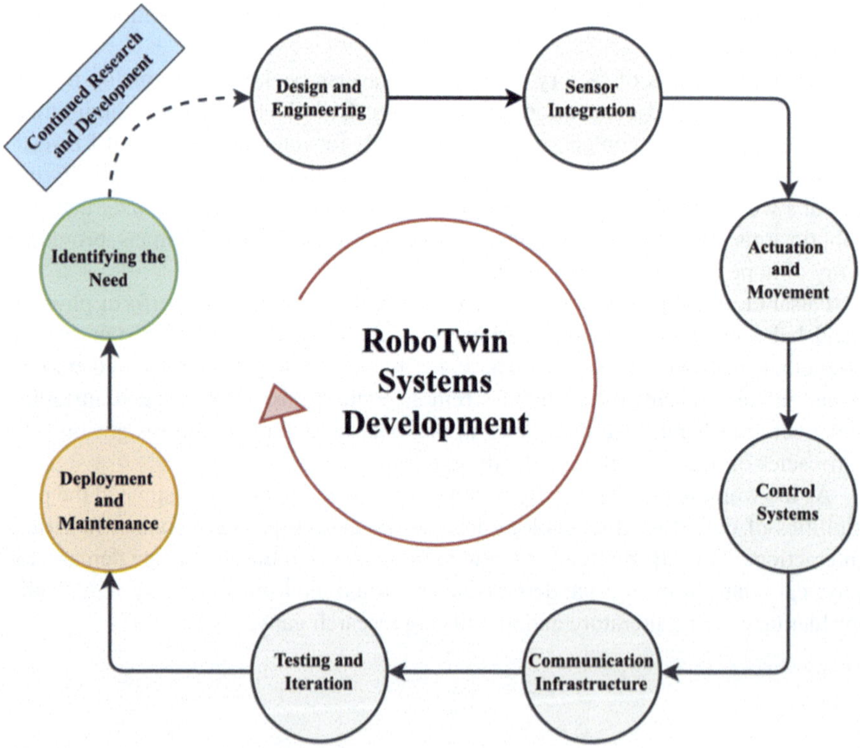

Fig. 9.6 Development steps of RoboTwin systems

In [32], a detailed description of the RoboTwin model system is provided, consisting of four critical subsystems:

1. Physics engines: These engines incorporate physics models to calculate real-world reactions of workpieces during physical manipulation or interaction with other objects.
2. Modeling and auto AI annotation tool: This tool facilitates the mapping of real-world textures onto the 3D CAD model of workpieces and automates the creation of datasets, annotating pickable and unpickable points for each workpiece.
3. Motion control: This aspect encompasses robotic models with motion control algorithms governing robotic arm movements and gripper reactions.
4. VR modules: These modules offer operators an immersive experience, enabling them to immerse themselves in a realistic virtual factory environment. This facilitates testing and refining the robotic picking system.

9.4 Metaverse Data Security, Privacy, and Trust in Robotic Environments

The use of robotics in our daily lives is become more common in the rapidly changing world of technology. Robots are taking on various jobs, from home chores to industrial production, and they frequently can gather and analyze enormous volumes of data [15, 16, 20]. Simultaneously, the notion of the metaverse, which refers to a shared communal virtual environment resulting from the combination of permanent online virtual spaces and digitally augmented physical reality, is gaining traction, providing novel avenues for interaction and involvement. However, amid this technical development, worries about data security in robotic settings have emerged, leading to a critical analysis of the kinds of data gathered, how sensitive personal data is, and the dangers and vulnerabilities that may arise from using them. As robots increasingly integrate into this metaverse, data security, privacy, and trust issues must be addressed [42, 43]. When people spend more time in virtual worlds where robots live, securing private data and building trust is more important than ever. To protect user interests, this article explores the complexities of metaverse data security, privacy, and trust in robotic environments. It also looks at potential solutions to these problems. Figure 9.7 shows the security, confidentiality, and trust in metaverse robotics.

The networked nature of people and artificial entities in the vast metaverse reveals a cyberattack-prone environment. Malicious actors can take advantage of various weaknesses in the virtual environment, such as hacking attempts, data breaches, and malware assaults. Strong cybersecurity measures are necessary since these threats compromise personal data security and can influence robotic systems for malicious intent. Trust in the metaverse is based on data integrity, which protects the validity and dependability of information shared between humans and robots from manipulation or tampering. Data transfers must be kept sacred since any

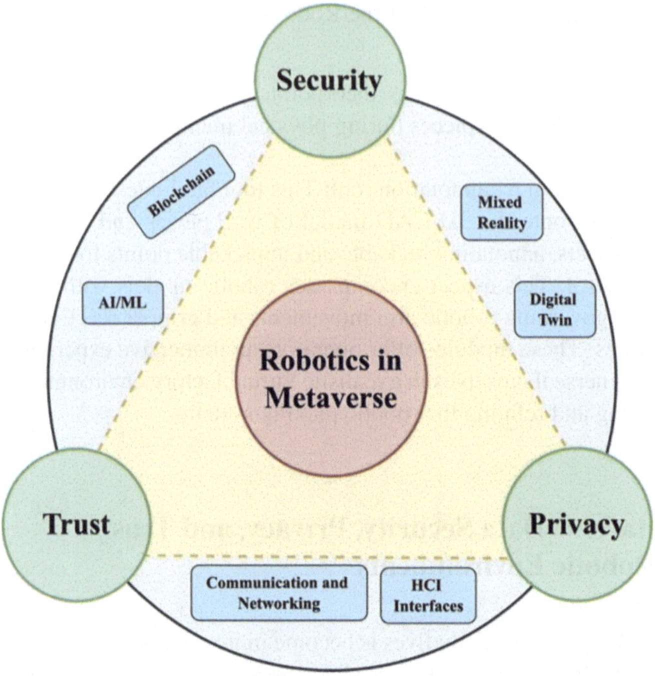

Fig. 9.7 Security, privacy, and trust in metaverse robotics

compromise to data integrity might spread false information, warp virtual experiences, or allow robotic entities to do unapproved acts. In the vast digital space of the metaverse, users entrust a plethora of personal information, including biometric data, communication logs, and behavioral patterns, raising privacy issues. Biometric data refer to unique physical or behavioral characteristics used for identification or authentication. Examples include fingerprint scans, facial recognition, iris scans, voiceprints, and gait analysis. In robotics and the metaverse, biometric data can enhance security by unlocking a robot with a fingerprint. Still, it also raises privacy concerns, necessitating careful handling of collected biometrics. Maintaining user privacy becomes critical to avoid unapproved monitoring, stop data collection, and avoid profiling without specific user permission, all of which create an environment of transparency and trust in the virtual world [44]. Strict identity verification procedures are necessary to determine the legitimacy of individuals and robotic entities while navigating the metaverse. Building trust is contingent upon verifying the authenticity of stated identities since identity theft and impersonation are severe risks in the virtual sphere. Strengthening identity verification processes can help the metaverse create a safe space favorable to real-world communication and cooperative projects.

9.4.1 Types of Data Collected by Robots in Metaverse Environment

Robots in metaverse environments are intended to engage with people and their surroundings; thus, for them to perform their jobs efficiently, a variety of data must be gathered. This information may be divided into many primary categories:

1. Sensor Data: Robots are outfitted with various sensors, such as cameras, microphones, and motion detectors, to sense and navigate their surroundings. By capturing optical, aural, and spatial data, these sensors let robots recognize things, identify barriers, and carry out orders.
2. Behavioural Data: Robots collect behavioral data, such as preferences, routines, and engagement patterns, through constant user contact. These data enhance the effectiveness of robotic jobs and allow for the customization of user experiences.
3. Biometric Data: Certain robots employ biometric sensors to recognize and validate humans based on physiological traits such as speech patterns, fingerprints, or facial features. It is susceptible because biometric information is unique to each individual and can be used for identity theft or unauthorized access.
4. Location Data: Robots in metaverse environments frequently use indoor positioning technologies or GPS-based location tracking systems to locate and navigate around virtual and real-world locations. Location data raise privacy and surveillance concerns by providing insights into user movement and spatial interactions.
5. Interaction Data: During user engagements, robots capture interaction data such as emotional reactions, gesture commands, and transcripts of conversations. By analyzing this data, robots can learn to communicate better and adjust their behavior to suit user preferences.

9.4.2 Sensitivity of Personal Data and Its Implications for Privacy

Robots' acquisition and use of personal data give rise to serious privacy issues since they reveal private information about people's identities, habits, and preferences. Robots are different from inanimate objects in that they can sense, comprehend, and react to human inputs, making distinguishing between humans and machines harder. Because of this, people's privacy and autonomy may be seriously jeopardized if personal data kept in robotic systems is mishandled or gained unauthorized access [45, 46].

1. Privacy Risk: Robots' random gathering of personal information is a risk to privacy as it allows for the monitoring, profiling, and surveillance of people without their knowledge or agreement. In addition, combining various data sources can

reveal detailed information about people's lives, which could result in discriminatory practices and other intrusions.
2. Consent and Control: In metaverse environments, users may interact with numerous robots simultaneously, gathering and analyzing data for different purposes. It becomes challenging to ensure informed permission and precise control over the use of personal data, which raises questions about accountability, transparency, and user autonomy.
3. Data Minimization and Purpose Limitation: To reduce privacy hazards, it is essential to incorporate these concepts into the design and execution of robotic systems. Developers can lessen the possibility of privacy violations and harm to persons by gathering just the data required for specified objectives and limiting the length of time that data is retained.
4. Contextual Integrity: Maintaining the standards and expectations around gathering and using personal data within specific social, cultural, and institutional settings is necessary to protect the contextual integrity of that data. Upholding individuals' privacy rights and promoting confidence in technology in robotic settings necessitates clarity about data methods and adherence to ethical norms.

Data privacy poses special difficulties in robotics in the metaverse, calling for creative solutions. To successfully handle these complications, a comprehensive strategy that includes sophisticated encryption methods, safe authentication procedures, and privacy-improving technology designed especially for robotic interactions is necessary. Ensuring the security of data transmitted between users and robotic entities in the metaverse requires the implementation of strong end-to- end encryption mechanisms. This strategy guards against potential eavesdropping or interception by malevolent actors by encrypting communication channels to guarantee that critical information is protected throughout transmission.

Modern privacy-preserving technologies can improve robotic interaction privacy protection while facilitating essential data processing and analysis, such as homomorphic encryption and differential privacy. By balancing privacy and utility in the metaverse, these strategies allow data modification without compromising the secrecy of sensitive information. Building trust and transparency requires giving people fine-grained control over their personal information and robot interactions [45]. Users may control how their information is shared and used in the metaverse by having explicit permission processes and privacy settings. This helps guarantee that their privacy choices are honored while interacting with robots.

To maintain data utility and safeguard user privacy in robotic interactions, anonymization and pseudonymization techniques must be used. Platforms can protect user identities while keeping valuable data analysis and metaverse capabilities by anonymizing or pseudonymizing personally identifying information. This strategy maximizes the advantages of robotics-driven experiences for users while reducing the possibility of privacy violations. Platforms may establish a safe and private space where users can easily communicate with robotic beings by including these techniques in the metaverse conception and use of robotics. In addition to boosting user confidence and trust, this proactive strategy supports the moral and responsible use of data in the metaverse's changing environment.

9.4.3 Examination of Potential Risks and Vulnerabilities

Robots in metaverse environments have advantages, but they also have hazards and weaknesses. These might range from malevolent assaults to technological issues. Comprehending these risks is essential to implementing robust security protocols and preserving the accuracy and privacy of individual information.

1. Cybersecurity threats: Robotic systems are susceptible to cybersecurity risks such as ransomware attacks, malware infections, and data breaches. Malicious actors may use lax authentication procedures or software flaws to obtain unauthorized access to private information or take over robots and use them for evil intent.
2. Physical Security risks: The confidentiality and accessibility of data kept in robotic devices may be jeopardized by physical security threats, including theft, manipulation, or sabotage. Robots are exposed to physical assaults due to insufficient security measures against unauthorized access or manipulation of hardware components, endangering the integrity of data processing processes.
3. Social Engineering attacks: People who operate robotic systems or interact with them on a human level are vulnerable to social engineering assaults, in which attackers use psychological flaws to trick victims into disclosing private information or providing access to areas they shouldn't. In metaverse contexts, phishing schemes, pretexting, and impersonation techniques pose serious hazards to data security and privacy.
4. Insider Threats: A significant obstacle to data protection initiatives is insider threats, which result from malevolent or careless acts by people with privileged access to robotic systems. Workers, independent contractors, or outside vendors tasked with maintaining or managing robots might unintentionally reveal private information or purposefully abuse their authority for nefarious purposes.

Data protection issues must be addressed to secure people's privacy and maintain trust in technology as robots continue to spread in metaverse environments. By identifying the kinds of data that robots gather, comprehending personal data sensitivity, and mitigating possible risks and vulnerabilities, stakeholders may cultivate a safe and moral environment for robotic innovation. By cooperating across disciplines and adhering to privacy-by-design principles, we may utilize robots' disruptive potential while maintaining fundamental rights and values in the digital age.

9.4.4 Current Developments Discussion

One crucial yet challenging task is robotic grasping, which is manipulating objects autonomously with robotic arms using a variety of computer vision algorithms.

Acquiring vast and diverse RGBD datasets is still a challenge, even with the advances in ML. MetaGraspNet was inspired by how the virtual and physical domains are merged in the metaverse [47]. It presented an extensive benchmark dataset consisting of 100,000 images of 25 different sorts of objects, divided into five difficulty levels, and generated via physics-based metaverse simulations. This dataset made assessing segmentation and object identification algorithms for robotic grasping applications easier. Furthermore, a new layout-weighted performance metric designed for grab applications was suggested.

Musculoskeletal robots are becoming increasingly popular in the industrial sector because they are more flexible, durable, and adaptable than conventional stiff robots. Designed explicitly for essential lifting duties, a hybrid controller addresses these robots' control issues and offers a wide range of industrial applications. A simulated-reality platform that makes use of metaverse technology verified the proposed approach. The feedforward controller of the hybrid controller was a muscle-synergy-based radial basis function network, which can be described as both phasic and tonic muscle synergies. Furthermore, adaptive dynamic programming handled optimum control problems by acting as the feedback controller. The ADP-based controller optimized muscle excitations using an actor- critic framework. Experiments on an upper limb musculoskeletal system demonstrated the controller's convergence, stability, and efficacy, showing it to perform better than other controllers in lifting tasks [48].

Digital twin (DTs) technology, gaining popularity across industries, improves system efficiency by generating virtual representations of physical entities for real-time monitoring and decision-making. Human Digital Twins (HDTs), a subclass of DTs, replicate persons and have uses in healthcare, such as continuous patient monitoring. Despite their benefits, DTs and HDTs face cybersecurity threats and privacy concerns, particularly about the sensitive personal data they handle. Sirigu et al. [49] discussed the dangers to DTs and HDTs, emphasizing rising cybersecurity vulnerabilities and privacy concerns. It also suggests future research directions to solve these issues successfully.

The next section discusses a novel metaverse-based robot model for education, building on current research and technological advancements.

9.5 Proposing Novel MetRo-SP2 Concept Model

Definition of *MetRo-SP2*[3]

MetRo: *Met*averse *Ro*bot
SP2 or (SPP): *S*ecurity, *P*rivacy and *P*rotection

Robust privacy and data protection safeguards must be included throughout the system, especially with integrated robotics into the metaverse. This part presented the concept model for a metaverse-based robot system that enabled security,

[3] *MetRo-SP2* research is a part of an ongoing project at Computational Intelligence and Robotics Research Laboratories at Sri Lanka Technological Campus.

privacy, and data protection techniques. The MetRo platform (a metaverse-based robot) is introduced in the first subsection; the second subsection is the MetRO-SP2 (a metaverse-based robot with security, privacy, and data protection) concept model; and the third subsection discusses its sensor data protection concept model.

9.5.1 *MetRo* (*Meta*verse *Robots*) *Concept Model*

This section details this requirement and describes the proposed MetRo-SP2 platform that will make it easier for robots to operate remotely by utilizing metaverse technology. The proposed system architecture, consisting of several elements from the physical and metaverse domains, is shown in Fig. 9.8. Computational devices, storage units, sensors, actuators, and communication devices are among the physical world components seamlessly incorporated into the proposed architecture. Thus, the architecture's AI/ML, HCI/BCI, RobotTwin, and Blockchain facets are all included in the metaverse facet. This networked metaverse architecture communicates with virtual environments, virtual assets, robot avatars, and other relevant services to provide a unified ecosystem for remote robotic activities in the metaverse.

Figure 9.9 shows the proposed *MetRo* concept model for remote operation for education purposes. Unreal Engine 5 was our base engine in this architecture. We

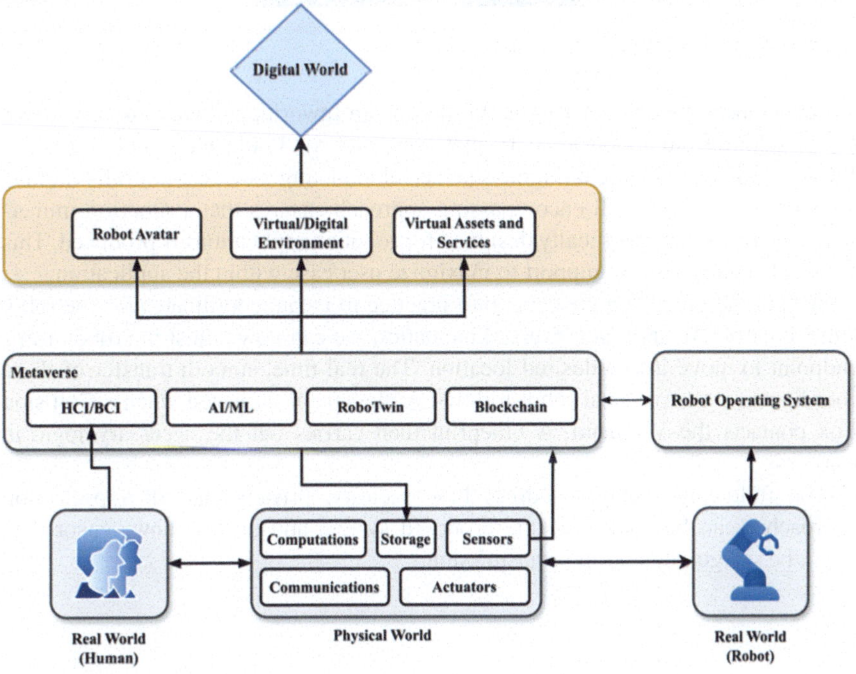

Fig. 9.8 Proposed human, robot-based metaverse interaction

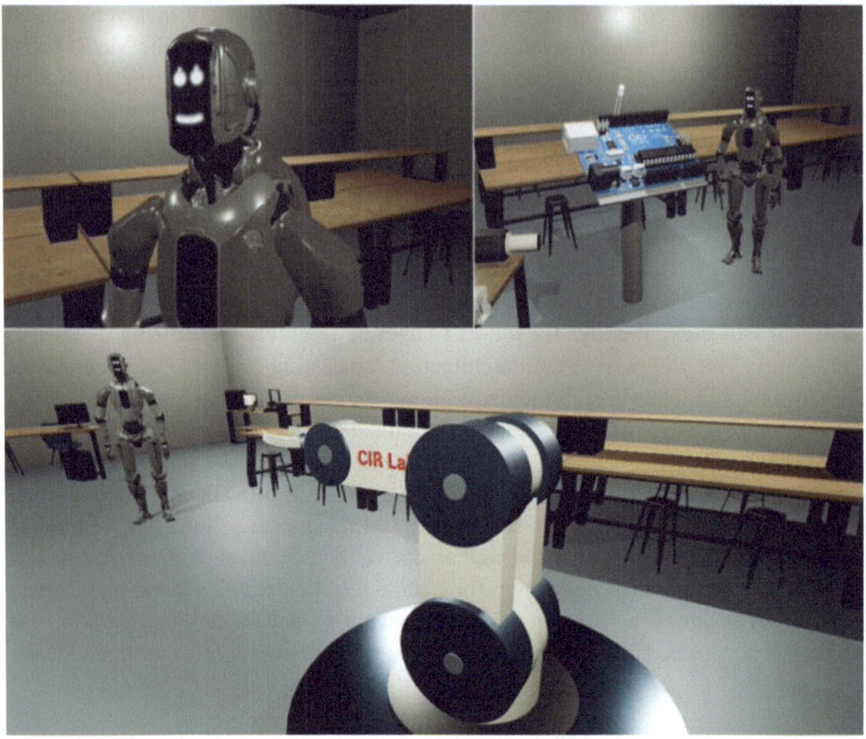

Fig. 9.9 Proposed *MetRo* concept model

presented the main character as an AI robot from Inworld, and we took inspiration for the setting from open-source 3D platforms such as "Poliheaven" and "Garbcat." Taking advantage of Unreal Engine's powerful modeling tools, we carefully molded the virtual world. This all-encompassing approach ensures that a superior, immersive VR experience specifically designed for robotics applications is produced. This model also integrates AI support to maximize user ease within the application.

We put the preceding concepts into practice to create coordinates for the robot arm's control. By applying inverse kinematics, we can now adjust the robot arm's endpoint to move it to a desired location. The real-time, smooth transfer of these coordinates to the physical robot follows. A message is activated when a Collision Box contacts the VR hand. A blueprint then carries out the necessary logic in response.

The following sections address how security, privacy, and data protection approaches can be applied to the proposed system and discuss how sensor data might be safeguarded during transmission.

9.5.2 MetRo-SP2 (Security, Privacy and Protection Integrated Metaverse Robot) Concept Model

Secure communication between the main components of the robot manipulator system, the metaverse platform, and the robot manipulator is ensured by the secure communication layer, which plays a crucial role in the system architecture. This layer is protective by utilizing the HTTPS/TLS protocol for data encryption and secure transfer. These strong security features guard against unwanted access and interception of private data transferred between the robot manipulator and the metaverse platform. Figure 9.10 illustrates the proposed system's security, privacy, and data protection approaches.

Moreover, the system's security architecture goes beyond encryption to include a range of security protocols and techniques applicable to different functional areas. The sections on authentication and user authorization, privacy and data protection, and security monitoring are essential to provide thorough security coverage.

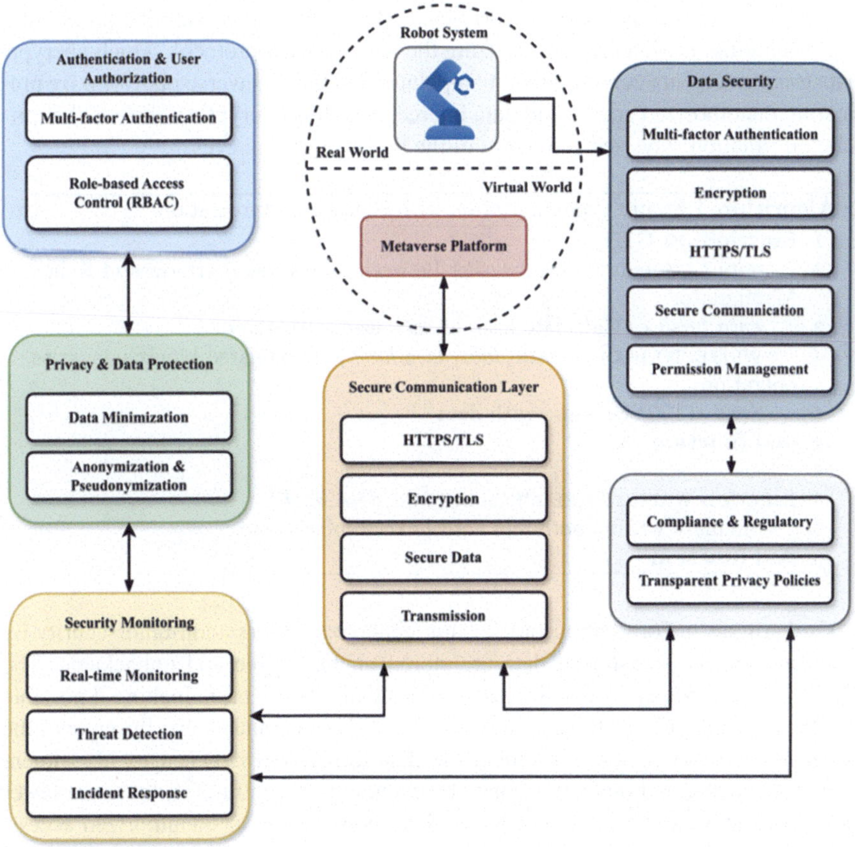

Fig. 9.10 Proposed security, privacy, and data protection protocol for metaverse-based robot environment

Various security mechanisms, including multi-factor authentication, role-based access control, data minimization, anonymization, real-time monitoring, and incident response, are implemented in these areas according to their specific roles. The diagram's arrows represent information flow and system component interaction, highlighting the critical role secure communication protocols play across the system architecture. The arrows illustrate the smooth transfer of data while emphasizing the importance of preserving data confidentiality, privacy, and protection during communication. The diagram shows how each metaverse-based robot manipulator system component has been methodically designed to incorporate security standards. All levels and features of the system are equipped with solid security measures that guarantee data availability, integrity, and confidentiality. This protects user privacy and creates a safe environment for easy interaction within the metaverse.

Algorithm 1 shows the proposed secure communication with blockchain integration between the metaverse and robot manipulator. Here, we suggest utilizing the HTTPS/TLS protocol to provide secure communication between the key parts of a robot manipulator system. To further improve security, we also propose utilizing blockchain technology. This combination provides a strong foundation for guaranteeing the system's data security, privacy, and safe data flow. We also proposed a secure communication channel utilizing the HTTPS/TLS protocol, which encrypts data transferred between the robot manipulator and the metaverse platform. By preventing unauthorized access and data interception, this encryption ensures that private information stays secret and out of the hands of prying eyes.

Algorithm 1 Secure Communication with Blockchain Integration

function DO_GET
    data ← Simulate data transfer between Metaverse Platform and Robot Manipulator
    data_hash ← Calculate hash of *data* using SHA-256
    WRITE_TO_BLOCKCHAIN(*data*, *data_hash*) ▷ Simulated blockchain write operation
    Send HTTP response with *data*
end function

function WRITE_TO_BLOCKCHAIN(*data*, *data_hash*)
    Simulate writing *data* and *data_hash* to blockchain
end function

Furthermore, incorporating blockchain technology offers additional security by ensuring that the transferred data are stored unchangeable and unhackable. The blockchain records every data transaction together with its hash, making it possible to verify the integrity of the data transparently. This guarantees that throughout the robot manipulator system's lifetime, the data transferred between its constituent parts are reliable and unaltered. Since blockchain is decentralized, there are fewer points of failure and a lower chance of data modification or unauthorized access, which improves the system's overall security posture.

It is recommended that a strong and secure foundation for communication within the robot manipulator system be established by fusing blockchain technology with the HTTPS/TLS protocol. This ensures data availability, confidentiality, and integrity while reducing potential security risks and weaknesses.

9.5.3 Sensor Data Protection Concept Model

The methods described in this section present a systematic approach to handling sensor data gathered from robots. Sensor data are essential in robotics because it allows robots to see and communicate with their surroundings. The initialization phase of the suggested technique is where communication channels with sensors are created, and encryption keys are initialized to protect the transmitted data. By doing this step, the security of the sensor data is guaranteed throughout the pipeline. Algorithm 2 discusses the proposed sensor data security platform for the MetRo model.

Algorithm 2: Sensor Data Security Algorithm

```
Input  : Sensor data
Output: Encrypted data
Initialize communication with sensors;
Initialize encryption key;
Function main ():
    while true do
        SensorData, DataHash ← collectSensorData();
        EncryptedData ← encryptData(SensorData);
        TransmissionStatus ← transmitData(EncryptedData);
        if TransmissionStatus = "success" then
            storeData(EncryptedData);
            Print "Data transmitted and stored successfully.";
        end
        else
            Print "Transmission failed. Retrying...";
        end
    end
Function collectSensorData:
    SensorData ← readSensorData();
    DataHash ← calculateHash(SensorData);
    return SensorData, DataHash;
Function encryptData(SensorData):
    EncryptedData ← encrypt(SensorData);
    return EncryptedData;
Function transmitData(EncryptedData):
    TransmissionStatus ← sendData(EncryptedData);
    return TransmissionStatus;
Function storeData(EncryptedData):
    InsertIntoDatabase(EncryptedData);
```

The main loop execution, where sensor data are continuously gathered, encrypted, sent, and stored, forms the basis of the approach. The first step in the procedure is gathering sensor data, which entails collecting information from several sensors built into the robot. Every data collection has a hash value computed to ensure data integrity and provide a way to verify integrity. The gathered data are then transmitted using a predetermined encryption algorithm to prevent unwanted access. After that, the encrypted data are sent to its target location, and the transmission status is noted. The system is built to attempt the transmission procedure again to guarantee data delivery in the event of a transmission failure. The approach strongly emphasizes real-time data transmission and processing, crucial for robotics applications where prompt answers are essential. Additionally, the methodology strengthens the data pipeline's robustness by integrating error-handling mechanisms, which allows the system to recover from transmission failures and continue operating.

In robotics applications, the suggested approach offers an organized framework for handling sensor data, guaranteeing data integrity, confidentiality, and dependability throughout the data pipeline. Following this methodology, developers can create a reliable and safe system for managing sensor data in robot-centric environments.

9.6 Discussion

Exploring "*Metaverse Meets Robotics*" has produced vital insights into the junction of robotics, metaverse, and related technologies. Previous research studies have highlighted the rapid advancement and increased integration of robotics, AI/ML, and immersive technologies such as AR/VR/XR and metaverse. The convergence of these technologies creates exciting prospects and considerable concerns, notably regarding data security and privacy. As robots become more linked and capable of operating in virtual environments, the necessity to protect sensitive information and user privacy grows.

The extensive assessment outlined in Table 9.1 highlights a noteworthy lack of substantial developments and in-depth discussion in the field of metaverse-based robotics. Despite the proliferation of systems arising at the intersection of the metaverse and robotics, a significant fraction of these initiatives has unfortunately ignored critical imperatives such as privacy, security, and data protection.

While a growing number of systems combine features of AR, VR, and mixed reality (XR), their use of the metaverse to provide real-time virtual versus physical control is still mostly undeveloped. This deficit shows a wasted chance to fully realize the metaverse‘s potential as a platform for complex robotic interactions. Furthermore, while remarkable progress has been achieved in robots and robot twin systems, a significant gap exists in incorporating robust data protection methods inside these frameworks. Furthermore, maintaining the secure transfer of data to and from robot twin systems is an area that requires immediate attention and fortification.

Table 9.1 Summary of the literature review

Reference	Year	Robotics	Metaverse	AR/VR/ XR	Digital twin	AI/ML	Privacy/data protection
[35]	2021	✓	×	×	✓	×	×
[37]	2021	✓	×	×	✓	×	×
[47]	2021	✓	✓	×	×	×	×
[36]	2022	✓	×	✓	✓	✓	×
[31]	2022	✓	✓	×	✓	✓	×
[38]	2022	✓	✓	✓	✓	×	×
[49]	2022	✓	×	×	✓	×	✓
[32]	2023	✓	✓	×	×	×	×
[30]	2023	✓	✓	✓	×	✓	×
[50]	2023	✓	✓	✓	✓	×	×
[39]	2023	✓	✓	✓	✓	×	×
[40]	2023	✓	✓	×	✓	×	×
[51]	2023	✓	×	✓	✓	×	×
[52]	2023	✓	×	✓	✓	×	×
[53]	2023	✓	✓	×	×	×	×
[54]	2023	✓	✓	×	✓	✓	×
[48]	2024	✓	✓	×	×	✓	×
[55]	2024	✓	✓	×	✓	×	×

The lack of significant involvement with AI and ML in the context of robots and robot twin systems exacerbates current limitations. Integrating AI/ML technology has enormous promise for improving robotic systems' autonomy, adaptability, and efficacy, but this opportunity is mostly unexplored.

The findings highlight an alarming reality: existing metaverse-based robotics advancements cannot address privacy, security, and data protection concerns. Addressing these shortcomings necessitates a coordinated effort to promote innovation, collaboration, and regulatory frameworks prioritizing the responsible and ethical evolution of metaverse-centric robotics technology.

Ethical issues are critical in ensuring equity, transparency, and accountability in creating and implementing these systems, particularly their influence on diverse populations and cultural norms. Prioritizing ethical concerns allows stakeholders to reduce risks, uphold moral values, and create public trust in the responsible and ethical evolution of metaverse-centric robotics technologies. In the context of MetRo-SP2, addressing ethical issues is critical to ensure equity, openness, and responsibility in developing and deploying metaverse robotic systems. Integrating ethical concerns into MetRo-SP2's fundamental design principles allows stakeholders to reduce potential biases and discriminatory consequences, fostering justice and inclusivity across varied populations and cultural norms.

This proactive approach to ethics allows developers and users to prioritize moral ideals, mitigate risks, and build public trust in the responsible evolution of metaverse-centric robotics technology. MetRo-SP2 can serve as a model for navigating the complex ethical landscape of emerging technologies, resulting in a more equitable and ethically sound future for human–robot interaction in virtual environments.

9.7 Conclusion

This study explores the rapidly developing fields of robots and VR, emphasizing the critical necessity for solid data protection protocols. The intersection of these domains offers both enormous possibilities and formidable obstacles as VR technology develops, and robots are increasingly incorporated into virtual environments. Since this intersection is acknowledged as a critical factor in revolutionary change, the study highlights how crucial it is to put in place thorough frameworks to protect user privacy and promote trust.

Given the swift advances in VR and robotics, it is crucial to safeguard confidential information. The MetRo-SP2 concept model is viable for managing robotic settings' intricate range of hazards. The MetRo-SP2 concept takes a comprehensive approach to data protection in the metaverse by combining blockchain technology, advanced security protocols, and encryption approaches. In the future, more research projects should concentrate on improving and applying the MetRo-SP2 model in practical settings. As the VR and robotics industries develop, it will be necessary to validate its efficacy in reducing data security threats and modify it to take advantage of new opportunities and difficulties. Through regular updates on industry developments and technical breakthroughs, researchers can guarantee that the MetRo-SP2 model continues to be applicable and efficient in protecting user privacy and data integrity. In addition, the responsible and ethical development of VR and robotics technologies depends on promoting collaboration amongst many stakeholders, including industry players, policymakers, and academic researchers. Together, these stakeholders can address ethical issues, societal ramifications, and regulatory concerns, laying the foundation for a technical environment that is more reliable and secure. The MetRo-SP2 concept model is a notable advancement in addressing the intricate data security problems in robotic systems seen in the metaverse. In the future, VR and robotics can coexist peacefully under ethical conduct, privacy, and security norms, provided research, cooperation, and innovation continues.

Conflict of Interest The authors declare that they do not have any conflict of interest.

References

1. Mystakidis, S. (2022). Metaverse. *Encyclopedia, 2*(1), 486–497. https://doi.org/10.3390/encyclopedia2010031
2. Mayda, M. (2022). Digital footprint management in digital visual culture. *Erciyes İletişim Dergisi, 9*(2), 1031–1044. https://doi.org/10.17680/erciyesiletisim.1116477

3. Floyd, T., Grieco, M., & Reid, E. F. (2016). *Mining hospital data breach records: Cyber threats to U.S. hospitals*. https://doi.org/10.1109/isi.2016.7745441
4. Ahmed, Z., Danish, Syed Muhammad, Qureshi, Hassaan Khaliq, & Lestas, Marios. (2019). *Protecting IoTs from Mirai Botnet Attacks Using Blockchains*. https://doi.org/10.1109/camad.2019.8858484.
5. Hinds, J., Williams, E. J., & Joinson, A. N. (2020). "It wouldn't happen to me": Privacy concerns and perspectives following the Cambridge Analytica scandal. *International Journal of Human-Computer Studies, 143*, 102498. https://doi.org/10.1016/j.ijhcs.2020.102498
6. Daswani, N., & Elbayadi, M. (2021). *The target and JPMorgan chase breaches of 2013 and 2014* (pp. 171–191). Apress EBooks. https://doi.org/10.1007/978-1-4842-6655-7_8
7. Dwivedi, Y. K., Hughes, L., Baabdullah, A. M., Ribeiro-Navarrete, S., Giannakis, M., Al-Debei, M. M., Dennehy, D., Metri, B., Buhalis, D., Cheung, C. M. K., Conboy, K., Doyle, R., Dubey, R., Dutot, V., Felix, R., Goyal, D. P., Gustafsson, A., Hinsch, C., Jebabli, I., & Janssen, M. (2022). Metaverse beyond the hype: Multidisciplinary perspectives on emerging challenges, opportunities, and agenda for research, practice, and policy. *International Journal of Information Management, 66*, 102542–102542. https://doi.org/10.1016/j.ijinfomgt.2022.102542
8. Tran, N. C., Wang, J., Vu, T. H., Tai, T.-C., & Wang, J.-C. (2022). Anti-aliasing convolution neural network of finger vein recognition for virtual reality (VR) human– Robot equipment of metaverse. *The Journal of Supercomputing, 79*(3), 2767–2782. https://doi.org/10.1007/s11227-022-04680-4
9. Haptic and Biometric Sensor Technologies. (2022). Spatio-temporal fusion algorithms, and virtual navigation tools in the decentralized and interconnected metaverse. *Review of Contemporary Philosophy, 21*, 105–121.
10. Das, P., Munshi, N. H., & Maitra, S. (2022). New key-dependent S-box generation algorithm on AES. *Micro and Nanosystems, 14*(3), 263–271. https://doi.org/10.2174/1876402913666210726163822
11. Zhou, L., Yeh, K.-H., Hancke, G., Liu, Z., & Su, C. (2018). Security and privacy for the industrial internet of things: An overview of approaches to safeguarding endpoints. *IEEE Signal Processing Magazine, 35*(5), 76–87. https://doi.org/10.1109/msp.2018.2846297
12. AL-Khassawneh, Y. A. (2023). A review of artificial intelligence in security and privacy: Research advances, applications, opportunities, and challenges. *Indonesian Journal of Science and Technology, 8*(1), 79–96.
13. Oseni, A., Moustafa, N., Janicke, H., Liu, P., Tari, Z., & Vasilakos, A. (2021). *Security and privacy for artificial intelligence: Opportunities and challenges*. ArXiv.org.
14. Curzon, J., Kosa, T. A., Akalu, R., & El-Khatib, K. (2021). Privacy and artificial intelligence. *IEEE Transactions on Artificial Intelligence, 2*(2), 96–108. https://doi.org/10.1109/tai.2021.3088084
15. Uzun, M. M. (2023). Metaverse Governance. In *Studies in big data* (pp. 231–244). https://doi.org/10.1007/978-981-99-4641-9_16
16. Raptis, T. P., Passarella, A., & Conti, M. (2019). Data management in industry 4.0: State of the art and open challenges. *IEEE Access, 7*, 97052–97093. https://doi.org/10.1109/access.2019.2929296
17. Munirathinam, S. (2020). Industry 4.0: Industrial Internet of Things (IIOT). *Advances in Computers, 117*, 129–164. https://doi.org/10.1016/bs.adcom.2019.10.010
18. Kalsoom, T., Ramzan, N., Ahmed, S., & Ur-Rehman, M. (2020). Advances in sensor technologies in the era of smart factory and industry 4.0. *Sensors, 20*(23), 6783–6783. https://doi.org/10.3390/s20236783
19. Herath, B., Mittal, M., & Kataria, A. (2024). Navigating the metaverse: A technical review of emerging virtual worlds. *Data Mining and Knowledge Discovery/Wiley Interdisciplinary Reviews, 14*(4). https://doi.org/10.1002/widm.1538
20. Bibri, S. E. (2022). The social shaping of the metaverse as an alternative to the imaginaries of data-driven smart cities: A study in science, technology, and society. *Smart Cities, 5*(3), 832–874. https://doi.org/10.3390/smartcities5030043

21. Soliman, M. M., Darwish, A., & Hassanien, A. E. (2023). The threat of the digital human in the metaverse: Security and privacy. In *Studies in big data* (pp. 247–265). https://doi.org/10.1007/978-3-031-29132-6_14
22. Cannavo, A., & Lamberti, F. (2021). How blockchain, virtual reality, and augmented reality are converging, and why. *IEEE Consumer Electronics Magazine, 10*(5), 6–13. https://doi.org/10.1109/mce.2020.3025753
23. Nguyen, Q. K., & Dang, Q. V. (2018). Blockchain technology for the advancement of the future. https://doi.org/10.1109/gtsd.2018.8595577
24. Huang, Y., Li, Y. J., & Cai, Z. (2023). Security and privacy in metaverse: A comprehensive survey. *Big Data Mining and Analytics, 6*(2), 234–247. https://doi.org/10.26599/bdma.2022.9020047
25. Nisiotis, L., Alboul, L., & Beer, M. (2020). A prototype that fuses virtual reality, robots, and social networks to create a new cyber–physical–social eco-society system for cultural heritage. *Sustainability, 12*(2), 645–645. https://doi.org/10.3390/su12020645
26. Pine, D. W. (2022, July 18). The story behind TIME's metaverse cover. *TIME*. https://time.com/6196282/time-metaverse-cover/
27. Koohang, A., Nord, J. H., Ooi, K., Tan, G., Al-Emran, M., Aw, E. C.-X., Baabdullah, A., Buhalis, D., Cham, T., Dennis, C., Dutot, V., Dwivedi, Y. K., Hughes, L., Mogaji, E., Pandey, N., Phau, I., Raman, R., Sharma, A., Sigala, M., Ueno, A., & Wong, L.-W. (2023). Shaping the metaverse into reality: A holistic multidisciplinary understanding of opportunities, challenges, and avenues for future investigation. *Journal of Computer Information Systems., 63*(3), 1–31. https://doi.org/10.1080/08874417.2023.2165197
28. Truong, V. T., Le, L. B., & Niyato, D. (2023). Blockchain meets metaverse and digital asset management: A comprehensive survey. *IEEE Access, 11*, 26258–26288. https://doi.org/10.1109/access.2023.3257029
29. Nnamonu, O., Hammoudeh, M., & Dargahi, T. (2023). Digital forensic investigation of web-based virtual reality worlds: Decentraland as a case study. *IEEE Communications Magazine, 61*(9), 72–78. https://doi.org/10.1109/mcom.005.2200688
30. Narin, N. G. (2023). The role of artificial intelligence and robotic solution Technologies in Metaverse Design. In *Studies in big data* (pp. 45–63). https://doi.org/10.1007/978-981-99-4641-9_4
31. Wang, Y., Wang, C., & Zhang, H. (2022). *Industrial robotic intelligence simulation in metaverse scenes*. 2022 China Automation Congress (CAC). https://doi.org/10.1109/cac57257.2022.10054666
32. Tsai, C.-H., Hernandez, E. E., You, X.-W., Lin, H.-Y., & Chang, J.-Y. (2023). RoboTwin metaverse platform for robotic random bin picking. *Applied Sciences, 13*(15), 8779. https://doi.org/10.3390/app13158779
33. Balbo, F., Daoud, A., Muller, G., Juganaru-Mathieu, Mihaela, Badeig, Fabien, Alqasir, Hiba, & Batton-Hubert, M. (2023). *RoboTwin: Combining digital twin and artificial intelligence domains for controlling robots in industry 4.0*. Smart Innovation, Systems and Technologies (pp. 15–25). https://doi.org/10.1007/978-981-99-3068-5_2
34. Juneja, N., & Goldenberg, A. A. (2002). *Kinematic calibration of a re-configurable robot (RoboTwin)*. TSpace (University of Toronto). https://doi.org/10.1109/robot.1997.606772
35. Zong, X., Luan, Y., Wang, H., & Li, S. (2021). A multi-robot monitoring system based on digital twin. *Procedia Computer Science, 183*, 94–99. https://doi.org/10.1016/j.procs.2021.02.035
36. Li, C., Zheng, P., Li, S., Pang, Y., & Lee, C. K. M. (2022). AR-assisted digital twin-enabled robot collaborative manufacturing system with human-in-the-loop. *Robotics and Computer-Integrated Manufacturing (Print), 76*, 102321. https://doi.org/10.1016/j.rcim.2022.102321
37. Tipary, B., & Erdős, G. (2021). Generic development methodology for flexible robotic pick-and-place workcells based on Digital Twin. *Robotics and Computer- Integrated Manufacturing (Print), 71*, 102140. https://doi.org/10.1016/j.rcim.2021.102140
38. Faisal, M. (2022). *Haptic-enabled robotic arms to achieve handshakes in the metaverse*. In uO Research. University of Ottawa.

39. Faisal, M., Laamarti, F., & El Saddik, A. (2023). Digital twin haptic robotic arms: Towards handshakes in the metaverse. *Electronics, 12*(12), 2603. https://doi.org/10.3390/electronics12122603
40. Faisal, M., Alejandro, R., Laamarti, F., & El Saddik, A. (2023). *Underactuated digital twin's robotic hands with tactile sensing capabilities for well- being* (pp. 15–38). Elsevier EBooks. https://doi.org/10.1016/b978-0-32-399163-6.00007-x
41. Ogenyi, U. E., Liu, J., Yang, C., Ju, Z., & Liu, H. (2021). Physical human– Robot collaboration: Robotic systems, learning methods, collaborative strategies, sensors, and actuators. *IEEE Transactions on Cybernetics, 51*(4), 1888–1901. https://doi.org/10.1109/tcyb.2019.2947532
42. Gupta, A., Khan, H. U., Nazir, S., Shafiq, M., & Shabaz, M. (2023). Metaverse security: Issues, challenges and a viable ZTA model. *Electronics, 12*(2), 391–391. https://doi.org/10.3390/electronics12020391
43. Ali, S., Abdullah, N., Armand, T., Athar, A., Hussain, A., Ali, M., Yaseen, M., Joo, M. -I., & Kim, H.-C. (2023). Metaverse in healthcare integrated with explainable AI and blockchain: Enabling immersiveness, ensuring trust, and providing patient data security. *Sensors, 23*(2), 565. https://doi.org/10.3390/s23020565
44. Di Pietro, R., & Cresci, S. (2021). *Metaverse: Security and Privacy Issues*. 2021 Third IEEE international conference on trust, privacy and security in intelligent systems and applications (TPS-ISA), Atlanta, GA, USA, pp. 281–288. https://doi.org/10.1109/TPSISA52974.2021.00032
45. Chatzimichali, A. P., Harrison, R., & Chrysostomou, D. (2020). Toward privacy- sensitive human–robot interaction: Privacy terms and human–data interaction in the personal robot era. *Paladyn, 12*(1), 160–174. https://doi.org/10.1515/pjbr-2021-0013
46. Ishii, K. (2017). Comparative legal study on privacy and personal data protection for robots equipped with artificial intelligence: Looking at functional and technological aspects. *AI & SOCIETY, 34*(3), 509–533. https://doi.org/10.1007/s00146-017-0758-8
47. Chen, Y., Zhixuan, Z. E., Gilles, M., & Wong, A. (2021). MetaGraspNet_v0: A large- scale benchmark dataset for vision-driven robotic grasping via physics-based metaverse synthesis. ArXiv.org. https://arxiv.org/abs/2112.14663
48. Qin, S., Li, H., & Cheng, L. (2024). A hybrid controller for musculoskeletal robots targeting lifting tasks in industrial metaverse. *IEEE Transactions on Cybernetics (Print)*, 1–12. https://doi.org/10.1109/tcyb.2024.3358739
49. Sirigu, Giorgia, Carminati, B., & Ferrari, E. (2022). *Privacy and security issues for human Digital Twins*. https://doi.org/10.1109/tps-isa56441.2022.00011
50. Gonzalez-Aguirre, D. I., Perez-Ramirez, J., Felix-Rendon, J., Leon, J. F., Bourgault, J., Esquivel, J. Z., & Nachman, L. (2023). Robot-based uniform- coverage and high-resolution LIDAR mapping for physically-grounded metaverse applications. *IEEE Internet of Things Magazine, 6*(1), 40–45. https://doi.org/10.1109/iotm.001.2200262
51. Kaarlela, T., Padrao, P., Pitkäaho, T., Pieskä, S., & Bobadilla, L. (2022). Digital twins utilizing XR-technology as robotic training tools. *Machines, 11*(1), 13. https://doi.org/10.3390/machines11010013
52. Inamura, T. (2023). Digital Twin of experience for human–Robot collaboration through virtual reality. *International Journal of Automation Technology, 17*(3), 284–291. https://doi.org/10.20965/ijat.2023.p0284
53. Shi, H., Liu, G., Zhang, K., Zhou, Z., & Wang, J. (2023). MARL Sim2real transfer: Merging physical reality with digital virtuality in metaverse. *IEEE Transactions on Systems, Man, and Cybernetics. Systems, 53*(4), 2107–2117. https://doi.org/10.1109/tsmc.2022.3229213
54. Ha, M., Lee, J., Cho, Y., Lee, M., Baek, H., Lee, J., Seo, J., Chun, S., Kim, K., Kim, J., & Lee, W. G. (2023). A hybrid upper-arm-geared exoskeleton with anatomical digital twin for tangible metaverse feedback and communication. *Advanced Materials Technologies*. https://doi.org/10.1002/admt.202301404
55. Sai, S., Prasad, M., Upadhyay, A., Chamola, V., & Herencsar, N. (2024). Confluence of digital twins and metaverse for consumer electronics: Real world case studies. *IEEE Transactions on Consumer Electronics*, 1. https://doi.org/10.1109/tce.2024.3351441

Chapter 10
Detecting Deepfakes Through the Classification of Facial Active and Passive Features Using Machine Learning

G. M. K. B. Karunasena, H. M. K. K. M. B. Herath, R. M. P. M. D. Rathnayake, H. D. N. S. Priyankara, and B. G. D. A. Madhusanka

Abstract Recently, there has been a much public concern over fake photos and films that contain facial information created by digital modifications, especially with deepfakes techniques. The widely used term "Deepfakes" describes a deep learning technique that swaps faces to produce fake identities. The distinction between actual and deepfake videos has become crucial due to their detrimental effects on the world. An intriguing vulnerability within live deepfakes has largely gone unnoticed by the security research community thus far. This research may be attributed to real-time deepfakes during video calls and conferences, which have recently become a notable cause for concern. Developing and evaluating existing deepfake detection methods requires large-scale datasets. Still, the deepfake datasets available today have inadequate image quality and do not appear like the deepfake videos that have been making the rounds on the Internet. This research presents a novel way for recognizing synthetic media produced using deepfake: an active and passive facial features based end-to-end deep neural network classifier model (AFPF-DNNC model). The facial landmark detector recognizes and extracts the

G. M. K. B. Karunasena (✉) · R. M. P. M. D. Rathnayake · H. D. N. S. Priyankara
Faculty of Engineering Technology, The Open University of Sri Lanka, Nugegoda, Sri Lanka
e-mail: gmkar@ou.ac.lk; rmpra@ou.ac.lk; hdpri@ou.ac.lk

H. M. K. K. M. B. Herath
Faculty of Engineering, Sri Lanka Technological Campus, Padukka, Sri Lanka

Computational Intelligence and Robotics Research Lab, Sri Lanka Technological Campus, Padukka, Sri Lanka
e-mail: kasunkh@sltc.ac.lk

B. G. D. A. Madhusanka
Faculty of Engineering Technology, The Open University of Sri Lanka, Nugegoda, Sri Lanka

Computational Intelligence and Robotics Research Lab, Sri Lanka Technological Campus, Padukka, Sri Lanka
e-mail: bgmad@ou.ac.lk

C. Hewage et al. (eds.), *Data Protection*,
https://doi.org/10.1007/978-3-031-76473-8_10

area from each individual video frame that contains the human face, then extracted facial regions used to extract active features in the faces. Simultaneously, a convolutional neural network (CNN) extracts passive features from these faces. The retrieved features are inputs into the end-to-end deep neural network classifier (DNNC), which is the last recognition step in the detection system. The proposed algorithm achieves 90.32% of an area under the receiver operating characteristic curve (AUC), 90.6 % accuracy, 90 % specificity, 91.3 % recall, 87.5 % precision, and 89.3 % *F*-score on the Celeb-DF (v2) a new deepfake forensics dataset. Finally, the experimental investigation validates the suggested method's effectiveness over the most advanced techniques.

10.1 Introduction

The creation, sharing, and editing of films and photographs are now easier than ever because of the rising prevalence of social networks like TikTok, Facebook, and YouTube and the accessibility of competent graphical content editing software and devices. Concerns about public privacy have recently been raised by the widespread distribution of incredibly lifelike fake photos and movies produced using deepfake technology on various social media platforms. Deepfake is a deep learning and correlate training generative neural networks (GAN) [1] architecture that can construct a video of the target saying or doing things the source person says or does by substituting the target's facial images in the source's video. Although deepfake technology may be misused to fabricate videos of politicians, discredit famous people and produce falsified information that confuses shareholders and tricks people, it is harmful.

The significant problem of face manipulation in images and films threatens global security. In interactions between people and biometrics-based human verification and identity services, faces are crucial. Home security heroes' "2023 State of Deepfakes" research thoroughly examines the level of deepfake technology today. It is based on an analysis of 95,820 deepfake videos, 85 channels dedicated explicitly to deepfakes on different internet platforms, and over 100 websites that are part of the deepfake ecosystem. According to the estimate, there were 95,820 deepfake videos available online in 2023, a 550% increase from 2019 [2]. Between 2022 and 2023, there has been a significant surge in the use of deepfake media in North America, more than doubling in frequency, rising from 0.2% to 2.6% in the United States. Deepfake activity has also increased significantly in Canada, from 0.1% to 4.6% in the same time frame [3]. Figure 10.1 shows the total increment of deepfakes in the United States and Canada.

The study discovered that about one-third of the videos on deepfake sexually explicit material websites included non-Western people. Notably, a considerable portion, 25 percent of the victims targeted, were K-pop singers from South Korea. Figure 10.2 shows an overview of nations found in deepfake sexually explicit material.

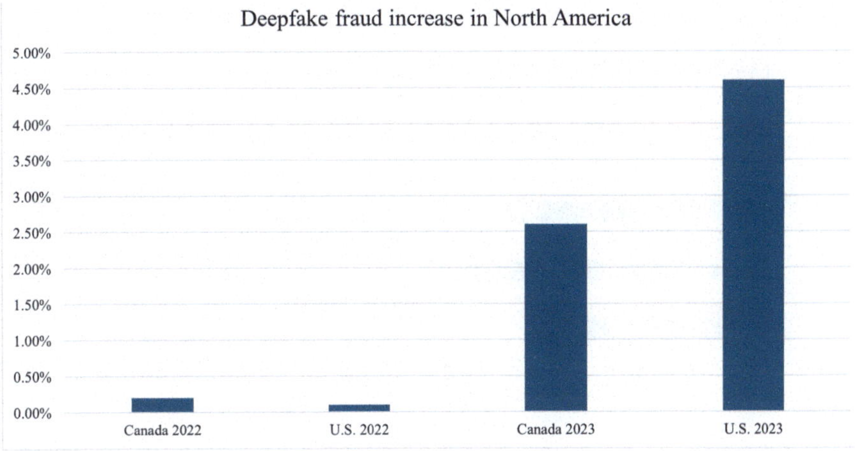

Fig. 10.1 Deepfake fraud increment comparison between 2022 and 2023, United States and Canada [2]

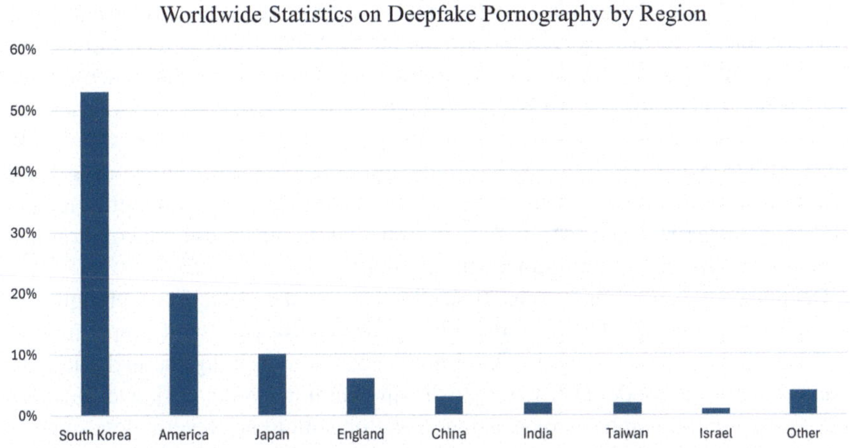

Fig. 10.2 Overview of nations found in deepfake sexually explicit material [2]

According to a survey conducted by iProov, globally, 71 % of respondents said they had no idea what a deepfake was, which is a very high percentage of ignorance. Less than one-third of customers worldwide said that they were aware of deepfakes. The number of respondents who indicate they know what a deepfake is varies by nation; the most significant percentages are in Mexico (40 %) and the United Kingdom (32 %). On the other hand, the least knowledgeable respondents are in Germany and Spain, where 75 % of those surveyed said they had yet to learn what a deepfake was [3]. Figure 10.3 shows the survey results.

In the past few years, deepfake's generating outputs have become more realistic, making it more difficult for human eyes to distinguish between the real and the fake.

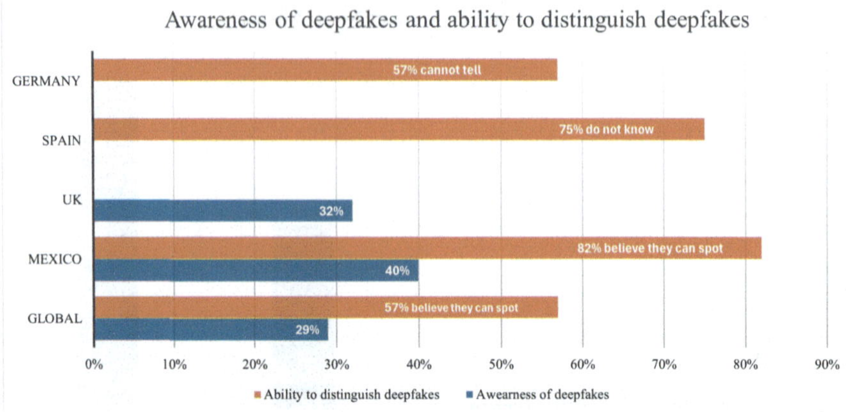

Fig. 10.3 Survey results on awareness of deepfakes and ability to distinguish deepfakes [2]

Therefore, credible face frame alterations can undermine confidence in online communication and security applications. Analyzing and identifying faces in images or videos is key to identifying fakes. In this field, numerous research articles have been provided: facial feature analyzing methods [4, 5], Viola-Jones detector algorithm [6], dlib framework [7], BlazeFace [8], RetinaFace [9], and convolution neural network, which has multiple processing capabilities [10], to name just a few. There are several methods available for identifying deepfake videos since the first deepfake video surfaced when a Reddit user inserted famous people's faces into pornographic videos [11]. Many of these techniques use deep neural networks to identify temporal discrepancies between face frames in films, while others use convolution networks to identify visual distortions within frames.

Deepfake detection algorithms typically use holistic detection techniques to assess if target material has been modified or artificially generated. Implementing AI-driven strategies to support algorithmic detection techniques, including the vision transformer model [12, 13], two-stream neural networks [14], and MesoNet [15], is a common emphasis of existing detection solutions. Nevertheless, manual image processing, which focuses on emphasizing the crucial areas of an image, needs more attention. Because of this, the model frequently has to process every video, which adds weight to the model.

The accuracy of the holistic methods directly depends on the dataset used to train the deepfake detection model. Deepfakes often involve manipulation of the video content, where facial expressions, movements, and lip synchronization evolve over time. Existing approaches may fail to capture temporal inconsistencies or artifacts from manipulating facial dynamics across multiple frames, due to analyzing only individual frames or images. These temporal factors can be crucial for detecting deepfakes in video sequences. This work focused on studying the characteristics of temporal factors such as eye blink period, eye blink frequency, and facial edge variations. The variation of these temporal features can be observed in a video sequence. This study focused on developing a detection model by combining these holistic features (passive features) and temporal features (active features). In this study, we employed an AI-based algorithm to extract passive features and manual image processing to extract active facial features to enhance the deepfake detection method. Before being integrated into deep learning frameworks, the most crucial data, areas, and characteristics will be meticulously filtered and processed. Finally, these features feed into the deep neural network to classify fake or authentic. The complete algorithm was identified as the active and passive facial features base end-to-end deep neural network classifier model (AFPF-DNN Model). Reducing the unnecessary learning processes on these networks and increasing the model accuracy are achieved by focusing on the most relevant aspects of learning.

This study will focus on utilizing some of the most popular and extensively used classification models to recognize and detect artificial video content produced using deepfake methods, and we will reveal how to reduce the complexity of the deepfake detection problem to a more manageable one. Since the current classification models are made to be highly accurate, choosing these models well would also improve the ability to address the difficulty of identifying deepfake content. Additionally, we provide processing techniques to change the deepfake detection input sequences into the inputs of a simple classification model with two classes: fake and authentic. Additionally, our proposal permits postprocessing synthesis and accounts for the processing steps to prevent the loss of important facial features. The paper continues with the following structure: In the second section, prominent deepfake datasets and methods for creating and detecting deepfake videos are reviewed. Section 10.3 shows a novel architecture for identifying deepfakes in video frames. The experimental findings and analysis are covered in Sect. 10.4. The conclusion and next steps are presented in Sect. 10.5.

10.2 Literature

Deepfake methods have been increasingly popular lately because of the excellent movies they produce and the fact that various consumers can utilize their apps. Based on GAN and autoencoder-decoder frameworks, a few well-known deepfake face apps are FakeApp, FaceSwap, ZAO, DeepFaceLab, Deep Art, and faceswap generative adversarial network (GAN).

The latent facial traits are extracted by the autoencoder component, and the face picture is synthesized from the encoded representations by the decoder module. To enable the transition between the source and target face images, two separate encoder–decoder combinations that have the same weights in each encoder part can be used. Every pair of encoders and decoders goes through a training phase using various collections of facial picture datasets. The features of the first face are then fed to the decoder of the second face in order to reconstruct the authentic face from the original face [19].

In GNAs, the discriminator and generator deep networks, which make up a GAN, train concurrently all over the learning stage. In order to mislead the discriminator during training, the generator component creates synthetic samples that are identical to real data distributions, making it difficult for the discriminator to correctly identify them that the photographs are not real, the discriminator is optimized to pick up on this distinction [20]. The most often used method is the StyleGAN algorithm [16], which allows scale-specific adjustments to the patterns to regulate the image generation. After developing the StyleGAN2 algorithm [17], predicated on data-driven unrestricted generative image reconstruction, it remains ability to create near-realistic, lifelike faces for fictious individuals typically impossible to distinguish from real people. This piece of software mentions a software program that uses StyleGAN2 to produce realistic images of imaginary individuals [18].

Since deepfake detection assesses the authenticity of face images through binary categorization, a sizable dataset of real and fake videos is required to train the detection model. The famous deepfake video datasets that can be accessed in the web are DeepFake-TIMIT [21], UADFV [22], FaceForensics++ (FF++) [23], Google/Jigsaw DeepFake Detection [24], Celeb-DeepFake (Celeb-DF(V1)) [25], Deepfake Detection Challenge (DFDC) [26], DeeperForensics-1.0 [27], WildDeep-fake [28], and Celeb-DeepFake (Celeb-DF(V2)) [29]. The previously mentioned datasets can all be utilized to create, segment, classify, and augment new data about images and videos. A comprehensive comparison of the above datasets is shown in Table 10.1. According to the generation method, UADFV, DF-TIMIT, and Face Forensics++ undergo 1st generation datasets, and DeepFake Detection, DFDC, Celeb-DF, WildDeepfake, and DeeperForensics belong to the 2nd generation datasets based on the amount and reliability of the produced data [25].

According to the research work in Li & Lyu [30], the DLIB detector extracts face regions from the video frames. Following that, the differences in pixel variations between the surrounding and distorted face region are used to identify the artifacts from face frames using four deep-learning architectures: ResNet50, VGG16, ResNet152, and ResNet101. Subsequently, this technique is tested on multiple

Table 10.1 Comparison of the datasets

Dataset	Real videos	Fake videos	Generating method	Generation
UADFV	40	49	FakeAPP	First
DF-TIMIT	320	320	Faceswap-GAN	First
FaceForensics++	1000	1000	DeepFakes, Face2Face, FaceSwap, NeuralTextures	First
DeepFake Detection	363	3068	Similar to FaceForen-sics++	Second
DFDC	1131	4113	Deepfake, GAN-based, and non-learned methods	Second
Celeb-DF	590	5639	Deepfake synthesis algorithm	Second
WildDeepfake	3805	3809	Deepfake synthesis algorithm	Second
DeeperForensics	50,000	10,000	End-to-end face swapping framework	Second

deepfake videos on social media and verified using two datasets of deepfake videos: UADFV and Deep-fake-TIMIT.

In the research conducted by Dang et al. [31], five facial landmarks are marked, and the bounding box of each face frame is cropped off using the insight face program. The VGG16 and XceptionNet network is used to create attention feature maps. Determining the realness of the video content aids in identifying the modified portions in the facial frames. This approach is tested on the UADFV and Celeb-DF datasets after being trained on the sizable DFFD dataset, which consists of videos produced by the Deep Face Lab and FaceForcies++ dataset.

The work done by Charitidis et al. [32] recommends a preliminary processing technique to eliminate many false positive face images following the face detection stage. They then use three distinct CNN models: MesoInception4, XceptionNet, and EfficientNet-B4. The Celeb-DF and FaceForensics++ datasets are used to analyze this algorithm once trained on the DFDC dataset. The work of Nguyen et al. [33] builds an autoencoder that utilizes a CNN to find the modified areas and identify the synthesis videos in the FaceForensics dataset. The detected face regions feed into an autoencoder comprising an encoder and decoder and use a partially supervised mode for learning.

Khalil et al. [34] propose a methodology that first detects the face region using the YOLO v3 model. The spatial information is then extracted using two feature extraction techniques: an enhanced high-resolution network based on CNN and local binary patterns based on texture. Following that, capsule networks are fed these spatial characteristics to identify deepfakes. DFDC-preview is used to train this approach, and Celeb-DF and DFDC-preview are used to test it.

Research done by Mehra [35] proposed a Mobilenet SSD face detector that uses the frame selection approach to identify the facial region in video frames. To find the deepfakes in the DFDC video dataset, the VGG19 model, capsule, and LSTM networks are used. To discover the temporal inconsistencies between frames, LSTM is utilized to capture these series of features and apply feature extraction techniques such as capsule networks to learn the spatial discrepancies inside frames. Singh et al. [36] apply the MobileNet-SSD detector to video frames to retrieve the facial regions. Next, four deep learning architectures, EfficientNet-B1, InceptionV3, XceptionNet, and EfficientNet-B3, and an LSTM model and a time-distributed layer are employed. This recognizes the legitimacy of the DFDC dataset videos and learns its spatial-temporal properties. According to Li et al. [37], the dlib detector extracts the face frames, and the landmarks on the face are labeled. Following the alignment of the face frames with an algorithm, the eye regions are chopped into a series of eye frames and sent into the VGG16-LSTM algorithm. VGG16-LSTM is more familiar with eye blinking patterns. Finally, 49 sample videos and the associated deepfake videos created using this technology are analyzed.

In Masi et al. [38], following the alignment of the face frames, the features from the color and frequency domains are combined using two DenseBlock models. Bi-LSTM learns temporal information and identifies deepfake videos by following these blocks. The algorithm is trained by using the FF++ dataset, and the testing is done by using Celeb-DF dataset. Zhao et al. [39] proposed a novel multi-attentional deepfake detection architecture that could be more successful in obtaining significant local features. Deepfake detection was designed to address a specific type of classification problem. According to the literature, some researchers use facial expressions to detect deepfake content. Researchers Mazaheri et al. proposed an approach that recognizes changes and localization in facial expression by combining techniques for image manipulation with facial expression recognition [40].

The research results reveal that many facial regions interact and coincide when making facial expressions or speaking. Deepfake systems are trained to extract areas of the face, nose, mouth, eyes, etc., instead of considering how a smile or any other facial expression might work in a synergistic effect. For example, this implies that a person's digitally modified image can have "laughing lips" but not "laughing eyes."

Juan et al. [41] provide a unique method for detecting deepfakes that makes use of the inherent weaknesses in deepfake videos, specific facial part inconsistencies. Figure 10.4 shows an example of the inconsistency of facial parts in reconstructed facial images. Numerous investigations have been carried out with an emphasis on identifying artifacts within the domain of deepfake detection. A work done by Li et al. [30] employs CNNs to find unique artifacts (FWA and DSP-FWA) in deepfake videos. Face warping artifacts in deepfake contents are intended to be detected using FWA; DSP-FWA is a more recent development that expands on this method. By including a spatial pyramid pooling module, it can handle differences in the initial target faces' resolutions. Additionally, self-collected facial photographs are used in the training of DSP-FWA. Again, Li et al. [42] demonstrated how to detect deepfakes by revealing the blending borders in photos that have been altered. Nevertheless, their method requires a sophisticated trace generator, which increases the computing burden. In contrast, Yang et al. [43] investigated possible evidence of

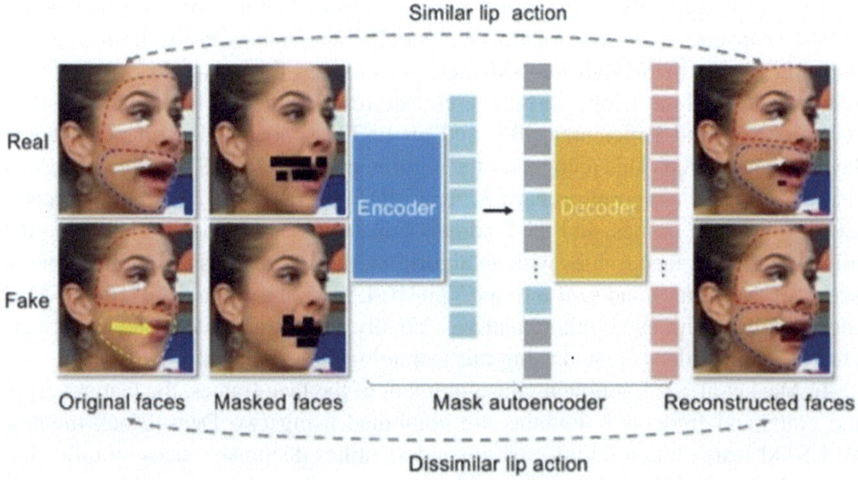

Fig. 10.4 Inconsistency of facial parts in reconstructed faces [41]

forgery utilizing attention mechanisms and replicated the forging process using picture-generating tools.

To identify deepfake materials, some research projects employed frequency domain analysis. The techniques used in the frequency domain to extract information from the face regions are fed into machine learning algorithms that categorize faces as either synthetic or real. Durall et al. [44] suggested a simple deepfake detection technique that used conventional frequency domain analysis and a simple classifier algorithm. They suggest a simple method of identifying fake face photos by employing DFT to transform the images into the frequency domain. After that, they use support vector machines (SVM) and logistic regression to extract features from a one-dimensional power spectrum representation of the fast Fourier transform.

Woo et al. [45] utilize frequency domain features to distill knowledge to enhance the identification of low-quality compressed deepfake images. For high-quality deepfake photos, their algorithms show an issue in detection efficiency. Research conducted by Giudice et al. [46] suggested a technique to identify deepfake videos by utilizing the discrete cosine transform (DCT) to detect aberrant frequencies in the frequency domain. Several shortcomings have been identified by media forensics researchers as telltale signs of fake media, including these factors: The signs and features that deepfake detection algorithms utilize include glitter, wobbly face and distortion, waviness in a facial movements, absent facial characteristics such as a recognized mole on a cheek, smoothness, and heaviness of clothing and hair, excessively soft skin, absent specifics such as teeth, wrinkles and hair, misalignment in face symmetry, abnormal movements of fixed objects, inconsistent mouth and speaking movements, inconsistencies in pixel levels, disparities in illumination, shadows as well as reflections, distorted edges, viewpoints and blurring of facial characteristics, lack of respiration, odd eye contact, and odd eye blinking [47] Figure 10.5 shows some clues and features of deepfake detection algorithms use [41]. Due to deepfakes' ability to maintain the photos' lighting, position, and facial expression, forensics models now face more difficulty detecting deepfakes when using deep-learning methods like CNN and GAN.

Deepfake detection techniques can be divided into two types in literature: feature-based and holistic techniques. The holistic approaches, which usually employ deep learning techniques and are used to detect deepfake face photos, primarily examine the face. By creating a compact collection of linear patterns from the picture pixels, these techniques try to reduce the dimensionality of the data. These combinations are then fed into a binary classification algorithm, which can distinguish between authentic and false images. On the other hand, feature-based approaches divide the entire face into distinct regions of interest, such as lips, eyes, nose, and color mismatches in the skin and head. They are utilized for both deepfake video and picture identification [48]. Kim et al. [41] suggested a detection technique that mitigates the anguish caused by pixel value disparities by combining color distribution analysis of the area around the vertical edge of an altered image with pixel restoration techniques. Last, research findings from the literature indicate that as deepfake content-generating algorithms constantly evolve, researchers must employ deep image processing techniques to conduct a more thorough analysis of

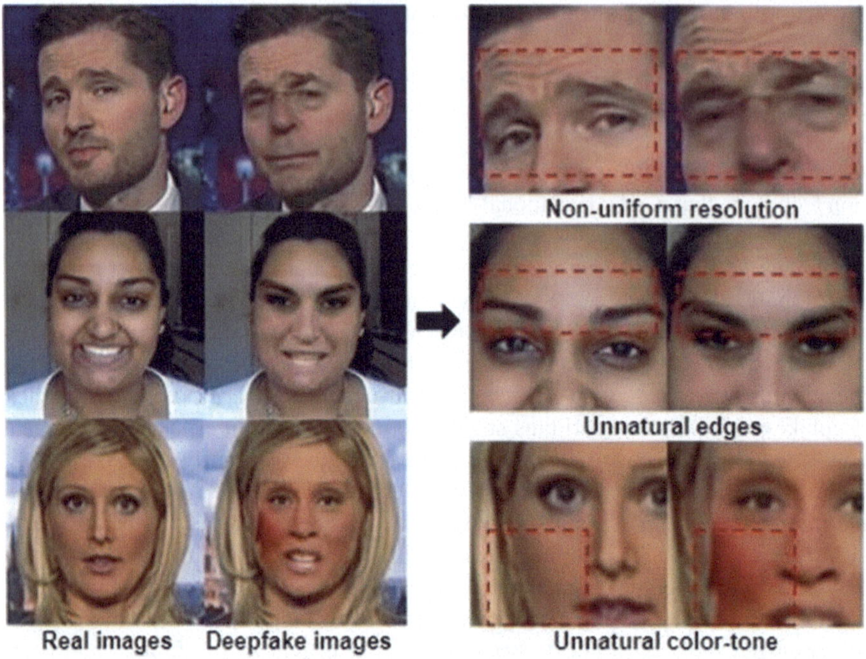

Fig. 10.5 Examples of forged patterns in the creation of deepfake images [41]

facial images to detect deepfake photos accurately. Hybrid deepfake detection methods that rely on feature-based and holistic methodologies still need more study.

10.3 Proposed Methodology

The suggested approach presents a practical technique for identifying deepfakes in videos. The overall architecture of the suggested synthesis video detection method is depicted in Fig. 10.6. The proposed methodology consists of three main stages: facial region extraction stage, feature extraction stage, feature classification stage, and culminating in testing. In the first stage, the face detection algorithm finds the face region and extracts the face ROI. Then extracted face region passes to the second stage; in the second stage, several image processing techniques and pretrained deep learning algorithms are employed to extract the required features for the classification process. Next, these extracted features feed into the pretrained deep neural network (DNN) algorithm for classification. Finally, the proposed methodology was validated using the publicly available dataset CFD-V2. The stages that have been proposed can be broadly explained as follows:

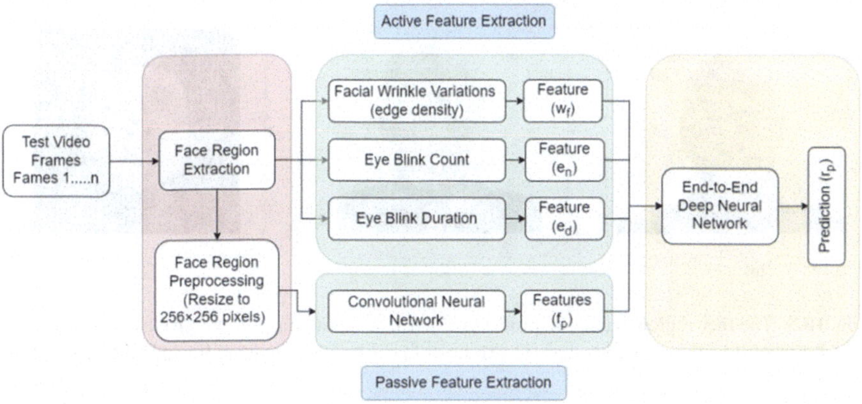

Fig. 10.6 Proposed architecture of the deepfake detection algorithm

Fig. 10.7 A graphical depiction showcasing the locations of 486 distinct points that define key features on a human face, as utilized by the facial landmark model

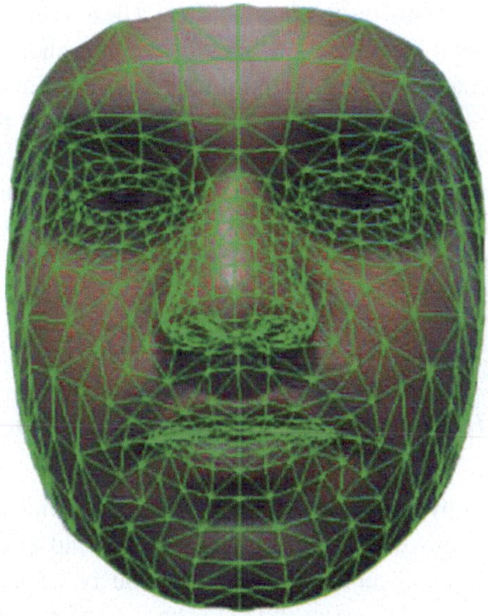

10.3.1 Face Region Extraction

Face detection is the initial stage in this procedure. With the open library "OpenCV" [49] library, frames are taken out of videos. The face region is then detected using the MediaPipe face detector. Regardless of the different facial directions, it uses detailed facial landmark data to guarantee efficacy. The precise placements of essential features on the facial region, including the forehead, jawline, lips, nose, ears, and eyes, are indicated by these facial landmarks, which are essential to this methodology. The MediaPipe face mesh is a crucial tool in this field that can precisely locate 468 critical spots on a face [50]. Figure 10.7 shows the detection of the

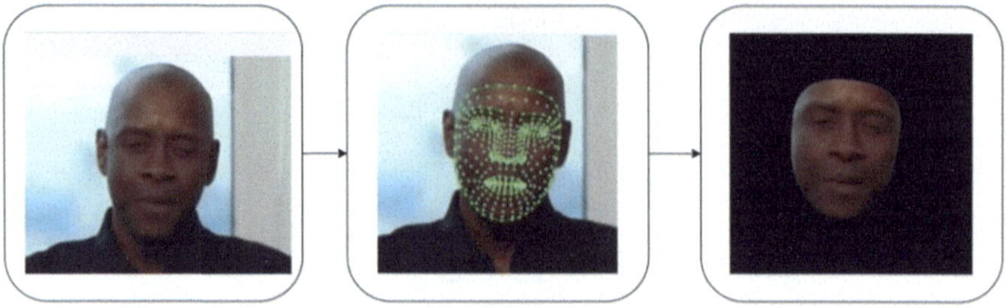

Fig. 10.8 The face region extraction process involves (**a**) input image frame, (**b**) face region detection by using facial landmarks, and (**c**) extracted facial region. Image extracted from the public dataset CFD-V2

face region of the face mesh algorithm. We can see the output of the MediaPipe facial landmarks algorithm, which successfully locates and maps 468 landmark locations on the human face. These meticulously labeled locations, representing a point from 0 to 468, provide a comprehensive topological understanding of the face characteristics essential for reconstruction and extraction.

During the initial preprocessing stage, the system detects whether a face is present in each video frame. If a face is found, the region containing the face is isolated, and sixty-eight specific facial landmark points are localized and annotated within that frame using a face detection model. Figure 10.8 shows a visualization of the face region extraction process. The extracted face region is then fed into further analysis.

10.3.2 Active Feature Extraction

Active facial features refer to the dynamic aspects of a person's face that can change or move during facial expressions and movements. These active facial features are dynamic and change continuously as a person expresses emotions, speaks, or responds to stimuli [51]. They are often used in facial expression analysis, emotion recognition, and other computer vision applications that track and interpret facial movements. The active feature extraction process consists of three subcategories: facial winkle variance extraction, eyeblink count extraction, and eyeblink duration extraction. In the following sections, we will elaborate on our three distinct detectors, each employing a different set of active feature spaces.

10.3.2.1 Facial Wrinkle Variance Extraction

Wrinkles represent the visible manifestation of aging on the face, emerging over time as one grows older. The formation of wrinkles on the face results from the changes in the facial skin's collagen and dermal tissues as an individual advances in age. Various elements, including excessive sun exposure and an unhealthy way of living, influence the development of wrinkles. Wrinkles are significant in face-based analytical techniques, a hallmark characteristic of normal physiological aging. Furthermore, wrinkles can be measured and quantified, making them a crucial marker for evaluating the condition and quality of facial skin. In this proposed method, we used facial wrinkle variation to determine factors for deepfake detection.

Deepfake generators may encounter difficulties replicating every detail on an actual face during the deepfake image generation process. This leads to cases where the produced deepfake images need more intricate details associated with facial wrinkles. Figure 10.9 shows the wrinkle variation of two frames in the same deepfake video. Deepfake videos can occasionally exhibit unintended artifacts or imperfections in the generated facial features. The two consecutive frames in Fig. 10.10 reveal an abrupt and unnatural manifestation of facial wrinkles within the same looping deepfake video sequence.

Our proposed methodology employed a specialized image filtering technique to quantify and analyze facial wrinkle variations (edge density) and fluctuations. We used a linear Gabor filter to extract these wrinkle features in this step. Image processing and analysis came to apply the Gabor features in many different fields after their emergence with Dennis Gabor in the optical scenario and transmission. However, these filters are the pearl for edge detection applications, surface evaluation, feature extraction, and object recognition [52–54]. The distinctive property of this class of sensors is their ability to best concentrate energy in the spatial and frequency domains, which enables their very distinguished use in mapping surfaces and textures. Gabor filters act in the same manner as band-pass filters, but with the only difference that the latter permits only the frequencies from a specific range to pass through while rejecting all the others.

Mathematically, a 2-D Gabor filter can be considered a sinusoidal wave analogous to a rippling pattern multiplied by a Gaussian function. This way, the filter responds to directions and at particular frequencies within the image. The sinusoidal

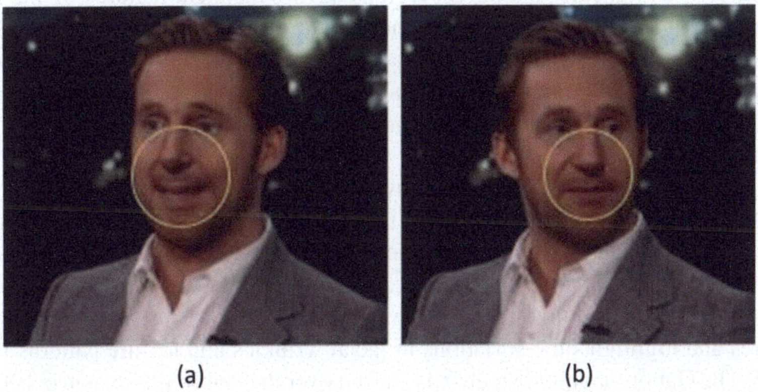

Fig. 10.9 Wrinkle variation of two frames in the same deepfake video loop. (**a**) both wrinkles are available on both sides of a person's nose, and (**b**) only a single wrinkle is present. Image extracted from the public dataset CFD-V2

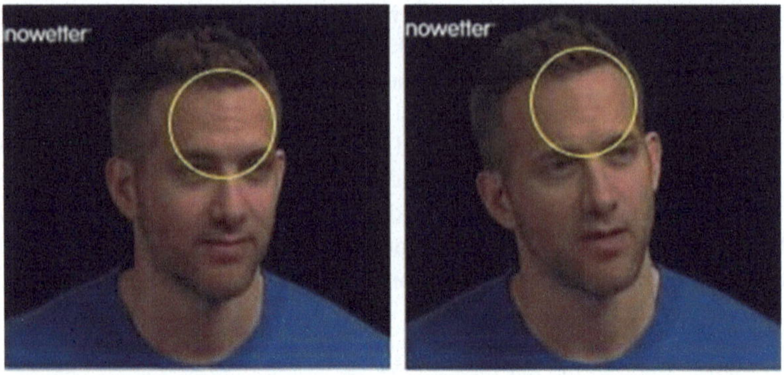

(a) (b)

Fig. 10.10 Two frames in the same deepfake video show wrinkle variations: (**a**) wrinkles available on the forehead and (**b**) a sudden disappearance of forehead wrinkles. (The image was extracted from the public dataset CFD-V2)

component conveys the oscillatory properties of the edges and texture features, and the Gaussian envelope localizes the filter's response, which is focused on a specific part of the image. Therefore, the Gabor filter can used to detect edge variations of images. The following equation represents the Gabor filter response for the test image:

$$G_{(x,;,y,;,\sigma,;,\vartheta)} = \frac{1}{2\pi\sigma^2} \exp\left(-\frac{u^2+v^2}{2\sigma^2}\right) \cos 2\pi \frac{u}{\lambda}$$

where, $\upsilon = -x \sin \vartheta + y \cos \vartheta$, $u = x \cos \vartheta + y \sin \vartheta$, ϑ is the orientation of the normal to parallel lines, λ is the wavelength of the complex exponential signal of the Gabor filter, and σ is the scale factor of the Gaussian envelope. Finally,

$G(x, y; \sigma, \vartheta)$ indicates maximum regions having identical features by the filter. Making use of this data, we extract the number of local feature points of the Gabor magnitude component to measure the rate of change of facial wrinkle variation—the Gabor magnitude component used as the feature related to facial wrinkle variation detection. Figure 10.11 shows the process of the Gabor magnitude component extraction process.

In this process, the first step is to convert the facial RGB image into a grayscale image. This grayscale image is then fed into the Gabor filter. The Gabor filter operates on the grayscale image and generates two separate image matrices: the Gabor magnitude matrix and the Gabor phase matrix. For the subsequent processing stages, we primarily utilize the Gabor magnitude matrix. This matrix effectively captures and highlights the variations in facial wrinkles and texture patterns in the image. The Gabor magnitude matrix is chosen over the Gabor phase matrix because it provides a more precise and more pronounced representation of the facial wrinkle details, which are crucial for the intended analysis. Furthermore, all the image

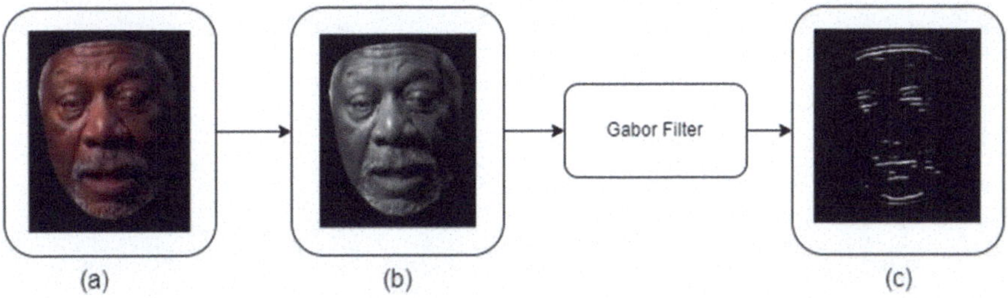

Fig. 10.11 Gabor magnitude component extraction process, (**a**) input image, (**b**) grayscale image, and (**c**) Gabor magnitude image. (Image extracted from the public dataset CFD-V2)

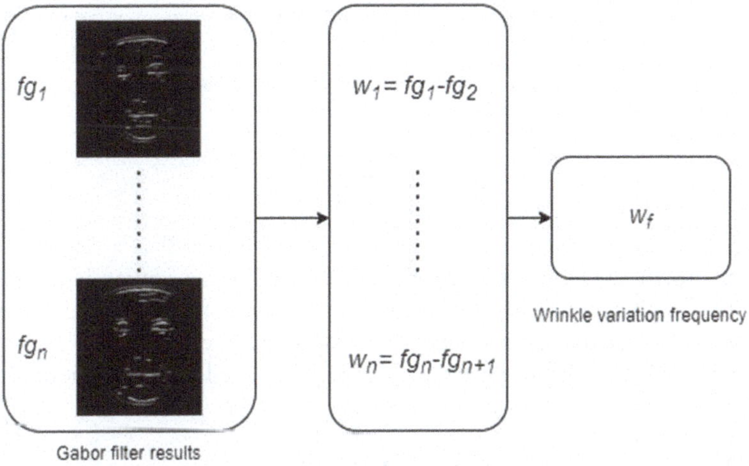

Fig. 10.12 Process of wrinkle variation frequency counting

frames in the sample video pass through the Gabor filter individually to measure the change in wrinkle variation w_f. Figure 10.12 shows the wrinkle variation frequency detection process. The wrinkle variation value w_f formulated by taking the tan sigmoid value of the average number of variations f_n values. The predefined threshold level is the threshold value defined in Fig. 10.13. The equation below represents the w_f formulation equation, where fg_n is the ratio of local feature points of the Gabor magnitude component in n^{th} frame.

$$f_n = \frac{\| fg_n - fg_{n+1} \|}{fg_n}$$

$$w_f^{'} = \left(\frac{\sum_{i=1}^{n} \left(\frac{fg_n - fg_{n+1}}{fg_n} \right)}{n} \right)$$

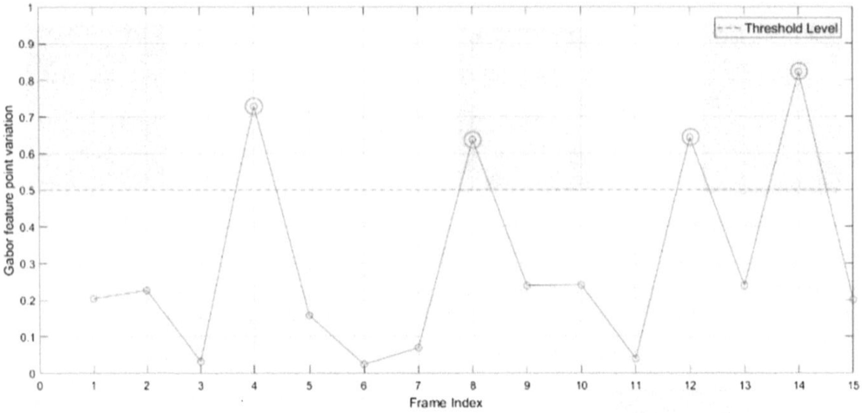

Fig. 10.13 The proposed method for appropriate threshold value finding

$$w_f = \left(\frac{2}{1 + e^{-2wf_f'}} \right) - 1$$

Figure 10.13 shows a proposed method for finding an appropriate threshold value. The graph was prepared for a deepfake video with a 20-second duration and 200 frames. The red line indicates the proper threshold level, 0.5. Values higher than threshold levels indicate unexpected wrinkle variation available between two frames.

10.3.2.2 Human Eye Blink Features Extraction

This step mainly focused on characteristics of eye blinking to detect deepfake detection. Deepfake contents can be detected through critical analysis of significant changes in the eye blinking patterns. Abnormal eye blinking behavior in deep-fake videos can be discovered by applying statistical methods such as duration of eye blinks, frequency of eye blinks, and time intervals between repeated blinks [55]. A human eye blinks for two to ten seconds on average, and the period of eye blinking takes a maximum time of 0.4 seconds. Most of the time, deepfake content creation algorithms cannot follow the above values. Almost all the time, the eye blinking speed or period is slower than the speed in deepfake videos [56]. In this proposed approach, the initial step involves isolating the regions containing the left and right eyes from the facial images within the video frames. This is achieved through the facial landmark detector MediaPipe, which accurately identifies and extracts these specific areas of interest. Figure 10.14 clearly shows the process of eye landmark extraction process.

After obtaining eye landmarks, we selected four landmark points p_i to calculate eye aspect-ratio (EAR). This EAR function calculates the overall area of the horizontal and vertical axes.

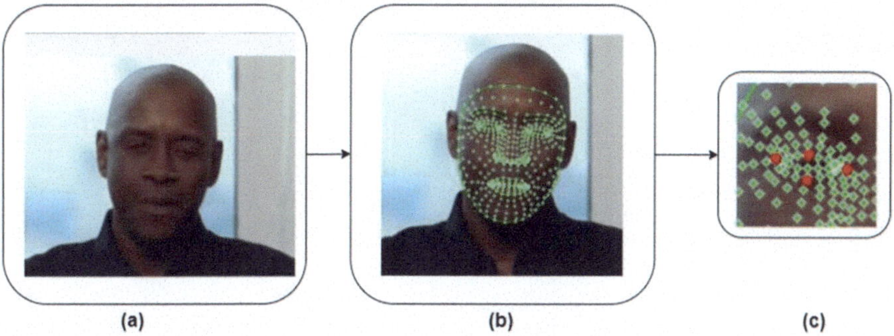

Fig. 10.14 Eye landmark extraction process. (**a**) Input image frame, (**b**) facial landmark detection, and (**c**) eye landmark points. Image extracted from the public dataset CFD-V2

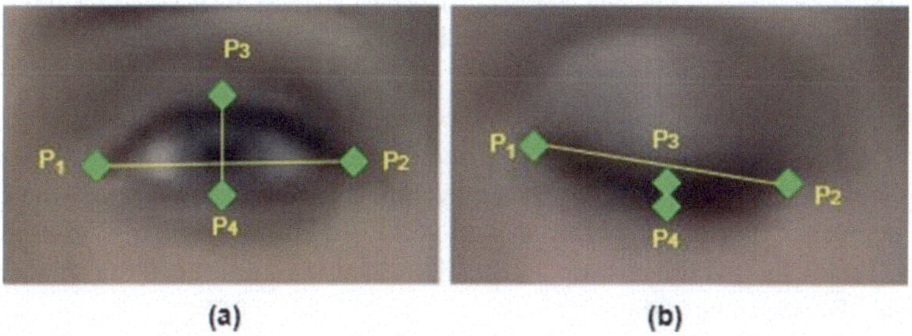

Fig. 10.15 The points around the eye are selected to calculate EAR (Eye Aspect-Ratio). (Image extracted from the public dataset CFD-V2)

Points p_1 and p_2 refer to the points of horizontal axis in the eye area, as shown in Fig. 10.15, and the points p_3 and p_4 refer to the points of vertical axis. In Fig. 10.15, the left image shows the landmarks' positions when the eye is opened, and the right image shows the landmarks' positions when the eye is closed. The following equation is used to calculate the EAR value.

$$\text{EAR} = \frac{\|p_3 - p_4\|}{\|p_1 - P_2\|}$$

When a person blinks, the eyelids of both the left and right eyes simultaneously close and reopen. Therefore, we utilize the mean EAR value (EAR_a). The right eye EAR value is EAR_r and the left eye EAR value is EAR_l. The value of EAR_a can detect an eyeblink that is less than a threshold level. Figure 10.16 shows the defined threshold level for eye blink detection.

$$\text{EAR}_a = \frac{\text{EAR}_r + \text{EAR}_l}{2}$$

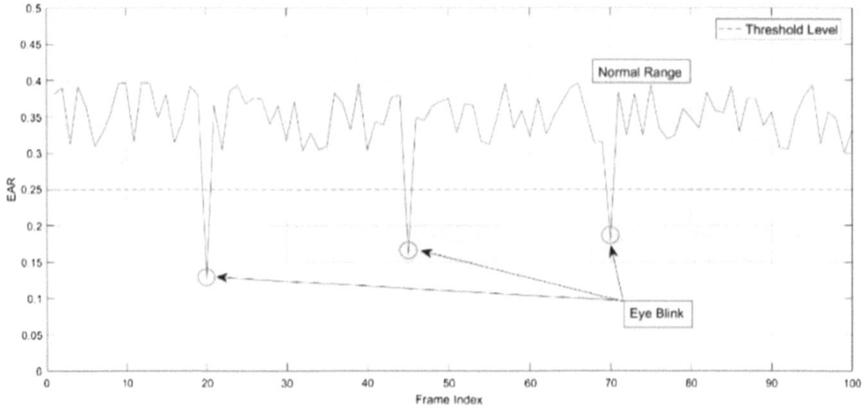

Fig. 10.16 Methodology for finding threshold. The graph indicates EAR values for 100 frames, and the red line indicates the appropriate threshold level

Once the eye EAR_a values are obtained, the algorithm calculates two distinct features: the average eye blinking period and the frequency of eye blinking observed throughout the image frames. When the eyelids are shut during a blink, the EAR_a value decreases and drops below a predetermined threshold across consecutive frames. When the eyelids are open, the EAR_a value returns to its normal range, reflecting the open-eye state.

Once the eye EAR_a values are obtained, the algorithm calculates two distinct features: the average eye blinking period and the frequency of eye blinking observed throughout the image frames.

When the eyelids are shut during a blink, the EAR_a value decreases and drops below a predetermined threshold (EAR = 0.25) across consecutive frames. When the eyes are open, the EAR_a value returns to its normal range, reflecting the open-eye state. The average eye blinking period measures the typical duration of a single eye blink, as shown in Fig. 10.17, providing insights into the natural rhythm and timing of the subject's eye blinking behavior. On the other hand, the frequency of eye blinking quantifies the number of blinks occurring within a given time frame. Finally, in this step, we obtain two features: eye blinking duration e_t and the number of blinks e_n in the sample video, as shown in Fig. 10.18.

10.3.3 Passive Feature Extraction

In this step, we used a convolutional neural network (CNN) to extract the passive features. This network consists of four convolutions and pooling layers, a dense network with two hidden layers. The convolutional layers employ rectified linear unit (ReLU) activation functions to enhance the ability to generalize and handle complex patterns, introducing nonlinear transformations to the input data. To

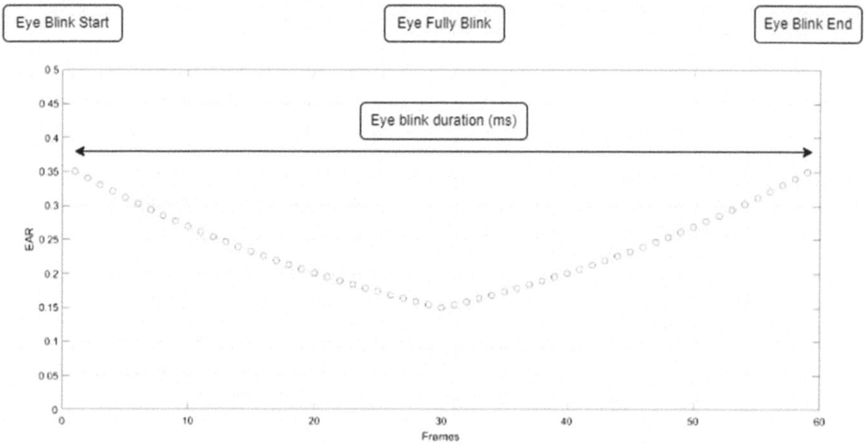

Fig. 10.17 The proposed method measures the duration of eye blinking and the time it takes to blink

Fig. 10.18 Obtained features from eye blinks in the test videos

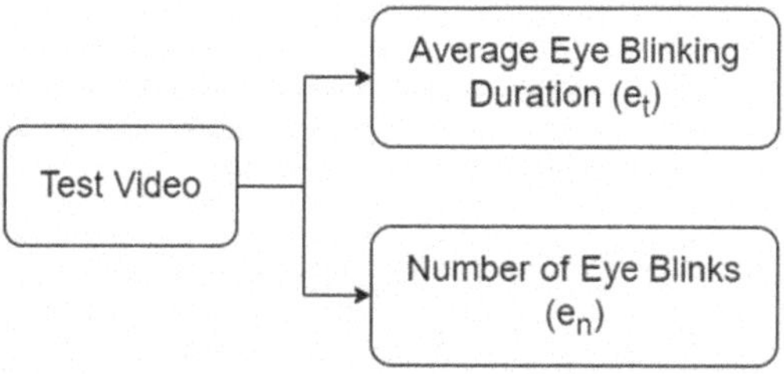

prevent overfitting and improve the reliability of the CNN predictions, the fully connected layers incorporate a dropout. Figure 10.19 shows the proposed CNN architecture to extract passive features.

The first convolution layer consists of 8 filters with a 3 × 3 kernel size and a max pooling layer with a 2 × 2 kernel size. The output feature map size of the first convolution is 128 × 128 × 3. The second convolution layer consists of 8 filters with 5 × 5 kernel size and a max pooling layer with 2 × 2 kernel size. The output feature map size of the second convolution is 64 × 64 × 8. The third convolution layer consists of 16 filters with a 5 × 5 kernel size and a max pooling layer with a 2 × 2 kernel size. The second convolution consists of a 32 × 32 × 16 output feature map. The fourth convolution layer consists of 16 filters with a 5 × 5 kernel size and a max pooling layer with a 4 × 4 kernel size. The output feature map size of the fourth convolution is 8 × 8 × 16. Then, these features feed into fully connected layer 1, consisting of 16 artificial neurons (nodes). It generates 16 features. Finally, these features feed into fully connected layer two, which consists of a single artificial neuron (node) to produce the last feature f_{p_i} (confidence value). The final f_p value is computed by calculating the average of the confidence values across all frames in the test video. The equation below represents the calculation of the average confidence value, where n denotes the total number of frames in the video.

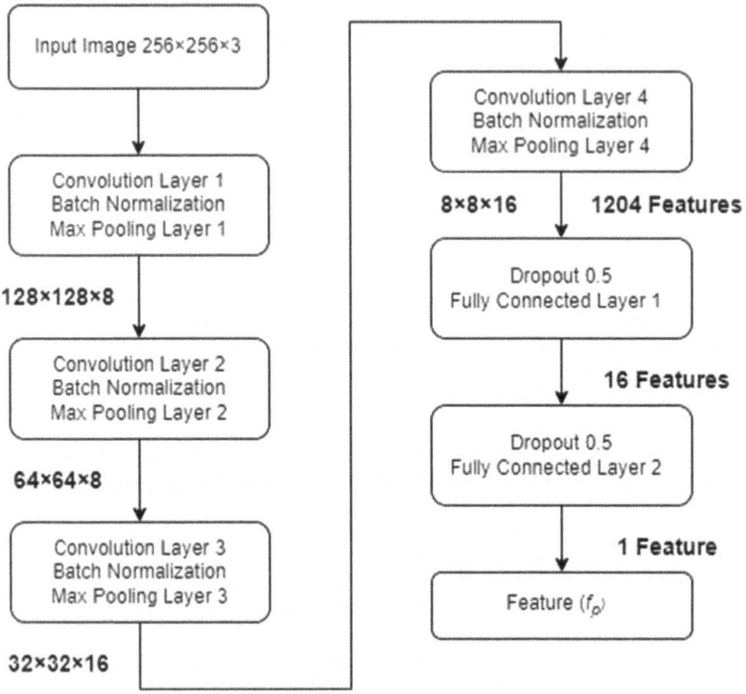

Fig. 10.19 Proposed CNN architecture

$$f_p = \frac{\sum_{i=1}^{n} f_{p\_i}}{n}$$

10.3.3.1 CNN Training Dataset and Training Parameters

In this research work, we used a public dataset called Celeb-DF-v2 [56]. The Celeb-DF-v2 dataset consists of high-quality synthesis videos with an average length of time 13 s and a frame rate of 30 fps and 590 actual videos with only a single face. We extracted 3000 authentic face images and 4000 deepfake images from the natural and deepfake videos to facilitate the training and validation. All these images are categorized into two classes: real (labeled as binary 1) and deepfake (labeled as binary 0) for binary classification. Figure 10.20 shows sample images of preprocessed mages. The extracted facial images resize to 256 × 256 pixels according to the CNN input layer size. The table below shows the training parameters selected for the CNN model training (Table 10.2).

The model was trained on a computer system with 8 GB of RAM (random access memory) and a CORE i7 CPU (central processing unit).

Fig. 10.20 Sample images from the preprocessed dataset. Image size 256 × 256 pixels. (Images extracted from the public dataset CFD-V2)

Table 10.2 Training parameters selected for the proposed CNN model

Parameter	Value
Optimizer	Adam
Learning rate	0.001
Batch size	1
Epoch	100
Class mode	Binary

10.3.4 Feature Classification Process

In previous topics, we explained all the methodologies behind the feature extraction processes. At the final stage, we classified test images as natural or synthesized by analyzing previously extracted features. We developed the end-to-end deep neural network (DNN) model to achieve this classification process. The proposed DNN architecture is shown in Fig. 10.21. The end-to-end deep learning approach uses a single deep neural network model to directly map raw input data to the desired output without preprocessing for tasks such as feature extraction.

The proposed architecture mainly consists of the input layer, four hidden layers, and an output layer. The input layer facilitates four inputs extracted from the previous steps (w_f, e_a, e_n, and f_p), and the output layer is made up of a single node (r_p). This network works with four hidden layers with a decreasing number of neurons (64, 32, 16, and 4) as a starting point. All the hidden layers utilize the ReLU activation function, and the output layer utilizes the Sigmoid transfer function. The

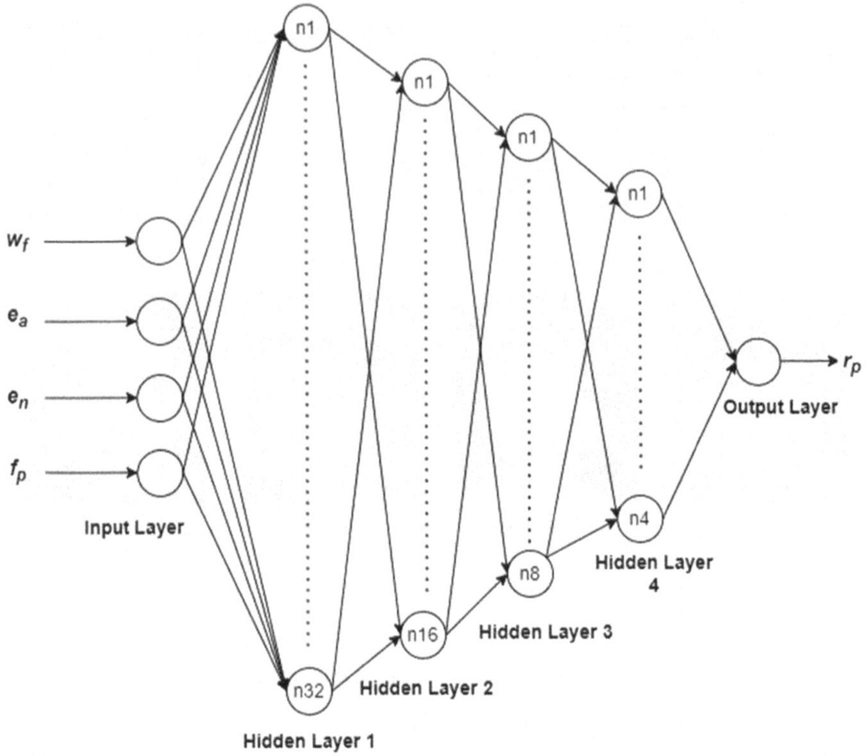

Fig. 10.21 Proposed multiple-input single-output (MISO) DNN architecture for binary classification

Sigmoid activation function is commonly used for binary classification tasks, where the output needs to be a probability value between 0 and 1.

10.3.4.1 DNN Training Dataset and Training Parameters

The proposed DNN architecture was trained by using a preprocessed numerical dataset. This preprocessed dataset consists of four input categories and two output classes. As already explained, this data state was generated using a publicly available dataset, the Celeb-DF-v2 dataset [25]. We select 100 real videos and 200 deep-fake videos to prepare the dataset. From each video, edge density variation (w_f), eye blink duration (e_a), eye blink count (e_n), and CNN output probability (f_p) were extracted. All the extracted features are categorized into two classes real (labeled as binary 1) and deep-fake (labeled as binary 0) for binary classification. Table 10.3 shows the properties of the extracted dataset, and Table 10.4 shows the selected training parameters.

The model was trained on a computer system with 8 GB of RAM (random access memory) and a CORE i7 CPU (central processing unit).

Table 10.3 Description of preprocessed dataset for training end-to-end DNN

Parameters	Range/Value
Edge density variation (w_f)	0–1
Eyeblink duration (e_a)	0.1–2 s
Eyeblink count (e_n)	n
CNN output probability (f_p)	0–1
Output labels (binary)	0/1
Dataset—real videos	150
Dataset–fake videos	200

Table 10.4 Selected training parameters

Parameters	Value
Optimizer	Adam
Learning rate	0.01
Epochs	1000
Training split	0.7
Validation split	0.15
Loss function	Binary cross entropy

10.4 Results and Analysis

This section focused on an explanation of our experimental results and an analysis of the proposed methodology. The results and analysis section consists of several subsections, each subjection allocated to explain results under facial wrinkle variation extraction (edge density variation) (w_f), eyeblink count (e_n), and eye blink duration extraction (e_a), results of the proposed CNN architecture (f_p) and finally, the results of the proposed deep neural network (DNN) architecture.

10.4.1 Facial Wrinkle Variation Extraction (Edge Density Variation)

In this section, we used the Gabor filter to detect edge density variation in the facial area. According to the Gabor filter results, the sudden disappearance or appearance of edges in the facial area can be visualized. Figure 10.22 shows edge density variation computed using a Gabor filter for two conscious frames in the same deepfake video.

Furthermore, according to the quantitative results, the edge density variation is essential in detecting deepfake videos because the edge density variation in the facial images depends on the facial wrinkle variations. Figure 10.23 shows the edge density variation of natural and deepfake videos. The duration of test videos is 15 seconds, and the frame rate is 30 fps.

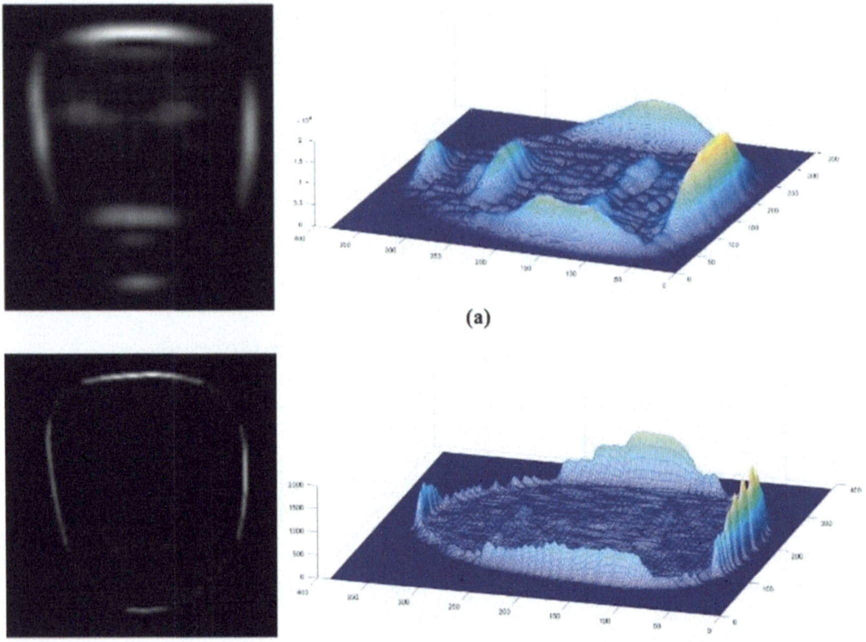

(a)

(b)

Fig. 10.22 Edge density variation is computed by using the Gabor filter. (**a**) Represents the edge variation of frame n and (**b**) represents the edge variation of the frame $n + 1$

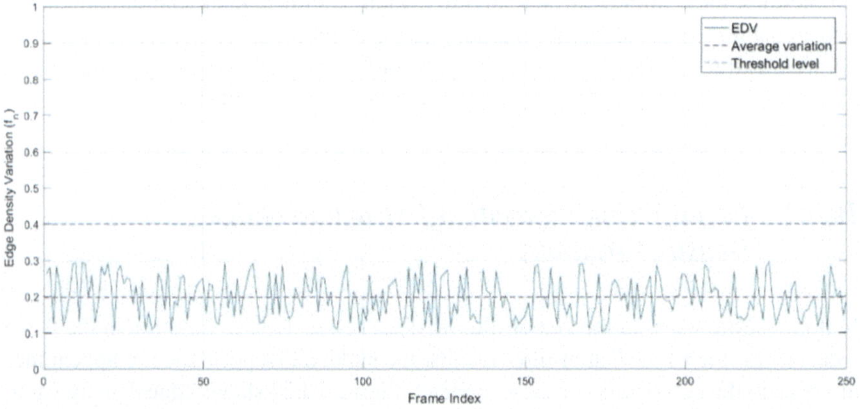

Fig. 10.23 Edge density variation for real video

According to Fig. 10.23, the facial wrinkle variation or facial edge density variation value (f_n) lies between 0.1 and 0.3, and the average value is 0.23, below the threshold value. Therefore, this real sample video has no considerable edge density variation.

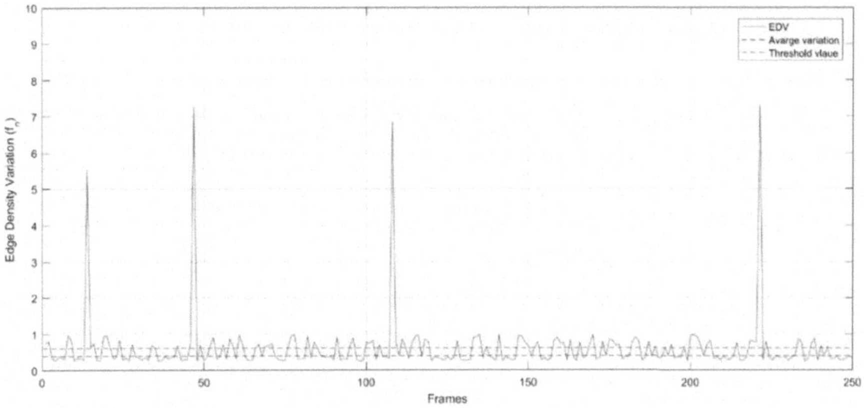

Fig. 10.24 Edge density variation for deepfake video

Table 10.5 Edge density variation analysis for real and deepfake videos

Sample video	Duration (s)	FPS	Average EDV
SR_01	15	30	0.25
SR_02	15	30	0.23
SR_03	15	30	0.36
SR_04	15	30	0.29
SR_05	15	30	0.21
SF_01	15	30	0.57
SF_02	15	30	0.49
SF_03	15	30	0.63
SF_04	15	30	0.58
SF_05	15	30	0.71

Furthermore, according to Fig. 10.24, the facial wrinkle variation or facial edge density variation value (f_n) lies between 0.25 and 7, and the average value is 0.61, over the threshold value. Therefore, there is considerable edge density variation in this deepfake sample video. Table 10.5 depicts edge density variations for ten sample videos, five for real and the remaining for deepfake. The table shows the difference in edge density variation between real and deepfake videos. In the table below, *SR* represents real sample videos, and *SK* represents deepfake sample videos.

10.4.2 Eye Blink Features Extraction

In this section, we analyze the characteristics of eye blinks, especially eye blink duration and number of blinks in a defined time duration. We used aspect ratio of eye (EAR) value for eye blink detection. According to the result analysis, we defined

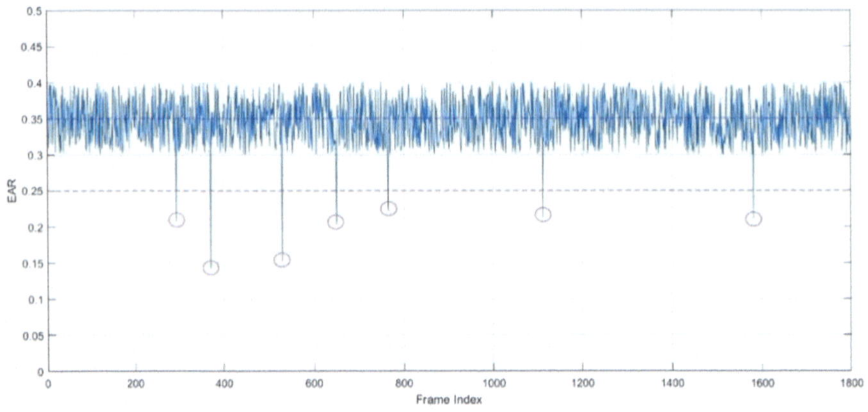

Fig. 10.25 EAR value analysis of 60 seconds deepfake video. The average EAR value is 0.35, and the number of eye blink detections is 7 (marked by red markers)

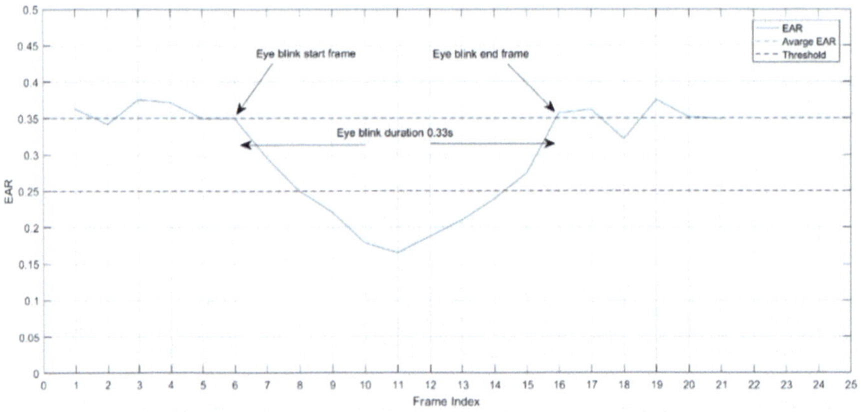

Fig. 10.26 Eyeblink time duration analysis of real scenario

the threshold value for the EAR. If the EAR value is lower than 0.25, the eye is in the closed state, and if the EAR value is higher than 0.25, the eye is in the open state. Considering this EAR threshold value, we could compute the blink rate. The blink rate indicates the number of blinks per minute. According to literature, the human eye blinks 17 to 22 times per minute [19]. Figure 10.25 shows the eye blink rate analysis of the deepfake video. The total duration of the test video is 60 seconds and 30 fps. However, we detected only 7 blinks for the entire video.

Figure 10.26 shows the eye blink time duration analysis of natural and deepfake scenarios. According to Fig. 10.26, the eye blink duration is 0.35 s, between 0.1 s and 0.4 s actual eye blink time. In Fig. 10.27, the eye blink duration is 0.75 s, an abnormal value.

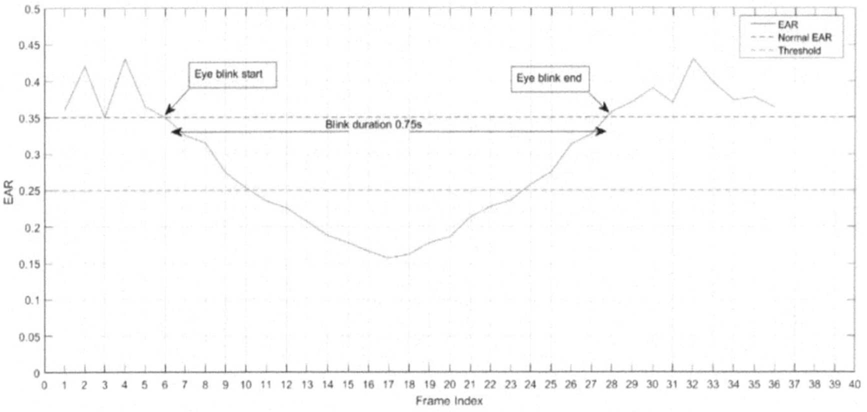

Fig. 10.27 Eyeblink time duration analysis of deepfake scenario

Table 10.6 Eyeblink analysis for real and deepfake videos

Sample video	Average EAR	Average eye blink duration (s)	Eye blink count
SR_01	0.351	0.23	5
SR_02	0.307	0.27	6
SR_03	0.333	0.31	4
SR_04	0.359	0.34	8
SR_05	0.309	0.26	9
SF_01	0.373	0.59	14
SF_02	0.391	0.66	23
SF_03	0.403	0.73	19
SF_02	0.382	0.69	21
SF_02	0.361	0.85	20

Table 10.6 shows results from ten sample videos, including five factual and five deepfake videos. The quantitative results confirmed that deepfake videos can be detected by critical analysis of eye blink features. In the table, SR represents real sample videos, and *SK* represents deepfake sample videos.

10.4.3 Passive Feature Extraction (CNN)

This section elaborates on the results and analysis of the passive feature extraction process using a convolutional neural network (CNN). Figure 10.28 shows the model accuracy curve during the training process. According to the graph, the training accuracy reached 92 %, and model validation accuracy reached 83 %. When the model is applied to training data, its performance shows better metrics than when it is applied to the validation dataset, and the validation set demonstrates slightly less accuracy. Figure 10.29 shows the model loss curve during the training process. The training loss curve reached 0.08, and the validation loss cover reached 0.18.

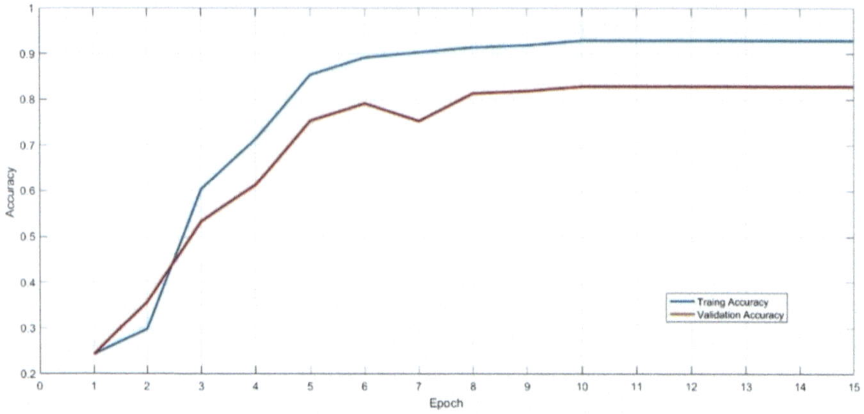

Fig. 10.28 The proposed CNN model's training performance: The accuracy curve; the blue represents the training performance, while the red represents the validation performance

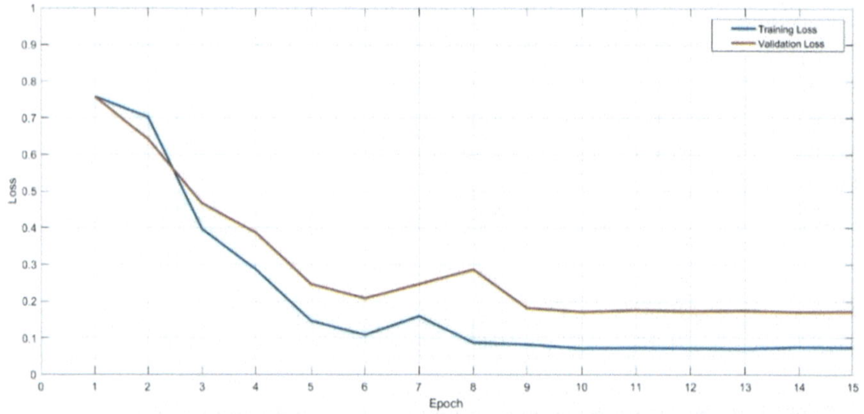

Fig. 10.29 The training performance of the proposed CNN model: Loss curve, the blue represents the training performance, while the red represents the validation performance

The CNN model performance was analyzed using ten test videos, five for real videos and five for deepfake videos. The test videos have a frame rate of 30 frames per second and a duration of 15 seconds. Table 10.7 shows the average confidence value predicted by the CNN model for each test sample. In the table, SR represents real sample videos, and SK represents deepfake sample videos. The results indicate that the CNN model generated average confidence values below 0.5 for the artificially generated videos, whereas it produced confidence values greater than 0.5 for the actual videos. The mean confidence value (f_p) is calculated by processing individual image frames in a test video using the proposed CNN. The confidence value is always between 0 and 1 because we used the Sigmoid activation transfer function for the final dense layer in CNN. Based on the obtained confidence values, we could

Table 10.7 Obtained average confidence values for the test videos

Test video	Duration (s)	FPS	Frames	Average confidence
SR_01	15	30	450	0.65
SR_02	15	30	450	0.53
SR_03	15	30	450	0.59
SR_04	15	30	450	0.71
SR_05	15	30	450	0.82
SF_01	15	30	450	0.43
SF_02	15	30	450	0.37
SF_03	15	30	450	0.31
SF_04	15	30	450	0.29
SF_05	15	30	450	0.41

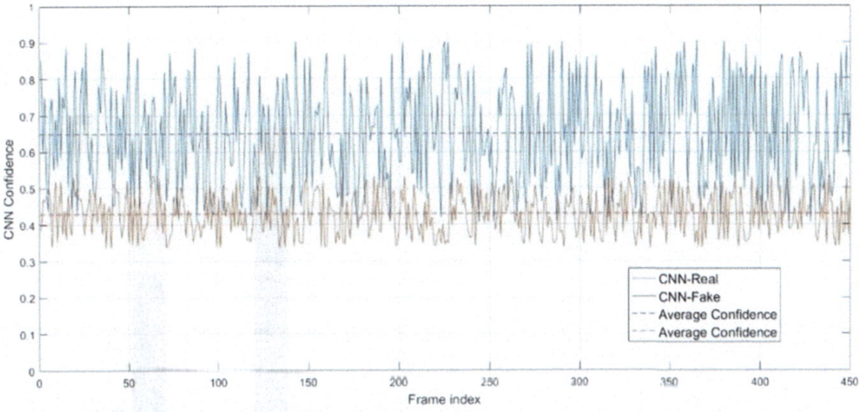

Fig. 10.30 CNN confidence comparison between actual and deepfake video. The blue line shows the confidence variation of real facial video. The orange line shows the confidence variation of the deepfake video. The purple and red horizontal lines indicate the average of CNN confidence

establish a suitable threshold value to categorize the test videos for further processing in the final stage.

Figure 10.30 compares the CNN confidence values predicted to individual frames from two test videos, one containing a real facial video and the other a deepfake. For these two sample videos, the mean confidence value (f_p) for the real facial video is 0.65, while the mean confidence value (f_p) for the deepfake video is 0.43.

10.4.4 Feature Classification Analysis

This section elaborates on the results and analysis of the final classification process by using an end-to-end deep neural network (DNN). Figure 10.31 shows the developed curve for the model training, testing, and validation data during the training

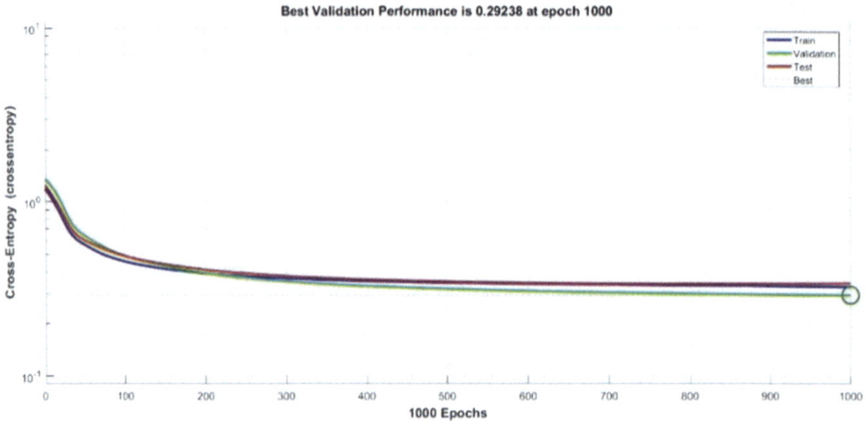

Fig. 10.31 Deep neural network training performance plot with best validation

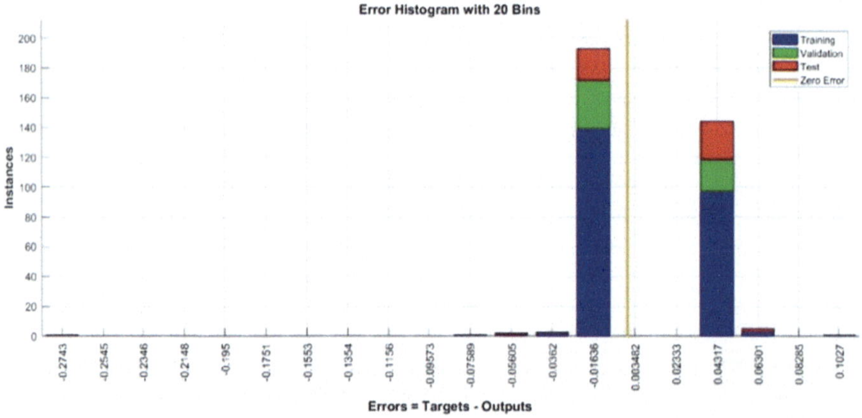

Fig. 10.32 Deep neural network error histogram plot for training

process. It indicates that the mean squared error (MSE) value decreases as the number of epochs increases. The MSE for the model training, testing, and validation decreases until epochs reach 1000. The epoch 1000 is where the best MSE occurs (MSE = 0.29). The lower MSE indicates better performance of the proposed predictive model.

Figure 10.32 shows the error histogram for the training process to assess the model performance. The blue, green, and red bars indicate training error, validation error, and test error values. According to the graph, most error values are near the zero-error line.

The reliability of the suggested model is evaluated by analyzing several important metrics, such as precision, accuracy, *F*-score, recall, and specificity. These evaluation measures were obtained by using the following standard equations.

$$\text{Accuracy} = \frac{\text{number of true negatives} + \text{number of true positives}}{\text{total number of samples}}$$

$$\text{Recall} = \frac{\text{number of true positives}}{\text{number of false negatives} + \text{number of true positives}}$$

$$\text{Specificity} = \frac{\text{number of true negatives}}{\text{number of false pocitive} + \text{number of true negatives}}$$

$$\text{Precision} = \frac{\text{number of true positives}}{\text{number of false positives} + \text{number of true positives}}$$

$$F \quad \text{score} = 2\left(\frac{\text{recall} \times \text{precision}}{\text{recall} + \text{precision}}\right)$$

Figure 10.33 shows the test data confusion matrix of the proposed model for detecting deepfakes. Using the confusion matrix, we calculated the above performance measurements. The table shows the quantitative values of the performance measurements. According to Table 10.8, the suggested model delivers 90.6 %

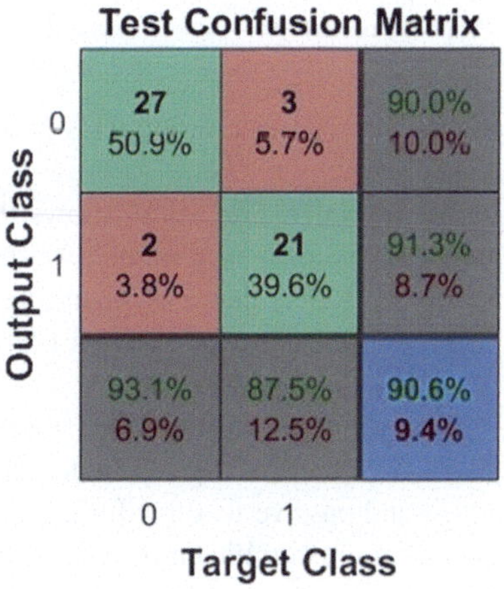

Fig. 10.33 The confusion matrix for the proposed model

Table 10.8 Performance of the proposed model

Performance measure	Value
Accuracy	90.6 %
Specificity	90.0 %
Recall	91.3 %
Precision	87.5 %
F-score	89.3 %

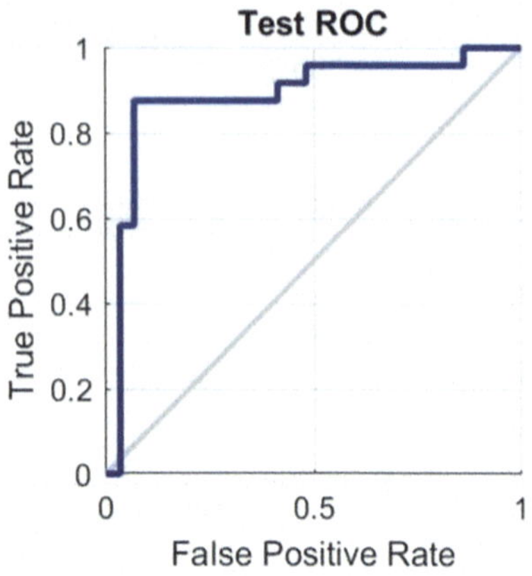

Fig. 10.34 ROC-AUC curve for the proposed model

Table 10.9 Comparison of the detection performance of the AUC values

	Celeb-DF v2	Number of parameters
Zhou et al. [14]	53.8%	24 M
Afchar et al. [15]	54.8%	27.9 K
Li et al. [30]	64.6%	–
Nguyen et al. [23]	57.5%	3.9 M
Rossler et al. [57]	48.2%	22.8 M
Chen et al. [58]	90.56%	42.8 K
Proposed model	90.32%	181.509 K

accuracy, 90% specificity, 91.3 % recall, 87.5 % precision, and 89.9 % *F*-score, respectively.

Additionally, Fig. 10.34 shows the area under the curve (AUC) for the performance of the proposed model. As shown in Fig. 10.34, the receiver operating characteristic (ROC) curve is positioned near the top-left corner, indicating that the proposed DNN model exhibits high-performance capabilities. The corresponding AUC score is 90.32 %.

According to the proposed method results, they finally achieved an AUC of 90.32 %, which is slightly better than theirs. Our fundamental model is based on active and passive features for classification; thus, the number of parameters is frequently more significant because this model consists of CNN and DNN. Table 10.9 shows a comparison of the area under the curve (AUC) values for the Celeb-DF v2 dataset; several studies [14, 15, 23, 30, 57] appear to have achieved relatively low AUC scores, indicating their methods may not perform as well in distinguishing deepfake and authentic images. The core of our model relies on both active and passive features for classification. Consequently, it often has more parameters since the model comprises CNN and DNN.

The experiments were performed on a Dell laptop running Windows 11 OS, equipped with an 11th Gen Intel(R) Core(TM) i7-11370H CPU with 8 GB RAM and an RTX 2060 GPU with 6 GB VRAM. The proposed model was implemented using

Table 10.10 Computational cost and memory usage of models in the training stage

Metrics	CNN	DNN
CPU-minutes	1.65	0.76
Megabytes (MBs)	0.53	0.32

Python version 3.7.4, leveraging various libraries such as Keras, TensorFlow, OpenCV, Scikit-learn, NumPy, and MediaPipe. MATLAB software was employed for result analysis and for generating technical graphs.

10.4.5 Potential Biases and Limitations of Dataset

The Celeb-DF (v2) dataset is a widely used and challenging dataset for deepfake forgery detection. However, it has certain limitations and potential biases that could impact the generalizability of the trained model. This dataset comprises videos of celebrities, which may introduce demographic biases regarding age, ethnicity, and gender representation. Therefore, this dataset may not perform well on deepfakes of individuals from underrepresented groups. The dataset may have a biased distribution of facial expressions and poses, as celebrities often appear in specific settings or contexts. This dataset may not generalize well to deepfakes with diverse facial expressions and poses outside the dataset's distribution. The Celeb-DF (v2) dataset primarily comprises frontal facial images; this may potentially lead to diminished accuracy and generalization capabilities for side-view or oblique facial profiles.

10.4.6 Computational Complexity

The experiments were performed on a high-performance computer, equipped with an 11th Gen Intel(R) Core (TM) i7-11370H CPU with 8 GB RAM and an RTX 2060 GPU with 6 GB VRAM. Table 10.10 shows the computational cost and memory usage of CNN and DNN classifiers in the training stage.

10.5 Conclusion

This research work proposed a novel methodology for identifying deepfake videos. This proposed methodology extracts facial features using image filtering and deep learning techniques. This proposed methodology uses active and passive facial features for deepfake detection. Under the active facial features, facial edge density variation, eye blink duration, and the number of eye blinks are deeply analyzed. The facial edge density variation ranges from 0 to 1, and 0.5 is the threshold value. The eye blink duration varies from 0.1 s to 0.4 s for real test samples and 0.5 s to 1.2 s for deepfake test samples. The eye blink count varies from 15 to 22 for real test samples and 6 to 13 for deepfake test samples. According to these active facial feature results, these features can be successfully used as features for deepfake content detection. Under passive feature extraction, this process uses a convolutional neural network (CNN) to analyze image frame's pixel levels and generate confidence value as a feature. The confidence value varies from 0 to 1, and the threshold value is 0.5 for separating real and deepfake videos. The CNN model training accuracy reached up to 92 %, and the model validation accuracy reached up to 83 %. Finally, all the features are used to analyze videos deeply and detect deepfake content. The final process was achieved using end-to-end deep neural network (DNN). The suggested method is quite effective as shown by the outstanding detection performance values obtained for all assessment criteria used. It depicts 90.32 % of the AUC score, 90.6 % accuracy, 90 % specificity, 91.3 % recall, 87.5 % precision, and 89.3 % F-score. The broad analyses of the performance proved that the proposed approach works as effectively as cutting-edge techniques. Since deepfake video generation algorithms continue to evolve, more work is required to enhance existing techniques for detecting such fabricated content. This work intends to combine features that showed exceptional performance in deepfake detection. The proposed model was trained by using the Celeb-DF (v2) dataset. This dataset primarily comprises frontal facial images, potentially leading to diminished accuracy, and generalization capabilities for side view or oblique facial profiles. Furthermore, the researchers plan to develop a more resilient deep learning-based approach for deepfake detection by leveraging facial color variations to stay ahead of advancements in the deepfake creation process.

References

1. Akhtar, Z., Dasgupta, D., & Banerjee, B. (2019, May). Face authenticity: An overview of face manipulation generation, detection and recognition. In *Proceedings of international conference on communication and information processing (ICCIP).*
2. Home Security Heroes. (2023). *2023 State of Deepfakes: Realities, threats, and impact.* [online] www.homesecurityheroes.com. Available at:. https://www.homesecurityheroes.com/state-of-deepfakes/
3. iProov. (2022). *Deepfake statistics & solutions* | Protect Against Deepfakes. [online] www.iproov.com. Available at: https://www.iproov.com/blog/deepfakes- statistics-solutions-biometric-protection
4. Vezzetti, E., Marcolin, F., Tornincasa, S., & Maroso, P. (2016). Application of geometry to rgb images for facial landmark localisation-a preliminary approach. *International Journal of Biometrics, 8*(3–4), 216–236.
5. Zhang, Z., Zhang, W., Liu, J., & Tang, X. (2014). Multiview facial landmark localization in RGB-D images via hierarchical regression with binary patterns. *IEEE Transactions on Circuits and Systems for Video Technology, 24*(9), 1475–1485.
6. Viola, P., & Jones, M. (2001, December). Rapid object detection using a boosted cascade of simple features. In *Proceedings of the 2001 IEEE computer society conference on computer vision and pattern recognition. CVPR 2001* (Vol. 1, p. I- I). IEEE.

7. King, D. E. (2009). Dlib-ml: A machine learning toolkit. *The Journal of Machine Learning Research, 10*, 1755–1758.
8. Bazarevsky, V., Kartynnik, Y., Vakunov, A., Raveendran, K., & Grundmann, M. Blazeface: Sub-millisecond neural face detection on mobile gpus. *arXiv 2019, arXiv:1907.05047.*
9. Deng, J., Guo, J., Zhou, Y., Yu, J., Kotsia, I., & Zafeiriou, S. Retinaface: Single-stage dense face localisation in the wild. *arXiv 2019, arXiv:1905.00641.*
10. Zhang, K., Zhang, Z., Li, Z., & Qiao, Y. (2016). Joint face detection and alignment using multitask cascaded convolutional networks. *IEEE Signal Processing Letters, 23*, 1499–1503.
11. Ismail, A., Elpeltagy, M., Zaki, S., & Eldahshan, K. (2021). A new deep learning-based methodology for video deepfake detection using XGBoost. *Sensors, 21*(16), 5413.
12. Wodajo, D., & Atnafu, S. (2021). Deepfake video detection using convolutional vision transformer. *arXiv preprint arXiv:2102.11126.*
13. Heo, Y. J., Choi, Y. J., Lee, Y. W., & Kim, B. G. Deepfake detection scheme based on vision transformer and distillation. *arXiv 2021. arXiv preprint arXiv:2104.01353.*
14. Zhou, P., Han, X., Morariu, V. I., & Davis, L. S. (2017, July). Two-stream neural networks for tampered face detection. In *2017 IEEE conference on computer vision and pattern recognition workshops (CVPRW)* (pp. 1831–1839). IEEE.
15. Afchar, D., Nozick, V., Yamagishi, J., & Echizen, I. (2018, December). Mesonet: A compact facial video forgery detection network. In *2018 IEEE international workshop on information forensics and security (WIFS)* (pp. 1–7). IEEE.
16. Karras, T., Laine, S., & Aila, T. (2019). A style-based generator architecture for generative adversarial networks. In *Proceedings of the IEEE/CVF conference on computer vision and pattern recognition* (pp. 4401–4410).
17. Karras, T., Laine, S., Aittala, M., Hellsten, J., Lehtinen, J., & Aila, T. (2020). Analyzing and improving the image quality of stylegan. In *Proceedings of the IEEE/CVF conference on computer vision and pattern recognition* (pp. 8110–8119).
18. Non-Existent Lifelike Face of Peoples Using StyleGAN2. Available online: https://www.thispersondoesnotexist.com/. Accessed on 11 December 2023.
19. Güera, D., & Delp, E. J. (2018, November). Deepfake video detection using recurrent neural networks. In *2018 15th IEEE international conference on advanced video and signal based surveillance (AVSS)* (pp. 1–6). IEEE.
20. Wang, Z., She, Q., & Ward, T. E. (2021). Generative adversarial networks in computer vision: A survey and taxonomy. *ACM Computing Surveys (CSUR), 54*(2), 1–38.
21. Korshunov, P., Marcel, S., Fakes, D. (2018). A new threat to face recognition? Assessment and detection.
22. Yang, X., Li, Y., & Lyu, S. (2019, May). Exposing deep fakes using inconsistent head poses. In *ICASSP 2019–2019 IEEE international conference on acoustics, speech and signal processing (ICASSP)* (pp. 8261–8265). IEEE.
23. Rossler, A., Cozzolino, D., Verdoliva, L., Riess, C., Thies, J., & Nießner, M. (2019). Faceforensics++: Learning to detect manipulated facial images. In *Proceedings of the IEEE/CVF international conference on computer vision* (pp. 1–11).
24. Dufour, N., & Gully, A. (2019). Contributing data to deepfake detection research. *Google AI Blog, 1*, 3.
25. Li, Y., Yang, X., Sun, P., Qi, H., & Lyu, S. C. D. (1909). A large-scale challenging dataset for deepfake forensics (2019). http://arxiv.org/abs/1909.12962, *35*, 36.
26. Dolhansky, B., Howes, R., Pflaum, B., Baram, N., & Ferrer, C. C. The deepfake detection challenge (dfdc) dataset. *arXiv 2020, arXiv:2006.07397.*
27. Jiang, L., Li, R., Wu, W., Qian, C., & Loy, C. C. (2020). Deeperforensics-1.0: A large-scale dataset for real-world face forgery detection. In *Proceedings of the IEEE/CVF conference on computer vision and pattern recognition* (pp. 2889–2898).
28. Zi, B., Chang, M., Chen, J., Ma, X., & Jiang, Y. G. (2020, October). Wilddeepfake: A challenging real-world dataset for deepfake detection. In *Proceedings of the 28th ACM international conference on multimedia* (pp. 2382–2390).

29. Li, Y., Yang, X., Sun, P., Qi, H., & Lyu, S. (2020). Celeb-DF: A large-scale challenging dataset for DeepFake forensics. In *2020 IEEE/CVF conference on computer vision and pattern recognition (CVPR), Seattle, WA, USA* (pp. 3204–3213). https://doi.org/10.1109/CVPR42600.2020.00327
30. Li, Y., & Lyu, S. (2018). Exposing deepfake videos by detecting face warping artifacts. *arXiv preprint arXiv:1811.00656.*
31. Dang, H., Liu, F., Stehouwer, J., Liu, X., & Jain, A. K. (2020). On the detection of digital face manipulation. In *Proceedings of the IEEE/CVF conference on computer vision and pattern recognition* (pp. 5781–5790).
32. Charitidis, P., Kordopatis-Zilos, G., Papadopoulos, S., & Kompatsiaris, I. A face preprocessing approach for improved deepfake detection. *arXiv 2020. arXiv preprint arXiv:2006.07084.*
33. Nguyen, H. H., Fang, F., Yamagishi, J., & Echizen, I. (2019, September). Multi- task learning for detecting and segmenting manipulated facial images and videos. In *2019 IEEE 10th international conference on biometrics theory, applications and systems (BTAS)* (pp. 1–8). IEEE.
34. Khalil, S. S., Youssef, S. M., & Saleh, S. N. (2021). iCaps-Dfake: An integrated capsule-based model for deepfake image and video detection. *Future Internet, 13*(4), 93.
35. Mehra, A. (2020). *Deepfake detection using capsule networks with long short-term memory networks.* Master's thesis. University of Twente.
36. Singh, A., Saimbhi, A. S., Singh, N., & Mittal, M. (2020). DeepFake video detection: A time-distributed approach. *SN Computer Science, 1*(4), 212.
37. Li, Y., Chang, M. C., & Lyu, S. (2018, December). In ictu oculi: Exposing ai created fake videos by detecting eye blinking. In *2018 IEEE international workshop on information forensics and security (WIFS)* (pp. 1–7). IEEE.
38. Masi, I., Killekar, A., Mascarenhas, R. M., Gurudatt, S. P., & AbdAlmageed, W. (2020). Two-branch recurrent network for isolating deepfakes in videos. In *Computer vision–ECCV 2020: 16th European conference, Glasgow, UK, August 23–28, 2020, proceedings, part VII 16* (pp. 667–684). Springer International Publishing.
39. Zhao, H., Zhou, W., Chen, D., Wei, T., Zhang, W., & Yu, N. (2021). Multi- attentional deepfake detection. In *Proceedings of the IEEE/CVF conference on computer vision and pattern recognition* (pp. 2185–2194).
40. Mazaheri, G., & Roy-Chowdhury, A. K. (2022). Detection and localization of facial expression manipulations. In *Proceedings of the IEEE/CVF winter conference on applications of computer vision* (pp. 1035–1045).
41. Hu, J., Liao, X., Gao, D., Tsutsui, S., Wang, Q., Qin, Z., & Shou, M. Z. (2023). Mover: Mask and recovery based facial part consistency aware method for deepfake video detection. *arXiv preprint arXiv:2303.01740.*
42. Li, L., Bao, J., Zhang, T., Yang, H., Chen, D., Wen, F., & Guo, B. (2020). Face x- ray for more general face forgery detection. In *Proceedings of the IEEE/CVF conference on computer vision and pattern recognition* (pp. 5001–5010).
43. Yang, J., Xiao, S., Li, A., Lu, W., Gao, X., & Li, Y. (2021). MSTA-Net: Forgery detection by generating manipulation trace based on multi-scale self-texture attention. *IEEE Transactions on Circuits and Systems for Video Technology, 32*(7), 4854–4866.
44. Durall, R., Keuper, M., Pfreundt, F. J., & Keuper, J. (2019). Unmasking deepfakes with simple features. *arXiv preprint arXiv:1911.00686.*
45. Woo, S. (2022, June). Add: Frequency attention and multi-view based knowledge distillation to detect low-quality compressed deepfake images. In *Proceedings of the AAAI conference on artificial intelligence* (Vol. 36, No. 1, pp. 122–130).
46. Giudice, O., Guarnera, L., & Battiato, S. (2021). Fighting deepfakes by detecting gan dct anomalies. *Journal of Imaging, 7*(8), 128.
47. Westerlund, M. (2019). The emergence of deepfake technology: A review. *Technology Innovation Management Review, 9*(11), 39.
48. Taeb, M., & Chi, H. (2022). Comparison of deepfake detection techniques through deep learning. *Journal of Cybersecurity and Privacy, 2*(1), 89–106.

49. Goyal, K., Agarwal, K., & Kumar, R. (2017, April). Face detection and tracking: Using OpenCV. In *2017 international conference of electronics, communication and aerospace technology (ICECA)* (Vol. 1, pp. 474–478). IEEE.
50. Alanazi, F., Ushaw, G., & Morgan, G. (2023). Improving detection of DeepFakes through facial region analysis in images. *Electronics, 13*(1), 126.
51. Siegel, D., Kraetzer, C., Seidlitz, S., & Dittmann, J. (2021). Media forensics considerations on deepfake detection with hand-crafted features. *Journal of Imaging, 7*(7), 108.
52. Huang, T., Yang, G. J. T. G. Y., & Tang, G. (1979). A fast two-dimensional median filtering algorithm. *IEEE Transactions on Acoustics, Speech, and Signal Processing, 27*(1), 13–18.
53. Lee, C. J., & Wang, S. D. (1999). Fingerprint feature extraction using Gabor filters. *Electronics Letters, 35*, 288.
54. Munawar, H. S., Qayyum, S., Ullah, F., & Sepasgozar, S. (2020). Big data and its applications in smart real estate and the disaster management life cycle: A systematic analysis. *Big Data and Cognitive Computing, 4*(2), 4.
55. Jung, T., Kim, S., & Kim, K. (2020). Deepvision: Deepfakes detection using human eye blinking pattern. *IEEE Access, 8*, 83144–83154.
56. Gupta, G., Raja, K., Gupta, M., Jan, T., Whiteside, S. T., & Prasad, M. (2023). A comprehensive review of DeepFake detection using advanced machine learning and fusion methods. *Electronics, 13*(1), 95.
57. Nguyen, H. H., Yamagishi, J., & Echizen, I. (2019). Use of a capsule network to detect fake images and videos. *arXiv preprint arXiv:1910.12467*.
58. Chen, H. S., Rouhsedaghat, M., Ghani, H., Hu, S., You, S., & Kuo, C. C. J. (2021, July). Defakehop: A light-weight high-performance deepfake detector. In *2021 IEEE international conference on multimedia and expo (ICME)* (pp. 1–6). IEEE.

Chapter 11
Smart Cities, Secure Data: Navigating the Evolving Landscape of Data Protection Challenges

N. D. G. T. Nanayakkara, G. A. S. Y. Nanayakkara, K. M. C. M. Karunanayaka, C. P. T. K. Kumara, and N. Karunanayake

Abstract The integration of technology and urban planning in the dynamic field of smart cities not only offers unprecedented opportunities but also poses significant challenges, especially in data protection.

In smart cities, the complex web of data from social dynamics to environmental indicators highlights the importance of sophisticated technology for data collection and analysis. Therefore, effective data visualization and communication are essential to transform complex data into actionable insights. Additionally, addressing the complexities of data ownership, management, and governance underscores the ethical obligations of responsible data governance.

Architectural strategies are explored to protect smart city infrastructures from dynamic threats, including network segmentation, encryption, and endpoint defense. In addition, challenges, such as cyberattacks, and vulnerabilities caused by Internet of Things (IoT) use are examined and preventative strategies to improve resilience are discussed. Insights from real case studies on incident response and crisis management are included. The importance of stakeholder collaboration and public awareness in strengthening cyber resilience is emphasized, as is the need to adapt data practices to evolving legal and ethical frameworks to ensure transparency and trust.

This study anticipates new trends and advocates for combining technological progress with ethical commitments, highlighting data protection as the key to sustainable urban planning. It integrates scientific and practical evidence and provides

N. D. G. T. Nanayakkara (✉)
College of Information Technology and Convergence, Pukyong National University, Busan, Republic of Korea
e-mail: thathsara@pukyong.ac.kr

G. A. S. Y. Nanayakkara · K. M. C. M. Karunanayaka · C. P. T. K. Kumara
Faculty of Information Technology, University of Moratuwa, Moratuwa, Sri Lanka

N. Karunanayake
Department of Radiology, Memorial Sloan Kettering Cancer center, New York, NY, USA
e-mail: karunan@mskcc.org

C. Hewage et al. (eds.), *Data Protection*,
https://doi.org/10.1007/978-3-031-76473-8_11

policymakers, practitioners, and scientists with a guide for developing resilient, ethically sound smart cities. By prioritizing technological advances as well as ethical considerations, we aim to create safe urban environments that increase public trust and ensure a better future.

11.1 Overview of Data in Smart Cities

11.1.1 Introduction

Smart cities represent a transformative approach to urban development, leveraging advanced technologies and data-driven systems to enhance various aspects of urban life.

11.1.2 Declaration of Smart Cities

A smart city is characterized by several key components that work together to create a holistic and interconnected urban ecosystem:

1. Digital Infrastructure: Smart cities are built upon robust digital infrastructure, including high-speed internet connectivity, sensor networks, and communication systems. These technological foundations enable the collection and analysis of vast amounts of data from various sources across the city [1, 2].
2. Data-Driven Decision-Making: Data serves as the lifeblood of smart cities, empowering decision-makers to gain valuable insights into urban operations and citizen needs. Through the use of advanced analytics, machine learning, and artificial intelligence, city officials can make informed decisions to optimize resource allocation, improve service delivery, and address emerging challenges.
3. Integrated Systems: Smart cities integrate various systems and technologies to streamline operations and enhance efficiency. This includes smart transportation networks, energy management systems, waste management solutions, and public safety initiatives. [1, 3] By connecting these systems, cities can achieve greater coordination and responsiveness across different sectors.
4. Citizen Engagement: Citizen Participation is fundamental to the success of smart cities. Through digital platforms and mobile applications, residents can actively engage with city authorities, provide feedback, and participate in decision-making processes. This two-way communication fosters a sense of community ownership and enables cities to better understand and address the needs of their citizens.
5. Sustainability and Resilience: Smart cities prioritize sustainability and resilience in their planning and development strategies. This includes initiatives to reduce

carbon emissions, improve air and water quality, enhance green spaces, and mitigate the impacts of climate change [1]. By adopting environmentally friendly practices and resilient infrastructure, smart cities aim to create healthier and more livable urban environments for current and future generations.

11.1.3 Overview of Data in Smart Cities

Data collected and utilized in smart cities span across various domains, including environmental, transportation, and social aspects:

1. Environmental Data:
    - Air Quality: Sensors measure pollutants such as nitrogen dioxide, ozone, and particulate matter to monitor air quality levels.
    - Weather: Weather stations collect data on temperature, humidity, precipitation, wind speed, and atmospheric pressure to support weather forecasting and climate analysis.
    - Water Quality: Sensors in water bodies monitor parameters like pH levels, dissolved oxygen, and pollutant concentrations to assess water quality and detect pollution events.
    - Energy Consumption: Smart meters track electricity, water, and gas consumption in buildings and infrastructure to optimize energy use and identify areas for efficiency improvements [1, 3].
2. Transportation Data:
    - Traffic Flow: Sensors embedded in roads, traffic lights, and vehicles gather real-time data on traffic congestion, vehicle speed, and road conditions.
    - Public Transit Usage: Automated fare collection systems and passenger counting sensors on buses and trains provide insights into public transit ridership patterns and demand.
    - Parking Availability: Smart parking meters and sensors detect available parking spaces and provide information to drivers to reduce congestion and emissions.
3. Social Data:
    - Citizen Behavior: Data from mobile apps, social media platforms, and Wi-Fi networks capture information on citizen activities, preferences, and interactions with urban services.
    - Demographic Information: Census data, demographic surveys, and administrative records provide demographic insights into the population composition, migration patterns, and socioeconomic characteristics of city residents.

11.1.4 Methods and Technologies for Data Collection

- Sensors and IoT Devices: Deployed throughout the city, sensors and Internet of Things (IoT) devices collect real-time data on various environmental and urban parameters. These devices can range from stationary air quality monitors to wearable health trackers.
- Mobile Apps: Citizens can contribute data through mobile applications that enable crowd-sourced information gathering, such as reporting potholes, tracking public transit usage, or participating in community surveys.
- Social Media Platforms: Analysis of social media data offers insights into citizen sentiment, opinions, and behavior patterns. However, this data collection method raises privacy concerns and requires careful consideration of ethical and legal implications.
- Advanced Analytics: Machine learning algorithms and data analytics platforms process and analyze large datasets to extract actionable insights and support data-driven decision-making in areas such as traffic management, urban planning, and public health [4].

11.1.5 Privacy and Security Implications

- Privacy Concerns: The collection of personal data through sensors, mobile apps, and social media platforms raises privacy concerns regarding data ownership, consent, and potential misuse. Smart cities must implement robust privacy policies and data protection measures to safeguard citizen privacy rights.
- Security Risks: IoT devices and sensor networks are susceptible to cyberattacks, hacking, and data breaches. Ensuring the security of smart city infrastructure requires implementing encryption protocols, access controls, and intrusion detection systems to mitigate cybersecurity risks and protect sensitive data from unauthorized access or manipulation [2, 3].

11.1.6 Data Analysis and Transformation

- Data Integration and Analysis:
  - Data from diverse sources such as sensors, IoT devices, mobile apps, and social media platforms are aggregated and processed using advanced analytics techniques.
  - Machine learning algorithms and data mining methods are applied to identify patterns, correlations, and trends within the data.
  - Statistical analysis and modeling help uncover relationships between different urban variables and predict future outcomes.

- Actionable Insights:
  - Urban Planning: Data analysis informs urban planners about population dynamics, land use patterns, transportation demand, and environmental factors. This insight guides the development of sustainable infrastructure, efficient land use zoning, and resilient urban design.
  - Resource Allocation: Decision-makers use data-driven insights to allocate resources effectively in areas such as transportation, energy, water management, and public safety. For example, real-time traffic data can inform traffic signal optimization or rerouting strategies to reduce congestion [2, 3].
  - Service Optimization: Analysis of citizen feedback, service usage data, and operational metrics helps optimize public services such as waste management, public transportation, healthcare delivery, and emergency response. This leads to improved service quality, reduced costs, and enhanced citizen satisfaction.

11.1.7 Importance of Effective Data Visualization and Communication

- Complex Data Simplification: Data visualization techniques transform complex datasets into intuitive visual representations such as charts, graphs, maps, and dashboards. This simplification makes it easier for decision-makers, stakeholders, and the general public to understand and interpret the information.
- Decision Support: Clear and visually appealing data visualizations provide decision-makers with actionable insights and facilitate evidence-based decision-making. Interactive visualization tools allow users to explore data from different perspectives and scenarios.
- Stakeholder Engagement: Engaging and informative data visualizations enhance communication and collaboration among stakeholders, fostering consensus-building and collective problem solving. Visual representations of urban data help stakeholders envision the impact of proposed policies, projects, and interventions [2].
- Public Awareness: Communicating urban data to the general public through accessible and engaging visualizations promotes transparency, accountability, and civic engagement. Public-facing dashboards, infographics, and storytelling techniques raise awareness about urban issues, solicit feedback, and empower citizens to participate in decision-making processes.

In summary, effective analysis and visualization of urban data play a crucial role in translating raw data into actionable insights that inform decision-making, drive policy formulation, and improve urban governance in smart cities. By leveraging data visualization and communication strategies, decision-makers can harness the power of data to address complex urban challenges and create more livable, sustainable, and resilient cities.

11.2 Data Governance and Data Ownership in Smart Cities

By definition, data governance refers to a collection of procedures, guidelines, standards, and measurements that assure the effective and efficient use of information [5]. In the context of smart cities, due to the exponential rise in the volume and complexity of data generated, a well-organized data governance structure is required. Governance procedures that guarantee the appropriate use and management of such large and complex datasets are necessary for their efficient management and utilization. In order to achieve the aforesaid objective, smart city data should be wealth in quality, integration, and interoperability. In addition, several other factors associated with data such as data security, consistency, accuracy, and accessibility also need to be paid attention. This diverse discipline further combines a number of aspects, such as risk management, data policies, data management, data quality, and regulatory compliance. To confine and discuss the broader subject of "data governance," it can be presented under three pillars: data quality, security, and privacy [5] (refer to Fig. 11.1).

Smart cities' data is ubiquitous in multifaceted applications such as transport, health care, education, fashion, and e-commerce, which is used in development, operations, decision-making, etc. Smart cities follow different strategies in accomplishing data quality, which shall include, but not be limited to accuracy, precision, relevance, validity, integrity, completeness, uniqueness, timeliness, consistency [9]. One means is to impose formal data quality frameworks [5, 6, 9, 11]. They cover protocols for data cleansing, validation, and routine audits, in addition to clearly defined metrics and standards. In the implementation, policies have to be introduced as rules and guidelines in data management [5]. In addition, a systematic approach can be established by standards and procedures working together [5]. There, data homogeneity, accuracy, and usability can be guaranteed by data standards.

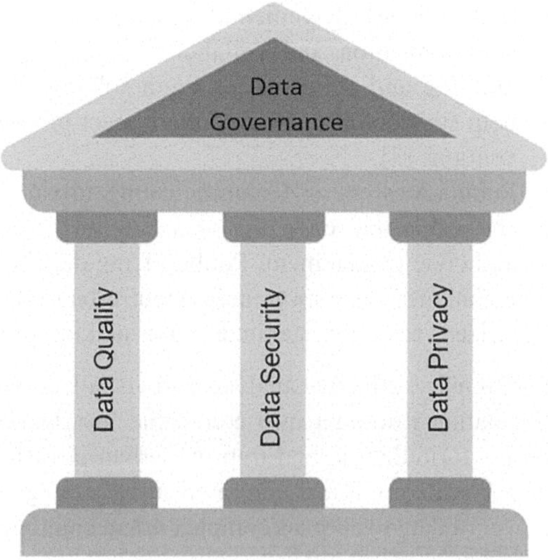

Fig. 11.1 Key pillars of data governance and data ownership in smart cities

Conversely, procedures can then outline the functional facets of data management, including gathering, storing, processing, and exchanging data. Roles and responsibilities also should be included in the data quality frameworks [5]. This means designating data owners, stewards, and a council or governing body to manage data governance operations. Accountability and ownership are ensured by a well-established governance framework. Data architecture and models also play a pivotal role in a framework [5]. This part includes metadata management, data warehousing, and data integration. It offers the framework required to support data governance procedures. Furthermore, frameworks should include risk assessment and mitigation techniques to protect data assets, manage data risks, and guarantee adherence to relevant data protection laws, such as the general data protection regulations (GDPR) [5, 8].

Moreover, the advancements in technology also can be utilized to ensure data quality [5, 9]. The ICT resources buried in the technical and informational levels need to be managed both inside and across the numerous organizations that make up a smart city ecosystem, in order to meet the sustainability and efficiency goals of smart cities. These technical levels shall include data processors, telecommunication infrastructure, data sources, etc. while information level includes usage support services, marketplaces, applications, and services. The usage of software and technologies also automates important steps in data quality management. This simplifies the upkeep of data integrity by including error detection, data cleansing, and reconciliation.

Besides, the public can effectively maintain good data quality in their smart cities [5, 9]. Therefore, it is critical to foster an individual culture that fully understands and values data quality. This means allocating funds for awareness campaigns, training programs, and the delineation of precise roles and duties for data stewardship. Mainly and at last, data has to be monitored and improved continuously to ensure its quality at all times [5]. This entails the regular evaluation of data quality measures and the creation of feedback systems to quickly detect and address problems.

However, several challenges are still making the assurance of data quality in smart cities difficult [5, 6]. A major example can be pointed out as its exponential growth, both in terms of volume and velocity. The probability of errors and inconsistencies can severely be increased with data complexity. Smart cities constantly struggle with data that comes in different formats and diversified sources. Because of this variety, it is more challenging to guarantee accuracy and consistency across different datasets [9]. Another major issue is unclear accountability or ownership of data which may eventually cause data standards to be neglected and deteriorate [7]. Therefore, all data has to be claimed with ownership to hold responsibility and in turn, provide value to the data shared [11].

Another pillar of data governance is "data privacy." [5] At several phases of the data analytics process, which may include data gathering, data sharing, data processing and archiving, data publication, and data utilization, any privacy threats should be recognized and addressed to confirm private data protection. Data privacy can be considered under several aspects such as the privacy of the individual, his actions and habits, used data and images, communication, thoughts and emotions, location, and association. Even though data privacy is an ethical matter to be concerned with, there are still challenges to accomplishing the goal [5, 8, 10]. Novel pathways should include spotting and reducing privacy threats; instead of drawing distinctions between private and public domains or between personal and non-personal data, improving user knowledge and comprehension above displaying business liability disclaimers, accountability is based on how data is used and the dangers that are posed to data subjects instead of establishing a formal division between data controllers and data processors, striking a reasonable balance between the need for data retention and individual rights, controlling the passage of data across borders according to accountability and continuing responsibility. To overcome these challenges, techniques are being introduced from time to time [8, 10]. One example is the anonymization. It refers to altering a data set's state such that it is impossible to identify the original source. This process would usually include complete de-identification while maintaining pseudo-identifiers in some way. Sensitive data masking and alteration are also a part of anonymization tactics. Another technique is the data perturbation. This process includes reassigning values in columns to break the links between features. The goal of this type of shuffling data is to remove connections between identifying traits and sensitive features without changing the entries in any other manner. Encryption tactics used to assure data privacy is also a widely applicable technology that can be used in every architectural level of a smart city. Information can be hidden via these encryption techniques, with the degree of concealment depending on the algorithm and cryptographic key used. Biometrics and homomorphy are two commonly used methods for data encryption.

The goal of data security, being the third pillar of data governance, in smart cities is to establish a reliable computing environment that facilitates secure cooperation both inside and across the involved institutions [5]. This is a difficult task, especially when one has to provide secure interactions between parties that may not have a common security infrastructure for reliably sharing information assets in the loosely coupled setting of smart. Smart city organizations have varied approaches to managing the security of their ICT infrastructures, use a range of devices and computing machines with variable processing and memory capacities, and, most importantly, have distinct goals and missions within a data ecosystem. Therefore, the data security is discussed more elaborately next sections of this chapter.

Apart from data governance, data ownership is another paradigm of data related to smart cities [7]. It speaks about the obligations and rights under the law pertaining to data. Data ownership gives a person or organization the power to decide how their data is gathered, utilized, kept, and shared. It determines the ultimately responsible personnel for the data and frequently includes legal agreements and intellectual property rights. In the context of smart cities, the variety of applications that share data, apart from individually owned data, has owners at the managerial levels. These data owners are essential to establish and protect the data environment in their respective domains. In addition, data owners have the power to decide what actions to take with reference to their data assets. They make decisions about quality, security, and strategic use of data. The access control to data assets is handled by the data owners by making sure the access complies with data governance guidelines and organizational requirements. They are also responsible for the security and accuracy of the data that falls within their jurisdiction. Moreover, data owners have the right to decision-making about data, such as those pertaining to usage guidelines, archiving, and storage.

11.3 Security Architecture and Infrastructure of Smart Cities for Data Protection

11.3.1 Design Principles of Secure Architecture for Smart City Systems

Security by Design Security should be a core component of the system design process, not an afterthought [12]. This necessitates evaluating potential threats and weaknesses during the design phase and incorporating appropriate security measures [13]. "Security by Design" is a crucial principle that underscores the significance of addressing security considerations from the outset of the system design process. This idea includes a several crucial steps.

The first phase is threat modeling, which entails identifying possible threats and vulnerabilities that may affect the system. These dangers might include illegal access, data breaches, and denial-of-service attacks. The second stage is to adhere to secure design principles. These principles guide the design of a secure system and include concepts such as least privilege (each module must have access to only the information and resources required for its legitimate purpose), fail-safe defaults (permission-based access choices as opposed to exclusion-based ones), and defense in depth. The third stage is to include security measures in the system architecture. These controls, which may be preventative (aimed at preventing a security incident), detective (aimed at identifying a security incident), or corrective (aimed at repairing the system following a security incident), are selected based on the identified risks and vulnerabilities.

In addition, the design should be examined and adjusted periodically to handle emerging threats and vulnerabilities. This is especially crucial considering the continuous evolution of cyberthreats. Another critical component of "Security by Design" is security testing. This includes thorough testing to guarantee that the security measures perform as intended.

Finally, extensive documentation is kept on the security design. This documentation contains information on the detected threats and vulnerabilities, the security measures that were applied, and the results of security testing.

The scope and complexity of smart city networks make "Security by Design" even more important. Insecure smart city systems can have major consequences for the social, environmental, and economic growth of cities. Therefore, smart city technologies must be designed with security in mind.

Privacy by Design Given the amount of data collected and processed in smart cities, data protection must be a top priority. This includes ensuring that data is anonymized wherever feasible and that strong data protection mechanisms are in place [12].

Privacy by design is an important topic in the context of smart cities. This is about creating systems and procedures that preserve data privacy from the start, rather than trying to add privacy protections later. This method ensures that the data is only used for legitimate purposes and no individuals are identifiable.

The complex and interconnected characteristics of smart cities present significant political, technological, and economical obstacles to the organizations, designers, and integrators tasked with overseeing these novel entities. A growing body of research examines the threats to security and privacy associated with smart cities. It highlights information security vulnerabilities and the infrastructure's issues in handling and processing personal data.

Smart cities are constructed with an infrastructure focused on information and communication technology (ICT) and Internet of Things (IoT) enabled sensor technologies to promote social and urban interconnectivity through enhanced citizen engagement and government efficiency. As a result, huge amounts of data are generated, examined, processed, and disseminated. Users and smart city services are continuously sharing information about where people go, what they visit, and what services they utilize.

This raises concerns not just about the scale, scope, and granularity of data collected and processed, but also about privacy and integrity. To address these concerns, a smart city infrastructure can include a distributed privacy-preserving identity management solution based on attribute-based credentials (p-ABC), a user-centric Consent Manager, and a General Data Protection Regulation (GDPR)-compliant Access Control mechanism [13].

Thus, the architecture makes it easier to define a specific aim, collect data, restrict access to personal data, and get user consent, all while ensuring selective and restricted exposure of personal information and user unlikability among service and identity providers. This comprehensive approach sheds light on many of the fundamental difficulties and suggests important directions for future research.

Resilience Smart city systems should be resilient against interruptions such as cyberattacks, natural disasters, and system breakdowns [14].

Resilience in smart cities is defined as the ability to sustain or restore balance in the face of disruptions and shocks. This equilibrium refers to the condition of balance throughout the smart city's domains, which include the natural environment, people, government, economics, lifestyle, and infrastructure. As cities become increasingly technologically reliant, their ICT attack surface expands significantly. Cybersecurity is critical for reducing shocks and stresses by assuring the confidentiality, integrity, and availability of data and data-driven infrastructure. However, security alone isn't enough. Cyber resilience takes a step further by guaranteeing that ICT systems continue to deliver services in the event of a security breach. Cities' cyber resilience may be characterized by their capacity to prepare, respond, and innovate. Efforts to improve cyber resilience are critical for surviving and, perhaps, profiting in the face of cyberattacks or physical calamities.

In smart city systems, resilience is anticipating problems, acting quickly to return to normal, and taking note of and adjusting to new situations as they arise. While municipal officials often have a thorough awareness of the physical hazards they face, their knowledge of how to manage cyberrisk is frequently limited. As a result, local officials must prioritize not only balancing investments but also ensuring that they are making the appropriate ones.

Interoperability A smart city's different technologies and systems need to interact seamlessly together. This necessitates the usage of established protocols and interfaces [15].

Interoperability is an important factor in the development of smart city technologies. It refers to the seamless integration of various systems, devices, and applications inside a smart city. This is crucial because, in order to connect to the physical and social spaces of cities, smart cities need a vast variety of hardware and software applications, creating intricate global ecosystems spanning several knowledge and activity categories.

One of the challenges facing the ecosystems of smart cities is the inherent diversity of IoT devices and platforms. Different IoT and software systems exchange data streams using different architectures and communication protocols. The proliferation of these platforms makes it difficult to establish comprehensive, universal access models and hinders the growth of smart city ecosystems.

Flexible smart systems and complex designs are recommended to stimulate and facilitate the development of new applications. These systems can connect domains from different smart city ecosystems, avoiding single vertical implementations that limit solution and infrastructure scalability.

In this context, interoperability refers to the capacity to mix and share data from many sources for use in a variety of applications. Existing decentralized smart city ecosystems may be linked by information-sharing pipelines, resulting in a horizontal layer that reduces load while simultaneously supporting a cohesive design.

Scalability As the city expands and advances, so will its technology demands. As a result, systems should be built to scale effectively [15].

Scalability is an important feature of urban planning and development, particularly in the context of a city's technology requirements. As cities develop and evolve, their technology infrastructure must be adaptable and expandable to suit the changing needs of its residents.

The first facet of scalability is the ability of the city's technological systems to accommodate an increasing number of people or tasks. For example, when the population of a city grows, so does the demand for utilities such as power, water, and internet access. The city's infrastructure must be constructed to support this expansion, ensuring that services remain dependable and efficient as demand grows.

Second, scalability refers to a system's capacity to enhance its functionality. As a city develops, new technical requirements may emerge. For example, the introduction of electric cars may entail the establishment of a citywide charging network. A scalable city is one that can incorporate these new technologies into its current infrastructure rather than creating whole new systems from the start.

Finally, scalability involves future-proofing. This entails developing systems in a way that accounts for future development and technology improvements. It entails making strategic infrastructure construction decisions that will allow for easy future upgrades and expansions. This might entail picking modular designs that can be readily expanded or selecting technologies that are widely supported and likely to stay important for many years.

11.3.2 Components of Secure Architecture for Smart City Systems

Secure Network Infrastructure This comprises secure communication protocols, network segmentation, and firewalls to safeguard data in transit [15, 16].

Secure network infrastructure is a vital component of smart city systems, ensuring that data is sent safely and efficiently across multiple devices and systems. This includes several crucial elements:

Secure Communication Protocols: These are the rules that govern how data is sent and received over a network. In the context of smart cities, these protocols must be secure to ensure integrity and the confidentiality of data exchanged. Secure communication protocols include the transport layer security (TLS) and the secure sockets layer (SSL), which enable secure internet communication. Furthermore, protocols like Zigbee, Z-Wave, and Thread are widely utilized for IoT devices in smart cities.

Network Segmentation This entails partitioning a network into numerous segments or subnetworks, each representing a network segment or network layer. Network segmentation improves security by isolating possible cyberattacks from a single segment. For example, if a cyberattack happens in one section, the negative impact can be limited to that segment, preventing it from spreading to other areas of the network. Furthermore, network segmentation improves network performance by lowering congestion.

Firewalls Firewalls act as a barrier between trustworthy and untrusted networks, monitoring and filtering incoming and outgoing network traffic according to specified security rules. They are an essential component of a secure network architecture, defending the network against unwanted access and many sorts of cyberattacks.

Endpoint Security Given the large number of devices linked to the network in a smart city, each one offers a potential security threat. Therefore, comprehensive endpoint security measures are crucial [17].

Endpoint security is an important component of cybersecurity, particularly in relation to smart cities. Smart cities use ICT, community-wide data, and intelligent solutions to digitally alter infrastructure and optimize government in response to public demands. They integrate operational technology (OT) that manages physical infrastructure to networks and applications that gather and analyze data utilizing ICT components including IoT devices, cloud computing, artificial intelligence (AI), and 5G.

In a smart city, every item linked to the network, from traffic lights and sensors to public utility systems, poses a potential security risk. These devices, also known as endpoints, are appealing targets for hostile cyber actors due to the data gathered, sent, stored, and processed, which can contain large volumes of sensitive information from governments, corporations, and individuals.

Endpoint security entails protecting these devices and ensuring they do not become entry points for security threats. It entails frequent software updates and patches, secure settings, access restrictions, and the use of security applications to prevent, detect, and respond to attacks.

However, the intricacy of smart city systems, which frequently include artificial intelligence-powered software, might bring new risks. As a result, cybersecurity technology, processes, and standards must be included in the portfolio from the start of the project. This involves safeguarding sensors, monitors, and other smart city assets from physical threats such as vandalism or environmental damage, as well as limiting remote access to susceptible equipment.

Cloud Services Cloud services offer scalable and adaptable computer resources. However, they must be appropriately protected using measures like as encryption and access control [18].

A range of on-demand services are offered by cloud providers to consumers and companies over the Internet. They consist of platform as a service (PaaS), infrastructure as a service (IaaS), and software as a service (SaaS). A variety of data storage solutions, including file, blob, disk, and table storage, are offered by well-known cloud service providers like Google Cloud and Microsoft Azure. Additionally, they offer encryption for Azure Data Lake, Azure Cosmos DB, and Azure SQL Database.

When it comes to protecting data in cloud services, there are a few recommended practices to follow. The first step is to select a reputable cloud service provider with safe data storage, encryption, and access controls. The supplier should adhere to appropriate security standards and laws, such as ISO 27001, Health Insurance Portability and Accountability Act (HIPAA), and Payment Card Industry Data Security Standard (PCI DSS).

Understanding your security duties is also important. In most circumstances, the cloud provider is in charge of safeguarding the infrastructure, while the client is responsible for protecting the data housed on that infrastructure. This is sometimes referred to as the shared responsibility model.

Data breaches may be significantly reduced with the use of robust authentication solutions like multifactor authentication. To get access to the cloud environment, users must provide several forms of authentication, such as a password and a code sent to a mobile app.

One essential component of cloud security is encryption. It comprises encoding information so that only individuals with permission may access it. Azure provides many encryption solutions, such as customer-managed keys in Key Vault, customer-managed keys on customer-controlled hardware, and server-side encryption utilizing service-managed keys. It is possible to manage and store keys on-site or in a different safe location using client-side encryption.

Edge Computing Edge computing can assist minimize latency and network congestion by processing data closer to where it was created. However, these edge devices must also be protected [19].

What Is Edge Computing? A distributed computing idea called "edge computing" brings processing and storage closer to data sources. This is intended to reduce bandwidth use and improve response times. It is a form of distributed computing that depends on topology and location and is an architecture rather than a single technology. Content distribution networks, which were set up in the late 1990s to provide web and video content from edge servers close to users, are where edge computing got its start. These networks grew to host applications and application components on edge servers in the early 2000s, leading to the first edge computing products that were sold commercially. One use of edge computing is the Internet of Things. There is a common misunderstanding that edge computing and the Internet of Things (IoT) are interchangeable. By utilizing the growing in-device computing capabilities, the edge computing infrastructure provides near-real-time predictive analysis and deep insights.

Edge computing may provide significant corporate benefits, including quicker insights, faster reaction times, and increased bandwidth availability. Since data is not sent across a network to a cloud or data center for processing, latency is significantly reduced. Faster and more thorough data processing made possible by edge computing leads to deeper insights, shorter reaction times, and improved customer experiences.

Security Considerations Edge computing has numerous advantages, but it also introduces new security issues. Edge devices are subject to physical manipulation, unlawful access, and cyberattacks. Therefore, safeguarding these devices, as well as the data they process and store, is critical. This may entail establishing strong access restrictions, encryption, secure boot protocols, and intrusion detection systems. Additionally, regular software upgrades and patches are required to guard against known vulnerabilities.

Identity and Access Management (IAM) IAM solutions guarantee that only authorized users may access certain data or systems [13].

IAM is an important part of current IT architecture. It is a security discipline that oversees user identities and access rights to various systems and applications within the IT infrastructure. The basic purpose of IAM is to ensure that the appropriate people, machines, and software components have access to the necessary resources at the appropriate moment.

IAM has two components: identity management and access management. Identification management entails creating, preserving, and maintaining identification information. Employee names, job titles, managers, direct reports, mobile phone numbers, and personal email addresses are among the details disclosed. The identity management database maintains a continuous record of everyone who should have access. When a user attempts to log in, their login information, such as their username and password, is validated against this database through a process known as authentication. Access Management, on the other hand, guarantees that each authorized user has the appropriate level and kind of access to a tool. This comprises assigning user roles and rights, as well as implementing authentication and authorization procedures. Access controls govern how users get access to systems and data.

Many companies require users to authenticate their identities using a process known as multifactor authentication (MFA). MFA, also known as two-way verification or two-factor authentication (2FA), provides greater security than using a login and password alone. IAM systems additionally include identity federation, user provisioning and deprovisioning, and the generation of reports for platform activities (such as sign-in time, systems accessed, and authentication type) to verify compliance and identify security concerns.

Data Protection Measures These include encryption, anonymization, and safe data storage [20].

Data protection is an important part of handling sensitive information, and it entails many crucial measures:

Encryption is the process of transforming data into a code that prevents unwanted access. It is an important aspect of data protection since it assures that even if data slips into the wrong hands, it remains unreadable and worthless to the unauthorized person.

*Anonymization***,** also known as pseudonymization, is an essential measure. It entails substituting personally identifiable information with other comparable data. This strategy is highly effective for safeguarding individuals' privacy, especially when working with enormous datasets.

Secure data storage solutions are also important. These include employing secure and accredited data centers to hold essential and sensitive information. Data maintained within an organization should be trustworthy, accurate, and not vulnerable to unauthorized alterations.

Furthermore, data protection methods frequently center on three critical areas: data security, data availability, and access control. Datasecurity is safeguarding data from deliberate or inadvertent harm. Data availability guarantees that data can be restored quickly if it is damaged or lost. Access control guarantees that data is only available to those who need it, not to anybody else.

In addition to these safeguards, companies must follow data protection principles such as lawfulness, fairness, transparency, integrity, and secrecy. These principles serve as the foundation for sound data protection practice and are essential for meeting data privacy requirements and legislation.

Incident Response Plan In the case of a security breach, a well-defined response strategy may assist reduce damage and recovery time [21].

An incident response plan (IRP) is a collection of guidelines for IT professionals to detect, respond to, and recover from network security problems. These strategies may address a variety of security catastrophes, such as data breaches and service disruptions, and help the firm recover.

The first step in developing an IRP is to identify incidents. This includes monitoring systems for odd behavior, examining system logs, and staying on top of any warnings or notifications from intrusion detection systems. Once an incident has been recognized, it is critical to respond swiftly to prevent more harm.

The next phase is incident containment. This might include isolating impacted systems or networks to keep the problem from spreading. It might also entail changing passwords, denying access, or installing extra firewalls or filters.

Following containment, the attention switches to incident elimination. This includes determining the underlying cause of the occurrence and eliminating it from the system. This might include uninstalling harmful code, erasing infected files, or correcting software vulnerabilities.

After the situation has been resolved, the following phase is recuperation. This includes recovering systems or data from backups, testing systems to ensure normal functioning, and keeping an eye out for any signals of aberrant behavior. The objective is to resume regular operations as fast and seamlessly as possible.

The post-event activity, the last stage of an IRP, comprises a thorough analysis of the incident, the effectiveness of the reaction, and any lessons discovered. This might involve a detailed investigation into how the incident happened, what could have been done to stop it, and what can be done to keep similar circumstances from happening in the future.

Regular Audits and Updates Regular security audits can help discover possible vulnerabilities, while regular upgrades safeguard systems from known threats [21].

Regular audits and upgrades are critical components of ensuring a safe digital environment. Regular security audits are methodical and quantitative technical evaluations of a system. Audits can be performed by internal or external personnel, but the purpose remains the same: to uncover any weaknesses that attackers could exploit [22].

During an audit, many parts of the system are evaluated. This involves examining user access limits, analyzing system configurations, and ensuring that the security software is up to date. Audits usually include a detailed assessment of information handling procedures to guarantee that data is safely transported and kept.

The results of these audits give significant insights into an organization's security posture. They identify areas of vulnerability and give a road map for remedial activities. Organizations may keep up with cyber risks by undertaking security audits on a regular basis.

Regular updates, often known as patch management, involve upgrading systems with the most recent security patches and updates. Software providers offer these updates to resolve security vulnerabilities in their products [23].

When a security vulnerability is found, the race starts. Hackers will try to exploit the weakness, while suppliers hurry to create, test, and deliver fixes. Organizations may defend themselves from known risks by upgrading their systems on a regular basis.

However, handling updates isn't always simple. It is necessary to strike a careful balance between preserving security, assuring system compatibility, and reducing downtime. To avoid any interruptions, fixes for essential systems should be evaluated in a controlled environment prior to deployment [24].

11.4 Threats and Challenges of Data in Smart Cities

Smart cities, with their interconnected networks of sensors, devices, and data streams, hold immense promise for improving efficiency, sustainability, and quality of life. However, this data-driven future comes with a hidden cost: an amplified vulnerability to cyberattacks and data breaches. This essay will explore the various threats and challenges associated with data security in smart cities, focusing on key areas like cyber security, internet of things (IoT) security, radio frequency (RF) security, infrastructure security, and data management in critical systems like transportation and energy grids.

One of the most prominent threats is cybersecurity. Smart city systems are complex ecosystems of interconnected devices, each with its own vulnerabilities. Hackers targeting critical infrastructure can gain access to a single device and then exploit its connection to cripple entire systems. A successful cyberattack could disrupt traffic management, manipulate energy grids, or even compromise public safety systems, causing widespread chaos and potential loss of life. The sheer volume and interconnectedness of data in smart cities make them prime targets for cybercriminals [25].

IoT security presents another significant challenge. Smart cities rely heavily on a multitude of IoT devices, from traffic sensors to smart meters. These devices often have weak security protocols, limited processing power, and short lifespans, making them easy targets for hackers. A compromised IoT device can become a gateway into a larger network, allowing malicious actors to steal data, disrupt operations, or launch denial-of-service attacks.

RF security adds another layer of complexity. Smart city systems often utilize wireless communication protocols like Wi-Fi and Bluetooth. These protocols, if not properly secured, can be intercepted by attackers using readily available tools. This can allow them to steal sensitive data, manipulate device readings, or even inject malware into the network.

Infrastructure security is also critical. Physical access to key infrastructure components, like control centers or data servers, can provide attackers with a backdoor into the system. Additionally, outdated or poorly maintained infrastructure can be more susceptible to physical attacks. Robust physical security measures are essential to prevent unauthorized access and safeguard critical data.

Managing device identity and access (IAM) is another crucial aspect of data security in smart cities. With a vast array of devices constantly connecting and disconnecting from the network, it's vital to ensure that only authorized devices have access to the systems and data they need. Implementing strong authentication and authorization protocols can help mitigate the risk of unauthorized access and data breaches [26].

Data security in specific sectors of smart city infrastructure presents unique challenges. Transportation systems that rely on real-time data for traffic management and public safety are particularly vulnerable. Hackers could manipulate traffic flows, disrupt emergency response systems, or even gain control of autonomous vehicles, leading to potential accidents and loss of life. Data in energy grids needs robust protection. A compromised grid could cause widespread blackouts, disrupt vital services, and even create a national security threat. Smart cities also rely on a variety of critical systems, ranging from waste management to water treatment. Breaches in these systems could have severe environmental and public health consequences.

Despite these efforts, smart cities face an increased attack surface, with interconnected devices creating multiple entry points for cyberthreats. Data breaches pose a significant risk, exposing personal and critical infrastructure information to malicious actors, while malware and ransomware infections could disrupt operations and endanger public safety.

Privacy concerns arise from extensive data collection and surveillance in smart cities, prompting the need for transparency, ethical data use, and responsible data governance. Additionally, data integration and interoperability issues, coupled with data accuracy and reliability concerns, necessitate standardized protocols and efficient data management practices to ensure seamless operation and decision-making.

As smart cities continue to evolve, addressing these multifaceted threats and challenges is imperative to unlock their full potential while safeguarding citizen's well-being and preserving privacy. By implementing comprehensive security measures, fostering collaboration, and prioritizing ethical data use, smart cities can navigate these obstacles and thrive in the digital age.

Several key strategies can be employed to address these challenges:

Prioritizing Cybersecurity Implementing robust security protocols, conducting regular penetration testing, and establishing clear incident response plans are essential for mitigating cyberattacks.

Securing the IoT Employing strong authentication and encryption protocols, implementing device patching and updates, and using secure communication channels are crucial for safeguarding IoT devices.

Enhancing RF Security Utilizing encryption for wireless communications, deploying intrusion detection systems, and implementing access control measures can help protect against RF attacks.

Fortifying Infrastructure Security Implementing physical security measures to control access to critical infrastructure, investing in infrastructure upgrades, and conducting regular security audits are essential steps.

Implementing Rigorous IAM Enforcing strong authentication protocols, utilizing role-based access control, and continuously monitoring identities and access rights can help mitigate unauthorized access.

Data Minimization Collecting and storing only the data necessary for specific purposes can help reduce the attack surface and minimize the impact of data breaches.

Data Encryption Encrypting data in transit and at rest can significantly improve its security in the event of a breach.

Transparency and Public Awareness Educating citizens about the importance of data security and empowering them to protect their privacy are crucial elements of a comprehensive security strategy [27, 28, 29].

In conclusion, the journey towards smarter cities presents some landscape rich with promise and potential, yet fraught with complexities and challenges. As we embrace interconnected networks and data-driven systems to enhance urban living, we must remain vigilant in addressing the multifaceted threats to data security. By prioritizing robust cybersecurity measures, fortifying infrastructure, and implementing rigorous data management practices, smart cities can navigate the intricate terrain of digital transformation while safeguarding the well-being and privacy of their citizens. Through collaboration, innovation, and a commitment to ethical data use, we can realize the full benefits of smart city initiatives and shape a future where technology serves as a catalyst for progress and prosperity.

11.5 Incident Response and Crisis Management

The Internet of Things (IoT) and Industry 4.0 are driving the rapid rise of smart cities around the globe. These cities are being further enhanced using cutting-edge, developing technologies, which in turn are creating very complex and fragile physical, natural, and cyber ecosystems [30]. Effective incident response and crisis management in smart city environments require proactive planning, clear communication, and collaboration among stakeholders to safeguard critical infrastructure and protect citizen data.

It's critical to be prepared for cyberthreats in a smart city because all infrastructure is interconnected via technology. Consider the possibility of a data theft or hacking attempt against the city's systems. A strategy is required for when this occurs. The plan should clearly say who does what. When it comes to resolving technological issues, such as identifying and thwarting hacks and restoring normalcy, the IT security team takes the lead. Big choices are made by local authorities, such as the mayor or city manager, who also keep the public updated on events. To ensure that everyone is protected and to apprehend the perpetrators, the police may become engaged [31].

It is crucial to keep people informed during all of this. A communication professional provides assistance by ensuring that all parties are aware of the situation without inciting fear. They communicate with the media and the public to ensure that everyone is informed at the appropriate moment. Below is an outline of procedures and strategies for handling the cybersecurity incidents and emergencies in smart city environments:

11.5.1 Procedures and Strategies

Authorities in smart cities are required to follow a predetermined set of procedures and strategies in the case of a cybersecurity incident or data breach. These are the common procedures that need to be followed these procedures can be changed as per the level of crisis management. There are,

1. Detection: To quickly discover any security issues, make use of anomaly detection algorithms and sophisticated monitoring technologies.
2. Assessment: Conduct a comprehensive assessment of the incident to determine the scope, severity, and potential impact on critical systems and services.
3. Containment: Implement measures to contain the spread of the incident and prevent further damage to infrastructure and data.
4. Eradication: Take proactive steps to eliminate the root cause of the incident and restore affected systems to a secure state.
5. Recovery: Execute recovery plans to restore normal operations and minimize downtime, including data restoration and system reconfiguration.

Roles and duties among stakeholders must be clearly defined in order to provide an efficient incident response within a smart city framework. Administrators of smart cities are essential in supervising all reaction activities, guaranteeing a coordinated strategy and resource distribution as required. They serve as everyone's primary point of contact, promoting dialogue, and the making of decisions between all parties.

Another essential element is IT staff, who are responsible for looking into events, putting technical remedies in place to lessen their effects, and eventually getting damaged systems back to normal operation. Their knowledge is essential for quickly determining the problem's underlying cause and putting precautions in place to stop it from happening again.

Law enforcement organizations contribute their investigative expertise to support forensic analyses, evidence gathering, and the hunt for and capture of offenders. Their participation is crucial in situations where there may have been illegal behavior because it guarantees that those responsible will be held legally accountable [32].

In addition, incident response in smart cities is greatly influenced by commercial sector partners. They share threat intelligence and cybersecurity measure knowledge, among other critical insights and resources, through their engagement with government entities. Together, the public and private sectors may effectively manage and mitigate emerging hazards through the collaborative efforts of this partnership.

Fundamentally, the ability of these disparate parties to work together and share resources and specialized knowledge in order to protect the city's digital infrastructure and people's safety is what makes incident response in a smart city so successful.

The authorities of smart cities understand the criticality of close cooperation between law enforcement and intelligence services in order to efficiently investigate cybercrimes and capture cybercriminals. Creating explicit guidelines for the exchange of information and managing evidence guarantees compliance with privacy and regulatory requirements. Simultaneously, public engagement is given priority, and open channels of communication—such as social media, official websites, and mobile apps—are employed to promptly disseminate instructions and updates during emergency situations. This builds community awareness and public trust [31].

11.5.2 Public Awareness and Education

Smart cities may develop more cyber-aware citizens who can navigate the complexity of the digital world while protecting their personal data and privacy by giving public awareness and education efforts first priority. One of the most important lines of defense for smart city data protection is an informed populace. Public awareness of cybersecurity threats enables people to see unusual conduct and alert authorities to it. This includes being aware of malware signs, phishing scams, the value of two-factor authentication, and strong passwords.

Campaigns for public awareness that make use of social and traditional media channels can increase knowledge of these problems. It is possible to create age-appropriate educational programs in collaboration with local organizations and schools. Incorporating cybersecurity education into current curricula allows schools to guarantee early digital literacy. Citizen workshops and training programs can offer useful advice on effective practices for data protection.

The objective goes beyond merely increasing public awareness to include enabling residents to take an active role in protecting data in smart cities. Building a secure smart city environment requires promoting a culture of responsible data sharing and online conduct. Smart cities may establish trust and guarantee the appropriate use of data in their communities by placing a high priority on public education and citizen empowerment. Initiatives to raise public awareness and educate the public can help address data privacy issues in smart city situations.

11.5.3 Importance of Awareness

The hazards of living in a smart city, where networks of connected devices and systems gather enormous volumes of personal data, must be understood by the populace. Educating people about privacy issues and cybersecurity risks enables them to make well-informed choices regarding their online behavior and data-sharing habits.

11.5.4 Privacy Rights and Data Security Best Practices

To reduce the danger of data breaches and unauthorized access, it is crucial to educate the public on their rights regarding privacy and best practices for data security. People are empowered to safeguard their personal data when they receive advice on how to make strong passwords, enable two-factor authentication, and use caution when sharing sensitive information online.

11.5.5 Initiatives for Promoting Digital Literacy

Initiatives aimed at promoting digital literacy are essential in providing citizens with the information and abilities they need to securely traverse the digital world. To increase awareness of digital literacy and cybersecurity, government organizations, academic institutions, and industry partners can work together to provide training courses, online resources, and seminars.

11.5.6 Citizen Empowerment

The cornerstone of creating a robust smart city ecosystem is enabling individuals to actively participate in their online safety. People may defend their digital identities and privacy rights by being empowered to take part in cybersecurity awareness initiatives, stay up to date on developing cyberrisks, and lobby for improved data protection laws.

11.6 Case Studies and Lessons Learned in Smart City Security Vulnerabilities and Mitigation Strategies

It is necessary to closely examine real-world situations where vulnerabilities have been exploited and the effects have spread throughout urban landscapes in order to comprehend the complexity of security risks in smart cities. When we consider the data protection challenges in the smart cities, recent case studies provide an overview of the complex issues raised by the data protection risks in relation to the development of smart cities. These case studies function as warning tales, shedding light on the possible dangers associated with the combination of urban infrastructure and technology. They also offer insightful advice and lessons on how to improve resilience and minimize potential threats in the future.

Every case study that is provided here illuminates a different aspect of the smart city security environment, from social engineering and IoT vulnerabilities to infrastructure attacks and data breaches. We seek to the fundamental dynamics behind these security incidents and investigate the response strategies used by impacted cities through in-depth investigation. By analyzing these real-world instances, we intend to identify key lessons that might guide future security plans and raise public, practitioner, and policymaker awareness. The dynamic nature of cybersecurity threats and the ever-expanding technical landscape were considered in the selection of current cases from various locations in the world related to the data protection challenges in smart cities.

11.6.1 Case 01: Florida Water Treatment Plant Hack (2021)

One of the best examples of a municipality utilizing smart city technologies is the US state of Florida. Florida has established a complex network of Internet of Things (IoT) platforms, cyber-physical systems (CPS), and smart connected devices across multiple domains by putting into practice a complete smart city engineering framework [33].

The IoT network in Florida, which is based on fiber optics and wireless connectivity among other high-speed, resilient communications infrastructure, is one of the main pillars of the state's smart city. This network allows for smooth data transfer and communication between smart city systems and devices across key thoroughfares, buildings, urban districts, universities, and residential neighborhoods [33].

Florida has used Internet of Things (IoT) technologies to monitor and analyze environmental data, including water quality, in real time when it comes to water management. The primary water canals in the city are equipped with buoy sensors that gather information on salinity, resistivity, conductivity, temperature, depth, and other pertinent parameters. In addition to helping marine biologists and water scientists monitor the health of waterways, this data also helps emergency managers and city engineers make well-informed decisions on the detection and mitigation of flooding [33].

In February 2021, a malicious actor targeted a water treatment plant in Florida, the cybercriminal took advantage of a dormant remote access software platform that hadn't been touched in six months. This flaw allowed the hacker to get into the plant's human-machine interface (HMI) system, which manages the treatment procedure, without authorization [34].

The hacker raised the concentrations of Sodium Hydroxide in the water to possibly hazardous levels using the compromised system, posing a serious risk to public health. Luckily, an attentive operator saw the illegal behavior and acted quickly to restore the chemical levels to normal, protecting the water supply from any negative impacts. The incident demonstrated, in spite of the quick response, how susceptible vital infrastructure in smart cities is to cyberattacks, and how such breaches may cause bodily harm [34].

Local authorities and cybersecurity specialists acted quickly to safeguard the water treatment system and minimize the threat after the attack was discovered. The public was not harmed because of the water treatment plant operator's alertness and prompt action and it is important to highlight the need of staff education and knowledge of cybersecurity procedures. Law enforcement agencies, government representatives, and cybersecurity companies worked together to investigate and analyze the attack vector, which made it possible to find vulnerabilities and openings in the system's defenses [34].

11.6.1.1 Lessons Learned from the Florida Water Treatment Plant Hack

The Florida water treatment facility hack exposed critical vulnerabilities in smart city security. Outdated software, ineffective remote access controls, and a lack of staff understanding contributed to this problem. Nevertheless, this case study also highlighted the importance of security audits, strong access controls, thorough staff training, and cooperative efforts amongst stakeholders. Smart cities may strengthen their defenses against cyberattacks and protect vital infrastructure by putting these precautions into place.

11.6.1.2 Integration of IoT in Critical Infrastructure

The Florida Water Treatment Plant Hack serves as an excellent example of how data protection issues and smart city infrastructure meet. Florida has created a complex network of cyber-physical systems (CPS) and Internet of Things (IoT) platforms to monitor and control vital services like water treatment within the framework of smart cities. Nonetheless, the hack revealed weaknesses in this system, emphasizing how crucial data security protocols are. The event emphasizes how important it is to have strong cybersecurity procedures in place to protect vital infrastructure from illegal access and guarantee the security and integrity of the data required to run urban services.

11.6.2 Case 02: Estonia Ransomware Attack (2021)

Estonia, a country known for its progressive digital governance and innovation, stands as a smart city development in the Baltic region. Estonia's capital, Tallinn, embodies the country's dedication to using technology to improve sustainability and urban life, even if it is not limited to just one city. Tallinn's journey toward becoming a smart city is notable for its focus on interoperability, accessibility, and user-friendliness all of which are vital pillars of the city's initiatives [35].

Estonia's smart city journey lies in a synchronized approach to urban and digital growth, aligning the city's development with the broader progress of its digital society. They have incorporated digital services into everyday life and meeting the demands of residents, visitors, and companies. Estonia has made use of shared infrastructures for authentication and data exchange, demonstrating resource optimization and efficiency in its digital ecosystem [35].

A notable cybersecurity threat case that occurred in Estonia in 2021 highlighted the susceptibility of its digital infrastructure to cyber assaults. An attacker coordinated this assault with an address in Estonia and targeted the Information System Authority (RIA), the government agency in charge of overseeing the nation's IT infrastructures and cybersecurity protocols. By taking advantage of a security flaw in the RIA's systems, the attacker was able to obtain about 300,000 picture documents from the documents database without authorization [36].

The responsible person was caught within a few days because different groups like the CERT-EE, the prosecutor's office, and the police worked together quickly. They also took the data that was taken. Even though they solved the problem, it reminded everyone how fast cyberthreats can change and why it's crucial to have good security to protect important government systems [37].

After the incident, Estonia's government has taken steps to strengthen its computer systems and improve their capacity to manage future crises in the wake of the ransomware outbreak and other cyber challenges. They set up systems to monitor for infections and automatically detect security vulnerabilities in order to identify and address issues more quickly. Also, they understand it's important to work with cybersecurity experts and good hackers. They have observed the creation of a plan to encourage these experts to find and report any weak spots in their systems. By working together and rewarding people who report problems, Estonia hopes to find and fix security issues before bad people can use them.

11.6.2.1 Lessons Learned from the Estonia Ransomware Attack

It's critical to identify and address cyberattacks promptly. Having systems that can monitor issues automatically and react quickly can be quite beneficial. The government, law enforcement, and cybersecurity professionals can all respond more quickly and effectively to cybercrime when they collaborate and exchange information.

Incentives for individuals who identify and report vulnerabilities in our computer systems before hackers can take advantage of them are also a good idea. In this manner, we are able to address problems before they escalate.

Overall, the Estonia ransomware attack shows us how important it is to stay alert against cyberthreats. By learning from what happened and putting in place ways to stay safe online, we can better protect our digital stuff and lower the chances of future attacks.

11.6.2.2 Digital Governance and Cybersecurity Resilience

The Information System Authority (RIA) in Estonia was the target of a ransomware attack, which highlights how vulnerable smart city digital infrastructure is to online attacks. When its systems were breached, Estonia a nation renowned for its forward-thinking digital government and smart city projects faced serious data protection issues. The incident demonstrated how crucial it is for government organizations, law enforcement, and cybersecurity experts to work together quickly in order to properly combat such risks. It also emphasized the necessity of proactive steps and ongoing monitoring to increase the resilience of digital infrastructure against changing cyberthreats.

11.6.3 Case Study 3: Tesla Car Hacked Remotely (2020)

Smart City planners are developing a more sustainable and effective transportation system, improving air quality, lessening traffic congestion, and increasing energy efficiency by incorporating electric vehicles (EVs) into smart cities. Using electric vehicles (EVs) instead of conventional cars can reduce pollution and improve the quality of the air in urban areas. EVs mostly run on electricity generated by solar and wind power, which is more environmentally friendly than the petrol and diesel used in conventional automobiles [37]. The integration of EVs into smart cities offers numerous benefits. Cities' carbon footprints are lowered by EVs. The International Energy Agency (IEA) estimates that transport contributes 24% of the world's CO2 emissions [38].

Two researchers have shown an automobile cybersecurity vulnerability in which they showed how a Tesla vehicle may be remotely compromised without requiring any user input. The ConnMan is a commonly used component in the automobile industry, the revelation raised concerns extending beyond Tesla vehicles, exposing vulnerabilities in ConnMan, an internet connection manager utilized in embedded devices like Tesla cars. This investigation was first carried out in the hacking competition, which offered substantial rewards for breaking into a Tesla car [39].

This attack leverages two vulnerabilities enabling attackers to gain full control of a Tesla's infotainment system remotely. Although the exploit does not provide direct control over the vehicle's driving capabilities, it does enable attackers to control a number of infotainment system features, including the ability to open doors, adjust seats, adjust the air conditioning, play music, and change the modes of steering and acceleration. Hackers can enter parked vehicles from up to 100 meters away by using a drone to carry out the attack over Wi-Fi.

Tesla quickly released a software upgrade in October 2020 to fix the vulnerabilities in response to this assault. The business allegedly stopped utilizing ConnMan in its cars as well, which reduced the danger associated with the vulnerabilities even further. Although the original ConnMan developer, Intel, was informed of the findings, the chipmaker stated that it was not in charge of fixing the flaws [39].

Two researchers who showcased this collaborated with Germany's national CERT to inform potentially impacted vendors in the automotive industry. It's yet unknown, though, if other manufacturers have acted on the researchers' conclusions. The discovery emphasizes the significance of continual awareness and aggressive measures to address evolving cybersecurity vulnerabilities in automotive systems, even with Tesla's quick mitigation efforts [39].

11.6.3.1 Lessons Learned from the Tesla Car Hacked

Continuous vulnerability assessment programs are essential for identifying and addressing vulnerabilities in automotive systems in smart cities proactively. Collaboration between researchers, cybersecurity experts, and industry stakeholders facilitates the timely disclosure and mitigation of cybesecurity vulnerabilities, enhancing overall resilience.

11.6.3.2 Cybersecurity in Smart Transportation Systems

The weaknesses in incorporating cutting-edge technologies like electric cars (EVs) into smart metropolitan transport networks are highlighted by the remote hack of a Tesla vehicle. Because of their connectivity and dependence on software systems, electric vehicles (EVs) present data protection challenges in addition to their many benefits for sustainability and urban mobility. In order to address cybersecurity concerns in smart city transportation, the incident emphasizes the significance of ongoing vulnerability evaluation and coordination among researchers, cybersecurity specialists, and industry players. Cities can take advantage of EVs while maintaining the security and privacy of data used to manage transportation infrastructure by tackling these issues.

11.7 Legal and Regulatory Frameworks

There is an increasing need to develop legal ways for collecting personal data as the smart city strategy gains traction to ensure data quality, privacy, security, management, etc. [8, 40–42]. Several legal and regularizing attempts have been made to standardize smart city data and assure privacy and data security. These regulations highlight the legal obligations surrounding data security and the consequences of failing to meet them [40]. Data governance frameworks also correlate with regulations [41]. The former provides a methodical and planned strategy to guarantee adherence, minimizing the monetary and legal ramifications of regulatory transgressions. Standardization improves an organization's capacity to manage the complicated regulatory landscape while also making compliance efforts simpler [8].

One such global approach introduced by the European Union (EU) to get more control over personal data is the General Data Protection Regulation (GDPR) [40]. Its primary objective is to standardize the legal framework across European countries. GDPR has prompted nongovernmental organizations (NGOs) to suggest that individuals ought to be aware that their data is being collected and that they have the ability to access and delete it. Additionally, they believe that all stakeholders of smart cities should be carried out in the public interest rather than the interests of the companies that supply cities with technological infrastructure. GDPR noncompliance may lead to significant fines for breaching. For example, GDPR allows for fines of up to €20 million or 4% of the annual global turnover, whichever is higher (GDPR.eu, 2020).

Similarly, the United States has introduced the "Consumers Bill of Rights" based on the "Fair Information Practice Principle (FIPP)." [40] It discloses the respect for context principle and individual control principle. In the former principle, the customers have the right to demand that businesses acquire, utilize, and disclose personal information in a way that is consistent with the circumstances surrounding the data's submission by the customers. In the latter principle, customers have the right to decide what personal information businesses collect about them and how they use it. Evolving with these legislations, the United States has imposed the "Smart Cities and Communities Act" in 2017 with the intention of innovative, cutting-edge and reliable energy, communication, and information technologies that are used to enhance people's health and quality of life, boost productivity, foster economic growth, enhance sustainability, safety, security, livability, and work-life balance. In addition, the United States introduced the "Internet of Things Cyber Security

Improvement Act" of 2017 creating the need that devices connected to the Internet to have a sufficient level of security. It places restrictions and obligations on contractors and developers that could expose them to risks.

Countries that have declared smart cities have imposed similar types of laws and regulations, which are more or less complementary with each other. For example, Mexico has introduced articles in their constitution on managing personal data lawfully [40]. Furthermore, the International Electrotechnical Commission (IEC) and the International Organization for Standardization (ISO) have introduced the "Privacy Framework" under ISO/IEC29100 [42]. In addition, the municipal and pal governing authorities of Switzerland have produced the "e-Governance Strategy Switzerland," which is primarily concerned with privacy laws in smart towns [41].

11.8 Future Trends and Challenges in Data-Driven Smart Cities

The area of data security and protection in smart cities is constantly changing, with new trends, technologies, and difficulties arising. This essay investigates some of these advances, with an emphasis on quantum computing risks, artificial intelligence (AI)-driven cyberattacks, The ethical concerns of smart city surveillance technologies, and the creation of ethical norms for the responsible application of developing technologies.

11.8.1 Quantum Computing Threats

Quantum computing provides a tremendous increase in processing capacity, but it also introduces new vulnerabilities to data security [43–46]. Quantum computing is a fast-expanding discipline that applies quantum physics concepts to computation. Quantum computers use quantum bits, or qubits, which may exist in several states at the same time due to a phenomenon known as superposition. This allows quantum computers to evaluate a huge number of possibilities simultaneously, resulting in a significant gain in computing capabilities.

Quantum computers have the potential to break many of the cryptographic methods in use today, rendering them worthless [43–46]. This is because quantum computers can execute algorithms capable of breaking the current public key encryption system. Public key encryption is an essential component of data security, safeguarding information as it travels across the internet. If quantum computers become widely available, they can decode sensitive data, exposing smart cities to new sorts of cyberattacks because they rely primarily on encryption for data security [43].

Despite these possible risks, the widespread availability of quantum computers is not likely [47]. The science of quantum computing is still in its infancy, with considerable technical obstacles to solve before large-scale quantum computers become a reality. These problems include regulating individual qubits at scale, managing cooling power and environmental management, and tackling manufacturability issues. As a result, while it is critical to begin planning for the quantum future, the immediate danger to data security is not as pressing.

11.8.2 Artificial Intelligence-Driven Cyberattacks

AI is rapidly being utilized to improve cyberattacks [48–51]. These AI-powered attacks use machine learning to analyze a human or machine target and discover the most likely ways to hack an organization. AI may be used, for instance, to automate the process of finding vulnerabilities in systems, thereby making it simpler for attackers to take advantage of them. This may be sending an email based on the social media profiles of your employees or utilizing bits of data to identify a target system's most likely weaknesses and initiate an attack.

Such attacks could be more intricate and challenging to identify than typical cyberattacks [48–51]. AI might be used, for instance, to automate the process of finding weaknesses in systems, which would make it simpler for hackers to take advantage of them [48–51]. Through deepfakes and AI-powered voice synthesis, fraudsters may quickly build ransomware, automate attacks, and improve the effectiveness of scams or social engineering attempts.

This is a huge difficulty for smart cities, since they frequently rely on complex networks of interconnected devices and systems, any of which might potentially be a source of vulnerability [48–51]. IoT-enabled sensor technologies and an ICT-based infrastructure are the foundation of smart cities, which are designed to foster social and urban interconnectedness through improved public participation and governmental efficiency. Designers, integrators, and organizations responsible for overseeing these new entities face significant political, technological, and financial challenges due to the complex and interconnected structure of smart cities. Studies on security, privacy, and risks in smart cities are becoming more prevalent; they highlight information security dangers and problems with smart city infrastructure related to personal data management.

11.8.3 Ethical Implications of Smart City Surveillance Technologies

Smart city technologies frequently include some kind of surveillance, which might pose ethical problems [52–55]. These technologies can collect massive amounts of data on individuals, possibly intruding on their privacy. For example, smart cities monitor their residents by placing sensors strategically across the city to gather information on many facets of urban life. Governments and other local authorities use these sensors to send, collect, and analyze data to get insight into the city's challenges related to garbage reduction, energy consumption, traffic management, and crime prevention.

There are also questions concerning how this information is utilized and who has access to it [52–55]. The widespread and constant flow of data from several sources into one government organization has raised concerns that these systems might develop into "electronic panopticons," where governments employ data-driven technology to optimize efficient population surveillance. Since information is transferred vertically between citizens and the government on a scale that calls into question the concept of urban anonymity, privacy concerns are the root of this argument.

Balancing the advantages of smart city technology with the requirement to safeguard individual privacy is a significant challenge [52–55]. To achieve these benefits while maintaining public confidence, towns and communities must strike a balance between innovation and privacy. This includes adhering to "privacy by design" principles, performing "at-the-edge" anonymization, and putting in place organizational and policy controls. By extending the uses of video surveillance beyond security with intelligent analytics, video content analytics technology converts video into a cross-functional resource that can be shared between enterprises and local governments for a variety of purposes.

11.8.4 Developing Ethical Guidelines for the Responsible Use of Emerging Technologies

Given these problems, there is a rising understanding of the importance of ethical principles for the responsible use of developing technologies [56–59]. These principles can assist in guaranteeing that the advantages of these technologies are achieved while limiting any negative consequences [56–59]. They can provide a framework for decision-making, assisting stakeholders in navigating the complex ethical concerns raised by new technologies [56–59]. The area of data security and protection in smart cities is encountering several new trends, technologies, and issues. Meeting these difficulties will need a combination of technological innovation, ethical thought, and regulatory control. By doing so, we can guarantee that smart cities continue to benefit their citizens while simultaneously preserving their safety and privacy.

11.9 Conclusion

The development of smart cities heralds a time when technology and urban surroundings are seamlessly combined to improve productivity, sustainability, and quality of life. However, there are a lot of obstacles in the way of realizing this goal, especially when it comes to protecting the security, privacy, and integrity of data.

We have explored the complex terrain of data protection issues in smart cities throughout this chapter. We have examined the challenges involved in guaranteeing appropriate data stewardship, from the gathering and examination of varied datasets to the governance of data. We have looked at the crucial roles that ethical issues, privacy-enhancing technology, and efficient data governance systems play in tackling these problems and striking a balance between the rights of the public and individual citizens.

Furthermore, in order to safeguard smart city systems from cyberattacks and vulnerabilities, strong are essential. This conversation has also expanded to the area of security architecture and infrastructure. We have emphasized the significance of incident response procedures and proactive cybersecurity measures in order to reduce risks and improve resistance to changing threats.

In addition, we have underlined how important stakeholder cooperation, public awareness, and education are to boosting cyber resilience and building confidence in smart city projects. We can build a more transparent and safer urban environment by encouraging digital literacy and giving citizens the tools they need to protect their personal information.

Future developments and technologies like artificial intelligence and quantum computing provide both possibilities and difficulties for data security and protection in smart cities. Establishing moral principles and legal frameworks that protect privacy and encourage responsible innovation in tandem with changing legal requirements is crucial.

To sum up, managing the always-changing data protection issues in smart cities calls for an all-encompassing strategy that considers ethical issues, technological innovation, legal compliance, and public involvement. We can fully utilize smart cities while defending residents' rights and interests in the digital era if we successfully handle these issues.

References

1. Deloitte Insights. (n.d.). *Forces of change: Smart cities*. Retrieved April 15, 2024, from https://www2.deloitte.com/us/en/insights/focus/smart-city/overview.html
2. Khan, M. A., Algarni, F., & Quasim, M. T. (Eds.). (2021). *Smart cities: A data analytics perspective*. Springer International Publishing. https://doi.org/10.1007/978-3-030-60922-1
3. OECD iLibrary. (n.d.). *Smart city data governance – e57ce301-en.pdf*. Retrieved April 15, 2024, from https://www.oecd-ilibrary.org/deliver/e57ce301-en.pdf?itemId=/content/publication/e57ce301-en&mimeType=pdf
4. OECD iLibrary. (n.d.). *Governing data in smart cities—A review of experiences | Smart city data governance: Challenges and the way forward*. Retrieved April 15, 2024, from https://www.oecd-ilibrary.org/sites/028eef3e-en/index.html?itemId=/content/component/028eef3e-en
5. Sargiotis, D. (2024). *Data governance in the digital age: Strategies, challenges, and best practices*. In Proceedings of the National Technical University of Athens
6. Liu, X., Heller, A., & Nielsen, P. S. (2017). CITIES data: A smart city data management framework. *Knowledge and Information Systems, 53*(3), 699–722. https://doi.org/10.1007/s10115-017-1051-3

7. Dalhi, A., Belghiti, I. D., & Guerouate, F. (2022). Big data management and smart city: From affordance technology perspective view size. *ITM Web of Conferences, 46*, 02002. https://doi.org/10.1051/itmconf/20224602002
8. Fabrègue, B. F. G., & Bogoni, A. (2023). Privacy and security concerns in the Smart City. *Smart Cities, 6*(1), 586–613. https://doi.org/10.3390/smartcities6010027
9. Choenni, S., Bargh, M. S., Busker, T., & Netten, N. (2022). Data governance in smart cities: Challenges and solution directions. *Journal of Smart Cities and Society, 1*(1), 31–51. https://doi.org/10.3233/scs-210119
10. Tene, O. (2010). Privacy: The new generations. *International Data Privacy Law, 1*(1), 15–27. https://doi.org/10.1093/idpl/ipq003
11. Osu, T., & Navarra, D. (2022). Development of A data governance framework for SMART cities. *The International Archives of the Photogrammetry, Remote Sensing and Spatial Information Sciences, XLVIII-4/W52022*, 129–136. https://doi.org/10.5194/isprs-archives-xlviii-4-w5-2022-129-2022
12. Tragos, E., Fragkiadakis, A., Angelakis, V., & Pöhls, H. C. (2016). Designing secure IoT architectures for Smart City applications. In *Designing, developing, and facilitating smart cities* (pp. 63–87). Springer. https://doi.org/10.1007/978-3-319-44924-1_5
13. Amir, M., Akhtar, K., Amit, M., & Upadhyay, K. (n.d.). *Architecture of security and data protection in smart cities. 18(6), 2021*. Retrieved April 13, 2024, from https://www.webology.org/data-cms/articles/20220213091114pmwebology%2018%20%286%29%20-%2076%20pdf.pdf
14. Brandon, Shopp. (2023). *Smart cities depend on edge computing, but securing it remains a challenge*. https://statetechmagazine.com/article/2023/05/smart-cities-depend-edge-computing-securing-it-remains-challenge
15. Jawhar, I., Mohamed, N., & Al-Jaroodi, J. (2018). Networking architectures and protocols for smart city systems. *Journal of Internet Services and Applications, 9*(1). https://doi.org/10.1186/s13174-018-0097-0
16. Cisco. (2019). *What is network segmentation? Cisco*. https://www.cisco.com/c/en/us/products/security/what-is-network-segmentation.html
17. Smart City Security & Solutions in Cyber Security Challenges in Smart Cities Safety Security and Privacy. (n.d.). *IEEE climate change*. https://climate-change.ieee.org/news/smart-city-security/
18. Phil, Goldstein. (2019). *How a secure cloud architecture can help smart cities*. Technology Solutions That Drive Government. https://statetechmagazine.com/article/2019/08/how-secure-cloud-architecture-can-help-smart-cities
19. Martin, Percival (2022, April 29). *Smart cities need edge computing*. TechRadar. https://www.techradar.com/features/smart-cities-need-edge-computing
20. Ismagilova, E., Hughes, L., Rana, N. P., & Dwivedi, Y. K. (2020). Security, privacy and risks within smart cities: Literature review and development of a smart city interaction framework. *Information Systems Frontiers, 24*(2), 393–414. https://doi.org/10.1007/s10796-020-10044-1
21. Cybersecurity Best Practices for Smart Cities. (n.d.). https://www.cisa.gov/sites/default/files/2023-04/cybersecurity-best-practices-for-smart-cities_508.pdf
22. Bhattacharya, S., & Sengupta, S. (2020). Security, privacy and risks within smart cities: Literature review and development of a smart city interaction framework. *Information Systems Frontiers, 22*(6), 1505–1522.
23. Alharbi, A., Almogren, A., & Alhazmi, O. (2020). *Opportunities and challenges of data-driven cybersecurity for smart cities*. 2020 IEEE International Conference on Big Data (Big Data) (pp. 3661-3668).
24. Kumar, A., & Kumar, A. (2021). The impact and challenges of big data applications in smart city management. *International Journal of Advanced Research in Computer Science, 12*(1), 1–6.
25. Kumar, A., & Kumar, A. (2021). IoT and big data applications in smart cities: Recent advances, challenges, and critical issues. *International Journal of Computer Applications, 179*(30), 1–5.

26. OECD. (2023). *Smart City data governance: Challenges and the way forward* (OECD urban studies). OECD Publishing.
27. ResearchGate. (n.d.). *Smart cities: Opportunities, challenges, and security threats*. Retrieved from https://www.researchgate.net/publication/342328737_Smart_Cities_Opportunities_Challenges_and_Security_Threats
28. Smart Cities World. (n.d.). *Infrastructure intelligence in one place*. Available at https://smartcitiesworld.net/AcuCustom/Sitename/DAM/012/Understanding_the_Challenges_and_Opportunities_of_Smart_Citi.pdf
29. Freya. (2023, October 16). *Smart cities: Challenges & security threats*. Government Business. Retrieved from https://www.governmentbusiness.co.uk/features/smart-cities-challenges-security-threats
30. Ahmadi-Assalemi, G., Al-Khateeb, H., Epiphaniou, G., & Maple, C. (2020). Cyber resilience and incident response in smart cities: A systematic literature review. *Smart Cities, 3*(3), 894–927. https://doi.org/10.3390/smartcities3030046
31. NIST. (2023). *Cybersecurity framework*. National Institute of Standards and Technology. https://www.nist.gov/cyberframework
32. Cybersecurity & Infrastructure Security Agency. (2020). Homepage | CISA. Cisa.gov. https://www.cisa.gov/
33. Rodulfo, R. (2020). Smart City case study: City of Coral Gables leverages the internet of things to improve quality of life. *IEEE Internet of Things Magazine, 3*(2), 74–81. https://doi.org/10.1109/iotm.0001.2000023
34. CNN, Marquardt, A., Levenson, E., & Tal, A. (2021, February 10). *Florida water treatment facility hack used a dormant remote access software, sheriff says*. CNN. https://edition.cnn.com/2021/02/10/us/florida-water-poison-cyber/index.html
35. Portal, L. (2023, September 22). *Tallin smart city: Estonia's flagship of innovation*. https://www.blue-europe.eu/analysis-en/short-analysis/tallin-smart-city-estonias-flagship-of-innovation/
36. Estonia and the world: cyber security 2021 in review—e-Estonia. (n.d.). E-Estonia.com. https://e-estonia.com/estonia-and-the-world-cyber-security-2021-in-review/
37. Apata, O., Bokoro, P. N., & Sharma, G. (2023). The risks and challenges of electric vehicle integration into smart cities. *Energies, 16*(14), 5274–5274. https://doi.org/10.3390/en16145274
38. Transport, Energy and CO2—Analysis. (n.d.). *IEA*. https://www.iea.org/reports/transport-energy-and-co2
39. Kovacs, E. (2021, May 3). *Tesla Car Hacked Remotely From Drone via Zero-Click Exploit*. SecurityWeek. https://www.securityweek.com/tesla-car-hacked-remotely-drone-zero-click-exploit/
40. Niebla Zatarain, J. M. (2019). Smart cities and personal data: Balancing innovation, technology and the law. *Revista Direito, Estado e Sociedade, 54*. https://doi.org/10.17808/des.54.1324
41. Tikhaleva, E. Y. (2023). 'Smart cities': Legal regulation and potential of development. *Journal of Digital Technologies and Law, 1*(3), 803–824. https://doi.org/10.21202/jdtl.2023.35
42. Europa.eu. (2015). *Privacy and data protection by design*. Retrieved from https://www.enisa.europa.eu/publications/privacy-and-data-protection-by-design
43. Tackling Smart City Problems with Quantum Computing | School of Computing and Information Systems. (n.d.). Computing.smu.edu.sg. https://computing.smu.edu.sg/news/2022/oct/04/tackling-smart-city-problems-quantum-computing
44. Organizations must protect against quantum threats. Here's how. (n.d.). World Economic Forum. https://www.weforum.org/agenda/2022/09/organizations-protect-quantum-computing-threat-cybersecurity/
45. News, I. (2020, May 22). App security quantum cryptography smart cities 5G investment industry leaders highlight priorities – ITU. ITU. https://www.itu.int/hub/2020/05/app-security-quantum-cryptography-smart-cities-5g-investment-industry-leaders-highlight-priorities/
46. Prisco, J. (n.d.). *Council post: The quantum computing threat: The road to quantum insecurity*. Forbes. Retrieved May 15, 2024, from https://www.forbes.com/sites/forbestechcouncil/2022/12/14/the-quantum-computing-threat-the-road-to-quantum-insecurity

47. Quantum Computing's Cyber-Threat to National Security. (n.d.). National Defense University Press. Retrieved May 15, 2024, from https://ndupress.ndu.edu/Media/News/News-Article-View/Article/2383102/quantum-computings-cyber-threat-to-national-security
48. Neshenko, N., Nader, C., Bou-Harb, E., & Furht, B. (2020). A survey of methods supporting cyber situational awareness in the context of smart cities. *Journal of Big Data, 7*(1). https://doi.org/10.1186/s40537-020-00363-0
49. Prabakar, D., Sundarrajan, M., Manikandan, R., Jhanjhi, N. Z., Masud, M., & Alqhatani, A. (2023). Energy analysis-based cyber attack detection by IoT with artificial intelligence in a sustainable Smart City. *Sustainability, 15*(7), 6031–6031. https://doi.org/10.3390/su15076031
50. Stanley, D. (2021, June 1). *What are the cybersecurity risks for smart cities?* Institute for Defense & Business. https://www.idb.org/what-are-the-cybersecurity-risks-for-smart-cities/
51. The Cybersecurity Risks of Smart City Technologies: What Do The Experts Think? (n.d.). *CLTC*. https://cltc.berkeley.edu/publication/smart-cities/
52. Ziosi, M., Hewitt, B., Juneja, P., Taddeo, M., & Floridi, L. (2022). Smart cities: Reviewing the debate about their ethical implications. *AI & SOCIETY*. https://doi.org/10.1007/s00146-022-01558-0
53. Ziosi, M., Hewitt, B., Juneja, P., Taddeo, M., & Floridi, L. (2022, January 5). *Smart cities: Reviewing the debate about their ethical implications*. Papers.ssrn.com. https://papers.ssrn.com/sol3/papers.cfm?abstract_id=4001761.
54. Moore, A. (2023, December 28). *Implications of smart city technologies*. Center for International Private Enterprise. https://www.cipe.org/blog/2023/12/28/implications-of-smart-city-technologies/
55. Kitchin, R. (2016). The ethics of smart cities and urban science. *Philosophical Transactions of the Royal Society A: Mathematical, Physical and Engineering Sciences, 374*(2083), 20160115. https://doi.org/10.1098/rsta.2016.0115
56. I T E P A P E R J U, W. (n.d.). *Governing smart cities: Policy benchmarks for ethical and responsible smart city development*. https://www3.weforum.org/docs/WEF_Governing_Smart_Cities_2021.pdf
57. *Governing smart cities: policy benchmarks for ethical and responsible smart city development*. (2021, July 13). World Economic Forum. https://www.weforum.org/publications/governing-smart-cities-policy-benchmarks-for-ethical-and-responsible-smart-city-development/
58. Responsible Use of Technology. (2019). https://www3.weforum.org/docs/WEF_Responsible_Use_of_Technology.pdf
59. Kendal, E. (2022). Ethical, legal and social implications of emerging technology (ELSIET) symposium. *Journal of Bioethical Inquiry, 19*(3), 363–370. https://doi.org/10.1007/s11673-022-10197-5

Chapter 12
Disinformation and the Impact on Democracy

Praahas Amin

Abstract The buildup of digital technology has led to the building up of false information, which has become an issue, particularly in the democratic process of elections and voting. The dissemination of misinformation and disinformation has been carried out to a significant degree through the use of social media. New developments in the field of generative artificial intelligence have significantly sped up the production and dissemination of false information, particularly by political actors for the purpose of manipulating public opinion, influencing voter behavior during elections, and having a detrimental effect on democratic processes. The transformation of social media from a beneficial uniting force to a polarizing instrument is the topic of discussion in this chapter. Additionally, the impact of artificial intelligence on the dissemination of false information and its ill effects on democratic processes is also discussed, as is the function of artificial intelligence in the increase in the production of fake content. At the same time that it can be used to protect democracy through the automated detection of false information, artificial intelligence can also be a threat to democracy through its use for mass surveillance, tools for generating false information, and suffering from bias introduced by developers or training data sources. This chapter examines how artificial intelligence can be a double-edged sword. The text examines a number of scenarios that have occurred all over the world in which social media and artificial intelligence have been utilized for the purpose of manipulating public opinion, resulting in the creation of a polarized society, a significant rise in partisanship, and a diminished capacity for rational thought. Additionally, some of the difficulties encountered throughout the process of constructing artificial intelligence models that have a high level of accuracy and a decreased level of bias are explored. The article presents a survey of the impact of social media and artificial intelligence technologies on democracy and discusses various solutions surveyed and challenges in implementation of the solutions.

P. Amin (✉)
Department of Computer Science and Design, Canara Engineering College, Mangalore, India
e-mail: praahas@canaraengineering.in

C. Hewage et al. (eds.), *Data Protection*,
https://doi.org/10.1007/978-3-031-76473-8_12

12.1 Introduction

Platforms such as X (which was then known as Twitter) and Meta (which was then known as Facebook) were developed with the intention of facilitating the quick distribution of information [1, 2]. Its purpose was to facilitate improved communication between friends and celebrities and the people who follow them. In the long run, the social media platforms could be utilized for the purpose of directly experiencing democracy in a participatory manner. This could be accomplished by serving as a platform that brings together protesters and rebellions against oppressive undemocratic governments, thereby providing citizens with a source of individual empowerment. Within a short period of time, social media platforms went from being a facilitator of democracy to becoming a foe of democracy. As a result of the Cambridge Analytica data scandal, the general public became aware of a world in which political parties and organizations had conspired to use the power of artificial intelligence and big data in order to undermine the democratic machinery. Cambridge Analytica exploited the data of Facebook users to manipulate the opinions of huge groups of people during events such as the presidential elections in the United States in 2016 and the Brexit referendum on the United Kingdom's membership in the European Union in 2016 [3]. The United States of America has a long history of employing disinformation and propaganda in order to target regimes in Latin America and the Middle East, as well as the Soviet Union [4–7]. In contrast to the present day, when the globe is connected by the Internet, artificial intelligence, and social media, the dissemination of misinformation during the Cold War was typically accomplished through the planting of false articles in print media, and the spread of this false information was much slower, less pervasive, and less convincing than it is now.

"Post-Truth" was selected by Oxford Dictionaries as the word of the year for the international community in the year 2016. It has been ten years since the adjective was first used, but it did not become well known until after the presidential election in the United States. The proliferation of misinformation and disinformation has become a frequent problem, particularly in the realm of politics and elections, in the interconnected world of the post-truth age, which is characterized by blurred boundaries between truth and untruth. In the wake of the proliferation of social media, the propagation of false information and disinformation has become not only more prevalent but also more straightforward. Disinformation refers to the deliberate dissemination of false information with the intention of malice, deceit, or manipulation, whereas misinformation refers to material that is erroneous or misleading and is disseminated without any malicious intent. Both kinds of incorrect information pose a risk to the legitimacy of democratic processes like elections, which are a fundamental component of democracy.

The computing technology of the time limited the researchers, academicians, and industry professionals who were working in the field of artificial intelligence during its early days. A certain degree of democratization of artificial intelligence has occurred as a result of technological advancements that have made computing power more accessible to the general public. The development of generative artificial intelligence-based technology, which has the capacity to generate realistic text, image, and video material, has further blurred the lines between fact and fiction, making it increasingly difficult for users to differentiate between fact and fiction. This has led to the development of technology that has the potential to generate realistic content. Falsified information has been used by political players for the purpose of subversion, the exploitation of pre-existing tensions in societies and communities, the deepening of the divide in vote banks, the manipulation of public opinion formations, the loss of trust in democratic institutions and the media, and ultimately influence the citizens' voting decisions. This chapter delves into the connections between the substantial influence that social media, artificial intelligence, misinformation, and disinformation have on election processes.

12.2 False Information and Creation of a Bias Bubble—The Problematic AI and Social Media Manipulation

The problem of fake news is a complicated and alarming one that has far-reaching consequences. Besides the dissemination of misleading information, it also has an effect on the stability of the economy, public safety, democratic processes, and the trust that people have in the media. Inaccurate information about important corporations or events that occur on a worldwide scale, for instance, can result in huge swings in the financial market, which can lead to losses for investors. It is possible for false information to impede efficient response efforts during times of emergency, such as when there is a natural disaster or a health crisis. A significant number of youths rely extensively on social media for their daily information consumption, and they frequently come across news by chance as they are scrolling through their feeds. Social media has become increasingly influential in the spread of hate speech, conflicts between professional and personal life, and the dissemination of online gossip or incorrect information, which poses a possible threat to the sustenance of enterprises [8, 9]. Users of social media platforms have the ability to create and distribute material anonymously or under pseudonyms and with this anonymity, individuals may feel more confident in spreading misleading information without worrying about the implications for themselves.

Additionally, social media platforms encourage users to create their own content. Despite the fact that this encourages creative thinking and involvement, it also makes it possible for people to post content without conducting detailed fact checks, which can lead to the rapid circulation of information that is either unverified or deceptive. To add more complexity to the problem, the rapid pace at which information is disseminated on social media platforms frequently allows little room for critical evaluation of the reliability of the material.

The breakout of the COVID-19 pandemic in 2019 was followed by a quick propagation of rumors and false information on social media. These rumors and fake news took advantage of the public's dread about the disease. For example, users of social media platforms described techniques of prevention and control that lacked scientific and medical basis in order to reap economic benefits. It was alleged by a World Health Organization specialist who specializes in social media platforms that this resulted in a surge of panic and worry that spread more quickly than the virus itself. As a result of the appearance of COVID-19 at the tail end of 2019, he indicated that a potentially catastrophic occurrence known as an infodemic had been brought about. This is characterized as the difficulty that individuals have in gaining access to legitimate sources and advice when they are in need of it, and this phenomenon, in turn, leads to the ease, and speed with which fake news can be spread, particularly on social media [10].

The identification of incorrect information using manual techniques is a task that is exceedingly arduous and is not a solution that can be scaled up. For the purpose of identifying false information, artificial intelligence has been widely considered as a solution that is both scalable and very successful. On the other hand, artificial intelligence, namely in the form of generative AI, has been extensively utilized in the production of false information. On the basis of enormous collections of information that was acquired by humans as big data, large language models (LLMs) are able to demonstrate the ability to recognize, predict, and generate text that is similar to natural language. The distinction between information generated by artificial intelligence and content created by humans is becoming increasingly difficult to discern.

The combination of artificial intelligence and social media has the potential to be a strong instrument that can be exploited for the purpose of manipulating the behavior of voters. Push notifications are utilized by media organizations through mobile applications in order to attract the attention of users, even in situations where the user has not consciously opted to see what is displayed on their mobile device. With the help of an underlying algorithm, social media platforms are able to maximize user engagement by delivering material to users in accordance with their profiles, which are generated on the basis of the users' actions on social media. It is possible to conduct user profiling in order to determine the requirements and necessities of an individual, and then microtargeting them in order to provide them with individualized content or advertisements. For instance, the algorithm will learn that a person who has a strong affinity for a particular political party will most likely enjoy watching information produced by that party. As a result, the algorithm will only display content that is relevant to that particular political party by default. It is possible that this will result in partisanship, an increase in the individual's prejudice toward that political party, and, in the long run, it will lead to the partisans believing that everyone is of the same opinion because they will not be exposed to other opinions. In addition, social media platforms enable targeted manipulation by delivering content to a particular group of individuals without providing the full picture. Consequently, this will result in a situation in which individuals are only provided with content that they are interested in hearing, thereby enclosing them in a bubble of their own opinion and rendering them blind to the perspectives or opinions of others. It is possible that artificial intelligence may be utilized in this scenario to detect groups of individuals who have contrasting viewpoints and then offer selective content to each of these groups, resulting in an information environment that is not balanced. The production of fake information is yet another instance in which artificial intelligence is being utilized. The application of artificial intelligence enables the production of a large quantity of misleading text, image, and video-based information. This content may then be disseminated by the campaign teams of political parties in order to disparage their opponents or communities in order to appeal to their vote banks. People will eventually come to accept it because they will be surrounded by a bias bubble that will prevent them from seeing any counterarguments. This will happen with increasing frequency as more content of this kind is presented. From a privacy and data protection point of view, the practice of microtargeting and profiling is problematic. Furthermore, the use of this technique for the purpose of manipulating voter behavior poses a threat to the democratic process. Furthermore, according to Article 22 of the General Data Protection Regulation (GDPR), which addresses automated decision making, which includes models of detection, and automated profiling, users have the right to resist being exposed to a decision that is exclusively based on automated processing, which includes profiling. Policymakers from the European Union have stated that personally identifiable information that has been processed by automated decision making cannot be utilized for political targeting since it violates the standards of the privacy law. Article 5 of the general data protection regulation (GDPR) mandates that companies must reduce the amount of data they collect and limit its use to the purpose for which it was initially intended. This can be a barrier to improving the accuracy of artificial intelligence models, as it may be necessary to use vast datasets in order to achieve the desired level of accuracy.

An emerging form of multimodal synthetic media that is generated using deep learning, known as deepfake, has added an additional layer of complexity to the situation. Deepfake can be used to produce false audio or video content in order to present something that never occurred as a fact, as was seen during the conflict between Russia and Ukraine [11]. Figure 12.1 shows a real image and the deepfake video of the Ukrainian President Volodymyr Zelenskyy. Deepfakes are created by superimposing a fake face that has been created with the help of artificial

Fig. 12.1 Video snapshot of the real (left) and the deepfake (right) version of the Ukrainian President Volodymyr Zelenskyy

intelligence on top of a genuine video. This creates a video that appears to authentically portray someone saying or doing something that they did not actually do. Deepfakes came to the attention of the public in 2018 when a community on reddit used artificial intelligence to transform pornographic films into videos featuring images of celebrities. Despite the fact that such digitally fabricated video is frequently far from perfect and exhibits artifacts, it is frequently utilized successfully for the purpose of spreading disinformation and poses a danger to the validity of political communication and news. Images and videos that are deepfakes are created by employing a pair of adversarial neural network models, which are referred to as the generator and detector. The data is produced by the generator, while the detector provides feedback on the data. The generator is responsible for producing data that has a higher level of fidelity. This process continues until the generator reaches a level of probabilistic confidence that it can outsmart the detector [12]. The generation of content using deepfake necessitates the accumulation of substantial amounts of input data, which is often readily accessible for public personalities. Two key causes for concern are the increasing accessibility of technology and the improvement in the quality of content that is created with deepfakes. The availability of computer power and the huge range of data that is currently available are both obstacles that are currently being faced. In spite of this, it is quite probable that these obstacles will be eliminated as a result of technical improvements and the availability of computing resources through the cloud. It is possible that this will eventually lead to a situation in which people will cease engaging in the democratic process of voting because they will have a difficult time distinguishing between actual material and fraudulent content, and they will also lose trust in the institutions that make up the democratic machinery [13].

WhatsApp is another platform that has been used widely by both the major political parties in India, the Indian National Congress and the Bharatiya Janata Party, for the spread of misinformation and disinformation especially in the run-up to the 2019 Lok Sabha Elections in India [14] during which there were instances of small, localized incidents and major killings and mob lynchings in which WhatsApp was used to spread rumors to serve a politically motivated agenda [15].

The social media platforms have been weaponized by political parties in order to facilitate the dissemination of coordinated disinformation campaigns in conjunction with major television news outlets simultaneously. There have been many cases of journalists who have been critical of the ruling party being exposed to harassment and attacks on social media. A few of the journalists have also been subjected to attacks in their offline lives. Furthermore, politicians on social media and news channels utilize provocative hashtags to justify and defend attacks of this nature. This type of attack is encouraged by these politicians. In recent years, there has been a consistent rise in political polarization all across the world, and this trend has been linked to the growing prevalence of content that is polarizing on the internet.

Table 12.1 lists various instances of misinformation and disinformation and their classification in different parts of the world.

Table 12.1 Region wise examples of false information and their classification

Region	Disinformation/misinformation type	Example
US	Political misinformation	Allegations that the presidential election in the United States was tainted by election fraud in mail-in the voting process
	Health misinformation	The dissemination of misleading information regarding the treatments and prevention methods for COVID-19 such as ingestion of bleach as a cure for COVID-19
	Social media rumors	Speculations that have gone viral on social media platforms regarding purported conspiracies by the government such as 5G Technology is responsible for the spread of COVID-19
Europe	Political disinformation	In an effort to influence elections in a number of European countries, the dissemination of fake narratives is taking place such as spreading false information about political opponents' criminal record to discredit them before elections
	Health misinformation	Spread of false information about COVID-19 vaccines such as the vaccine containing microchips for tracking individuals.
	Conspiracy theories	Propagation of conspiracy theories speculating COVID-19 to 5G technology and the widespread belief on social media that COVID-19 is a bioweapon engineered to enforce global control
Asia	Disinformation campaigns	Disinformation campaigns that are coordinated and directed at political opponents and ethnic minorities.
	Social media hoaxes	There have been hoaxes going viral on social media platforms about natural disasters and actions taken by the government such as false rumors on WhatsApp claiming that the government is planning to implement a nationwide lockdown with no evidence.
	Health misinformation	Spreading false information about traditional remedies as cures for COVID-19 such as drinking boiled garlic water can cure COVID-19
India	Social media disinformation	Viral disinformation campaigns on WhatsApp regarding communal tensions and religious issues such as messages on WhatsApp claiming that a particular religious group is planning attacks against another community
	Political disinformation	False narratives spread by political parties to manipulate public opinion during elections such as spreading false rumors about an opponent's personal life to tarnish their image and sway voters.
	Health misinformation	The dissemination of incorrect information regarding COVID-19 treatments and preventative measures, including treatments that have not been verified such as circulation of a fake news article on Facebook claiming that drinking cow urine can prevent and cure COVID-19.

(continued)

Table 12.1 (continued)

Region	Disinformation/misinformation type	Example
UK	Misinformation in mainstream media	The dissemination of incorrect information by tabloid publications concerning the rates of immigration and criminal activity, respectively, by publishing a sensationalized story falsely linking immigrants to an increase in crime rates.
	Political disinformation	Spreading of false narratives by political parties to sway public opinion on Brexit-related issues by using misleading statistics to exaggerate the economic benefits of Brexit in campaign materials.
	Health misinformation	Dissemination of false information about COVID-19 restrictions and government responses such as Spreading false information on social media platforms claiming that the government is planning to implement mandatory vaccination without evidence.

12.3 AI Based Supervision—The Problematic Surveillance

The technology of artificial intelligence has the potential to be utilized for mass surveillance of citizens, which may constitute an invasion of an individual's right to privacy. For example, the Chinese government collects massive amounts of information about its citizens, including information about their income, taxes, health, purchasing patterns, behavior on social media, and facial data acquired from surveillance cameras that are equipped with facial recognition systems all throughout the country. On the basis of this data, the citizens are then assigned a social credit score, which serves as an incentive for law-abiding citizens and citizens who behave responsibly. Actions such as donating blood, volunteering in community service, and making regular loan repayments are examples of actions that contribute to a rise in score. A citizen's score will decline if they engage in acts that are considered to be examples of negative behavior. Some examples of such behaviors include not paying back debts on time and littering, amongst others. What makes good conduct and poor behavior is a decision that is made by the government, and when citizens voice their ideas against the government or its policies, then such activities can be classified as the behavior that is considered to be bad. With a low social credit score, a citizen will be unable to obtain loans, board airplanes, or even find employment. This is because of the low credit score. Those governments that are not democratic have the ability to effectively use this as a tool for censorship in order to stifle opposition that is voiced by the citizens of the country against oppression. When it comes to situations like these, the employment of artificial intelligence presents a challenge because the success of the system is dependent on the quality and quantity of the data that is used to train it.

12.4 Trolling—The Cyberbullying and Intimidation Problem

Trolls are responsible for ensuring that the incorrect information that is generated on the internet lasts and reaches the greatest number of users, as well as instilling bias in certain groups. The main objective of these individuals is to instill fear in others who voice opinions that are contrary to their own, including critics, people who check the facts, and citizens. The term "troll" refers to those who have profiles on social media platforms that are generally fraudulent in character. They resort to using aggressive words and insulting others [16]. When a troll does not agree with an ideology, they will resort to targeting public personalities who support that ideology in order to damage the legitimacy of those public figures or individuals. Trolls are characterized by their inability to tolerate any viewpoint that is contrary to their own convictions. It is possible for trolls to inspire other users to resort to intimidating behavior toward users who hold dissimilar opinions. In order to get what they want; they resort to propagating false information that is in line with an ideology that they support. There is a presence of trolls across all social media platforms. On various social media platforms, they become members of various groups and earn the trust of the individuals who are a part of those groups. In order to undermine the credibility of the job that they have done, they resort to smear campaigns against their opponents. The trolls have not been subjected to any significant legal consequences of any kind for their actions. Through the use of major news outlets and social media platforms, trolling opponents has become an increasingly frequent practice in the realm of politics. The campaign teams of political parties are now even employing trolls to work in their information technology cells in order to propagate disinformation, discredit and damage opponents, and affect the behavior of voters. This poses a significant risk to democracy because individuals who are not well-versed in the media and information will be more likely to trust the fake information to be accurate and they will also be more likely to spread this incorrect information. Trolls empowered with artificial intelligence tools that can generate fake content such as toxic memes, videos, and posts that are a major threat to the democratic machinery. Having to deal with trolls can result in a loss of both time and productivity. Trolls have the potential to be responsible for the loss of potential clients as a result of the negative and incorrect information that they spread on social media platforms. These kinds of posts have the potential to tarnish the reputations of famous personalities or businesses, as well as to undermine their legitimacy in the eyes of the general public. It is possible for the reader to be misled into thinking that the meme in question is a fact because it presents itself as a believable fact. Ideas that are discriminatory on the basis of gender, race, class, religion, creed, caste, sexuality, beliefs, and ideologies can be propagated and reinforced through memes that are created in the name of humor.

12.5 Solutions for Tackling Disinformation

12.5.1 Algorithmic False Information Detection and Flagging

The verification of the authenticity of content has traditionally relied heavily on human action as the primary means of fact-checking. Since the introduction of technology that is based on artificial intelligence has resulted in a significant increase in the amount of misleading information that is generated, rendering manual fact-checking ineffective for checking every piece of content that is produced on the internet. After the elections in the United States in 2016, there has been a substantial surge in the number of innovations in automated fact-checking approaches for the purpose of evaluating content that is found online. The majority of the organizations that have been at the forefront of this development have been nonprofit and independent organizations.

Artificial intelligence solutions have been used effectively for the removal of undesirable content online by internet companies and social media platforms such as Google, Meta, and X. However, there is a need for improvement in the accuracy of these solutions. There is a possibility of both false positives and false negatives occurring in artificial intelligence models. It is possible for true data to be suppressed and labeled as misinformation as a result of false positives in a system that uses artificial intelligence for the elimination of misleading information. This has the potential to have a significant influence on the freedom of expression. Fact-checking systems that are powered by artificial intelligence have the potential to overlook implied statements that are buried inside complicated words, contextual or cultural cues, and sarcasm, all of which are easily identifiable by a human being. There is also the possibility that artificial intelligence models will be influenced by human prejudices derived from the data that they are trained with, which would result in outcomes that are skewed against particular communities or groups of people. It is possible for the developers who train the model to be the source of these biases. It is possible for biases of this kind to emerge as a result of training data that is either incomplete or not representative. When models are trained with historical datasets, they have a tendency to reveal the biases that were present in society at the time. It is possible for these biases to influence the decisions that are made by the AI in automated systems, such as recruitments. The artificial intelligence models that are being trained are susceptible to human biases, as the data itself contains human biases. As an illustration, in February of 2024, a post on X caused a stir in India when the artificial intelligence tool Gemini from Google was questioned about whether or not Prime Minister Narendra Modi was a fascist [17]. According to the response provided by the artificial intelligence tool, "he was accused of implementing policies that some experts have characterized as fascist." Furthermore, it was stated that "these accusations are based on a number of factors, including the Hindu nationalist ideology of the BJP, its crackdown on dissent, and its use of violence against religious minorities." In contrast, when questions of a similar nature were asked regarding former President of the United States Donald Trump and Chinese President Xi Jinping, it was stated that there was no definitive response.

Systems that are powered by artificial intelligence, particularly those that are powered by artificial neural networks, are fundamentally black box solutions. The reasoning behind these systems, which are based on self-teaching, is beyond the comprehension of the people who originally developed them. Another limitation of artificial intelligence is that it is difficult to see and understand the underlying mechanisms of the systems that are engaged in decision making in Artificial Neural Network-based systems. Despite the fact that it is more accurate, the creators themselves are unable to comprehend the rationale that may have been used by the artificial intelligence system to make a decision. Organizations such as the Defense Advanced Research Projects Agency (DARPA) and the United States Defense Research Agency have been working on explainable artificial intelligence (XAI) with high prediction accuracy [18]. The objective of this project is to provide human users with the ability to comprehend and explain the rationale behind the decision-making logic of artificial intelligence, as well as to characterize the strengths and weaknesses of these systems. Systems that are based on artificial intelligence may someday benefit from increased accountability and openness as a result of this situation. It is possible for human moderators to provide support for information verification systems that are based on artificial intelligence. On the other hand, the analysis of the content that is being moderated can be impacted by the previous experiences, personal philosophies, mental state on a particular day, or mood of the human moderator.

12.5.2 Artificial Intelligence Regulations—The EU AI Act and US Executive Order on AI

The European Commission proposed the first European Union Regulatory framework for artificial intelligence in April 2021. It states that artificial intelligence-based systems that find use in different applications are analyzed and classified differently based on the risk they pose to the users. The parliament's aim is to ensure that the artificial intelligence systems operational in the European Union are safe, transparent, nondiscriminatory, and unbiased.

The risk-based approach proposed by the commission and classifies artificial intelligence systems into several risk categories, with different degrees of regulation is shown in Fig. 12.2.

1. Unacceptable risk, artificial intelligence systems considered a threat to people. The manipulation of cognitive and behavioral processes of people, as well as specific vulnerable groups, social scoring, biometric profiling of individuals, and real-time and remote biometric identification technologies, such as facial recognition, are all included in this category. However, there are some exclusions

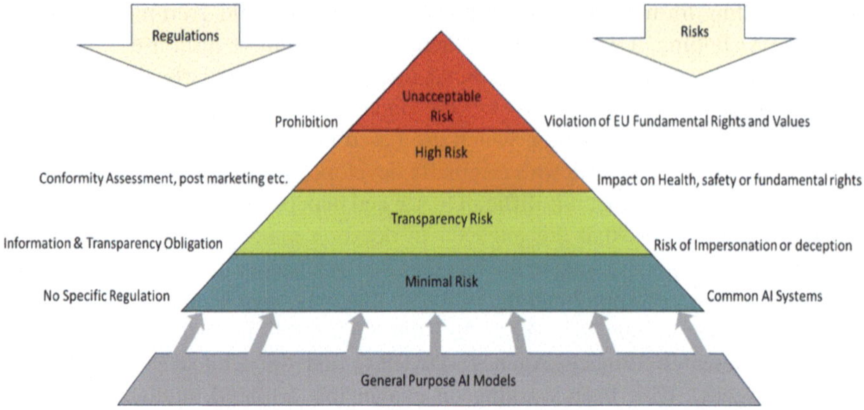

Fig. 12.2 EU AI Act—risk-based approach

made for specific law enforcement needs. It is forbidden to use them. They are prohibited.

2. High-risk artificial intelligence systems considered to be a threat to safety or fundamental rights. Safety components of items that are covered by sectoral EU's product safety legislation are included in this category. Some examples of these products are medical equipment, systems, toys, airplanes, automobiles, and elevators. If it is determined that artificial intelligence systems do the profiling of individuals and determine admission to educational and professional courses, then these systems should be categorized as high risk. Before their products can be launched into the market, providers of artificial intelligence systems are required to undergo a conformity assessment method. Additionally, they will be required to comply with a variety of standards pertaining to testing, data training, cybersecurity, and fundamental rights impact assessment.

   Immediately following the product's release, the providers are obligated to carry out post-market monitoring and, if necessary, take corrective actions accordingly.
3. Transparency risk artificial systems are those that may pose a risk to users in terms of impersonation or deception irrespective of whether they classify as high risk or not. In the event that content has been generated or modified by an artificial intelligence system, such as deepfakes, the producers of such systems have an obligation to reveal this information. It is necessary for these kinds of systems to create watermarks that are reliable and robust when they produce synthetic content in large volumes. An employer who uses artificial intelligence systems in the workplace is required to disclose this information to employees and representatives of those employees.
4. Minimal risk artificial intelligence systems are those that pose minimal risks to users, such as spam filters, recommender systems on OTT Platforms and music streaming platforms. Such systems will not be subject to further obligations beyond currently applicable legislation such as GDPR.

US President Joe Biden's Executive Order on AI is based on the following guiding principles:

1. Ensuring safety and security via comprehensive, repeatable, and reliable standardized AI system evaluations, policies, institutions, and risk mitigation methods before deployment. Testing, assessments, post-deployment performance testing, effective labeling, and provenance controls can assist mitigate AI's dangers without compromising its benefits.
2. Promoting responsible innovation and competition to allow the United States to lead in AI.
3. Focusing on the benefit of American workers by creating new employment and industries with a diverse workforce, providing employment training and education and guaranteeing that AI does not undermine rights, worsen job quality, encourage undue worker surveillance, lessen market competition, introduce new health and safety risks, or cause harmful labor-force disruptions.
4. Exploring how AI might unlawfully discriminate or facilitate the administration of federal programs and benefits to promote equity and civil rights.
5. Enforce consumer protection rules and prevent fraud, unintended prejudice, discrimination, privacy breaches, and other AI-related damages.
6. Ensuring data collection, usage, and retention is legal and secure reduces privacy and confidentiality threats.
7. Manage risks due to the use of artificial intelligence by the federal government and strengthen its internal ability for monitoring, overseeing, and encouraging responsible artificial intelligence.
8. Advance global societal, economical, and technological advancement by taking the lead in developing systems and ensuring responsible use of technology. This activity involves working with global allies to mitigate AI threats, and explore the potential of AI for the benefit of mankind.

12.5.3 Auditing of Artificial Intelligence

Bias in artificial intelligence-based decision-making systems can be countered by auditing such systems. This would ensure that the data that is used for training the models are thoroughly examined, which would result in an improvement in the accuracy of the models without adding any bias into the system. The government of the United States has proposed a law called the Algorithmic Accountability Act, which would require businesses to conduct audits of their artificial intelligence systems to check for bias, provide impact assessments, and take corrective actions. This might be underlined in the law.

12.5.4 Solution for Deepfakes

Law Professors Robert Chesney and Danielle Citron [19] offered 3 ways to detecting deepfakes. The dataset used for training the model for deepfakes, lacked faces with closed eyelids, so an approach for identifying anomalous eyelid movements could be employed for flagging deepfakes. However, the deepfakes evolved and adapted. Digital watermarking of the audio or video content at the source of media creation was another solution explored. This could be referenced for authenticity when compared with a deepfake. Another solution explored was Authenticated Alibi Services, where the user is lifelogged and every movement and action of an individual is stored. However, this may have major negative ramifications for personal privacy. Blockchain could be used to track the provenance of content and establish trust.

12.5.5 Regulation of Social Media Content

A number of proposals for the regulation of content on social media have been met with opposition from a variety of stakeholders, including the platforms themselves, civil rights organizations, and end-users. This is due to the fact that these proposals have either imposed additional liabilities on social media platforms or have enabled governments to exert a greater degree of censorship and control over content that is found online. The sole responsibility for controlling content on the internet should not be placed on governments. It is not appropriate for the platforms to bear the sole responsibility for this matter. Both of these things have the potential to be harmful to democracy, particularly when it comes to selective censorship that is conducted by governments or by platforms. It is not reasonable to expect governments to observe a neutral stance when it comes to the enforcement of censorship. Technologies typically advance at a quicker rate than official rules made by the governments, and therefore, the responsibility of regulating social media cannot be given solely to the governments. In addition, an increasing hold that governments have on user data could result in the violation of civil freedoms by governments that are undemocratic and oppressive in order to stifle opposition and repress dissent. Continuous feedback and contributions from all relevant stakeholders are required in order to accomplish the development of regulations for the regulation of a technology that is undergoing rapid evolution. There ought to be a process of collaboration between the government, the academic community, and the business sector and industries. According to the Information Technology (Guidelines for Intermediaries and Digital Media Ethics Code) Rules, 2021, the Indian government has just introduced, even multinational technology corporations based in other countries, such as Meta, X, Netflix, Amazon, and others, are required to comply with these rules. False news, fake bot accounts, and unlawful content monitoring are the primary areas of concern for social media networks and their respective platforms. The highest possible compliance standards are applied to platforms that have a minimum of fifty lakh registered users. The appointment of a Chief Compliance Officer, a Nodal Contact Person, and a Resident Grievance Officer, all of whom must be citizens or permanent residents of India, is a requirement that must be completed by all key social media intermediaries. Additionally, it is required of the companies that they have a physical contact address anywhere in India.

12.5.6 Educating and Creating Awareness in Citizens

A citizen's fundamental capacity to interact with the media in an efficient manner, along with the development of critical thinking skills that enable them to become active citizens, is what it means to be media and information literate. Being media and information literate would mean that a person would be able to comprehend the content that is presented in the media, as well as produce it and evaluate it critically. This would be an efficient approach for combating the spread of misleading information. It would be possible for citizens to engage in critical analysis of the content that they encounter on social media outlets. They would be skeptical of posts and messages on social media and would be careful about sharing a post on social media based solely on the amount of "likes" or "shares." They would also be critical of the content of social media messages. They would not limit themselves to a single channel or source of information and would instead have a more diverse selection of media outlets that they prefer to use for news and information sourcing. They would have a heightened awareness of the dissemination of incorrect information on the internet, and they would be able to bring it to the attention of those in their circle who are not aware of it, so bringing about a sense of awareness. An important component of the remedy is being aware of the situation and avoiding the dissemination or publication of misleading information. The person who is responsible for the dissemination of information and media must also take the initiative to prevent the dissemination of incorrect information and news. It would be necessary for them to have a solid understanding of the methodology involved in identifying bogus information.

European countries have launched anti-disinformation campaigns at schools that emphasize critical thinking and techniques of identifying trolls and bots by examining social media profiles. Young people especially in schools are vulnerable to propaganda, misinformation, and disinformation as they spend a significant time on social media platforms. They rely on content available online for their source of information. Abundance of false information available on social media platforms can significantly impact their perception of reality. Parents themselves may not be media and information literate, and therefore, schools have a responsibility of providing the youth with the tools and skills needed for being critical thinkers and for stopping the spread of misinformation and disinformation. This is essential for the survival of the democratic machinery of nations because if the youth is blinded by propaganda, then the future of the nations will be in jeopardy as the nations will have a highly polarized, divided, and a partisan society.

12.5.7 *Defeating Online Trolls Using Technology*

It is possible to combat trolls and the harmful content that they spread on social media platforms by utilizing solutions that are based on artificial intelligence. The model of artificial intelligence would need to be able to comprehend the sentiment of text postings or messages embedded in images, such as memes, and identify whether or not the post contains incorrect information that was made by a troll. To recognize text that is contained in a picture, the system would need to employ an optical character recognition (OCR) technique that is based on deep learning. After that, it would need to apply a sentiment analyzer that is based on machine learning to determine whether the text contains misleading information or content that is trolling in nature.

12.6 Conclusion

Demos, which means "people," and Kratos, which means "power," are the two terms that are the origin of the word democracy. A democratic form of government is one in which the power is vested in the people and exercised by them directly or indirectly through a representation system that typically involves free and fair elections being held on a periodic basis. The social media platforms can serve as a potent instrument that empowers democracy by giving a forum for individuals to voice their thoughts. Nevertheless, political parties make extensive use of social media in order to manipulate the behavior of voters in order to either maintain their power or win elections to gain power. Democracy is under danger as a result of this.

Extremists benefit from online forums and escape law enforcers by the cover of subjectivity. They use internet technologies to promote their opinionated ideologies. The internet provides extremists the avenue to access a potentially enormous audience. It facilitates communication among like-minded people across borders and oceans by anonymously and cheaply enhancing their ability to promote and recruit for their extremist causes. By providing more effective techniques of fact-checking, trolling detection, and deepfake detection, democratic processes can be strengthened through the utilization of tools that are powered by artificial intelligence. Artificial intelligence may have its own intrinsic biases due to the limited training data or the bias of the developer. "Is technology a boon or a bane?" is a question that has been asked for a very long time, and it is something that we see in both artificial intelligence and social media. It has come to our attention that technology is a double-edged sword. The technology in the hands of the right people has the potential to be utilized as a tool for strengthening democracy and giving people more power. It provides a platform for collectively voicing opinions and ensures it reaches the representatives so that they take decisions for the betterment of the citizens rather than betterment of the political parties. However, in the wrong hands, it has the potential to become a weapon that destroys democracy. In the hands of trolls, political parties, certain corporations, artificial intelligence, and social media technologies have the potential to widen the social divide, creating partisans, and destroying the fabric of civil society. In this society, questioning and criticizing those in power is discouraged, and technology is used for censorship and control. It is necessary to handle the various ambiguous gray areas that are present in technologies that are based on artificial intelligence. Regulation of technologies that are based on artificial intelligence is something that needs to be done by a body that includes academicians, researchers, scientists, policymakers, and industries. A single segment of society cannot hold the power to regulate such strong technologies. This power cannot be centralized in the hands of a single societal group. The creation of artificial intelligence models must be carried out in a responsible manner, and the field of artificial intelligence needs to be made more transparent. The development of artificial intelligence that is responsible is a necessity in order to ensure that the judgments made by the artificial intelligence models are held accountable. It is possible to construct artificial intelligence as gray box models rather than black box models for use in systems that have the potential to have a negative impact on the social fabric. Regulations on social media should be carried out in a just manner. A sufficient level of responsibility should lie on the platform in order to remove disinformation and misinformation to have better accountability. Taking down bot accounts that propagate misleading information and troll accounts should be done in an effective by the platforms and genuine information that may be critical of the governments that are currently in power should not be removed from platforms as this leads to unfair censorship. In order to ensure that political parties that hack the election process and use trolls and disinformation to gain political mileage and advantage are held accountable, these parties need to be held accountable through the development of policies. There must be strong regulations on political parties colluding with organizations for misusing the public big data for voter behavior manipulation. Social media and artificial intelligence technologies have the potential to be the strength of democracy, if regulated correctly and in a fair manner and without regulation, they can lead to the downfall of democracy.

It can be concluded that cutting-edge technology, such as blockchain-based solutions and machine learning algorithms for content verification, have the potential to reduce the spread of misinformation and improve electoral transparency. Additionally, programs that attempt to promote media literacy and critical thinking skills are crucial in order to provide citizens with the tools necessary to differentiate between information that is reliable and information that is not credible. When it comes to devising comprehensive plans to address the myriad of difficulties that are posed by artificial intelligence-driven misinformation and disinformation, collaboration between governments, technology corporations, and civil society players is absolutely necessary. It is necessary for regulatory frameworks that control the moderation of online material, transparency in political advertising, and data privacy to undergo evolution in order to keep up with the rapid growth of technology and the emergence of new risks. We are able to protect the fundamentals of democracy and make certain that elections accurately reflect the will of the people if we cultivate greater transparency, accountability, and resilience in our information ecosystems.

Conflict of Interest The author has no relevant financial or nonfinancial interests to disclose.

References

1. Kaplan, A. (2012). If you love something, let it go mobile: Mobile marketing and mobile social media 4x4. *Business Horizons, 55*(2), 129–139. https://doi.org/10.1016/j.bushor.2011.10.009
2. Kaplan, A., & Haenlein, M. (2011). The early bird catches the news: Nine things you should know about microblogging. *Business Horizons, 54*(2), 105–113.
3. Rehman, I. (2019). Facebook-Cambridge Analytica data harvesting: What you need to know. *Library Philosophy and Practice (e-journal)*, 2497.
4. Bittman, L. (1990). The use of disinformation by democracies. *International Journal of Intelligence and Counter Intelligence, 4*(2), 243–261. https://doi.org/10.1080/08850609008435142
5. Fabre, C. (2018). The case for foreign electoral subversion. *Ethics & International Affairs, 32*(3), 283–292. https://doi.org/10.1017/S0892679418000424
6. Johnson, L. K. (1989). Covert action and accountability: Decision-making for America's secret foreign policy. *International Studies Quarterly, 33*(1), 81–109. https://doi.org/10.2307/2600495
7. Levin, D. H. (2016). When the great powers get a vote: The effects of great power electoral interventions on election results. *International Studies Quarterly, 60*(2), 189–202. https://doi.org/10.1093/isq/sqv016
8. Allcott, H., & Gentzkow, M. (2017). Social media and fake news in the 2016 election. *Journal of Economic Perspectives, 31*(2), 211–236. https://doi.org/10.1257/jep.31.2.211
9. Rost, K., Stahel, L., & Frey, B. S. (2016). Digital social norm enforcement: Online fire- storms in social media. *PLoS One, 11*(6), e0155923. https://doi.org/10.1371/journal.pone.0155923
10. Al Zou'bi, R. M. (2022). The impact of media and information literacy on students' acquisition of the skills needed to detect fake news. *Journal of Media Literacy Education*. Pre-Prints. Retrieved from https://digitalcommons.uri.edu/jmle-preprints/28
11. Twomey, J., Ching, D., Aylett, M. P., Quayle, M., Linehan, C., & Murphy, G. (2023). Do deepfake videos undermine our epistemic trust? A thematic analysis of tweets that discuss deep- fakes in the Russian invasion of Ukraine. *PLoS One, 18*(10), e0291668. https://doi.org/10.1371/journal.pone.0291668
12. Shen, T., Liu, R., Bai, J., & Li, Z. (2018). *Deep Fakes using generative adversarial networks (GAN)*. University of California. http://noiselab.ucsd.edu/ECE228_2018/Reports/Report16.pdf
13. Mosk, M., Turner, T., & Faulders, K. (2018). Russian influence operation attempted to suppress black vote: Indictment. *ABC News*. Retrieved from https://abcnews.go.com/Politics/russian-influence-operation-attempted-suppress-black-vote-indict- ment/story?id=53185084
14. Campbell-Smith, U., & Bradshaw, S. (2019). *Global cyber troops country profile: India*. Project on Computational Propaganda, Oxford Internet Institute. https://demtech.oii.ox.ac.uk/wp-content/uploads/sites/12/2019/05/India-Profile.pdf
15. Aneez, Z., Neyazi, T. A., Kalogeropoulos, A., & Nielsen, R. K. (2019). *Reuters Institute India digital news report*. Reuters Institute for the Study of Journalism. https://reutersinsti- tute.politics.ox.ac.uk/our-research/india-digital-news-report

16. MacDermott, Á., Motylinski, M., Iqbal, F., Stamp, K., Hussain, M., & Marrington, A. (2022). Using deep learning to detect social media trolls. *Forensic Science International: Digital Investigation, 43*(Supplement), 301446., ISSN 2666-2817. https://doi.org/10.1016/j.fsidi.2022.301446
17. Mukherjee, M.(2024). *AI deepfakes, bad laws—and a big fat Indian election*. https://reu- tersinstitute.politics.ox.ac.uk/news/ai-deepfakes-bad-laws-and-big-fat-indian-election
18. Gunning, D., Vorm, E., & Wang, J. Y. (2021). DARPA's explainable AI (XAI) program: A retrospective. *Applied AI Letters, 2*(4, Special Issue). https://doi.org/10.1002/ail2.61
19. Citron, Danielle K., Chesney, Robert, & Deep Fakes. (2019). *A looming challenge for privacy, democracy, and national security, in 107 California Law Review 1753*. https://scholarship.law.bu.edu/faculty_scholarship/640

Index

C. Hewage et al. (eds.), *Data Protection*,
https://doi.org/10.1007/978-3-031-76473-8

SPRINGER NATURE

GPSR Compliance

The European Union's (EU) General Product Safety Regulation (GPSR) is a set of rules that requires consumer products to be safe and our obligations to ensure this.

If you have any concerns about our products, you can contact us on ProductSafety@springernature.com

In case Publisher is established outside the EU, the EU authorized representative is:

Springer Nature Customer Service Center GmbH
Europaplatz 3
69115 Heidelberg, Germany

The manufacturer's authorised representative in the EU is Springer Nature Customer Service Centre GmbH, Europaplatz 3, 69115 Heidelberg, Germany. If you have any concerns regarding our products, please contact ProductSafety@springernature.com

Printed and bound by CPI Group (UK) Ltd, Croydon, CR0 4YY
30/09/2026
02225569-0011